Communications and Control Engineering

Communications and Control Engineering is a high-level academic monograph series publishing research in control and systems theory, control engineering and communications. It has worldwide distribution to engineers, researchers, educators (several of the titles in this series find use as advanced textbooks although that is not their primary purpose), and libraries.

The series reflects the major technological and mathematical advances that have a great impact in the fields of communication and control. The range of areas to which control and systems theory is applied is broadening rapidly with particular growth being noticeable in the fields of finance and biologically-inspired control. Books in this series generally pull together many related research threads in more mature areas of the subject than the highly-specialised volumes of *Lecture Notes in Control and Information Sciences*. This series's mathematical and control-theoretic emphasis is complemented by *Advances in Industrial Control* which provides a much more applied, engineering-oriented outlook.

Indexed by SCOPUS and Engineering Index.

Publishing Ethics: Researchers should conduct their research from research proposal to publication in line with best practices and codes of conduct of relevant professional bodies and/or national and international regulatory bodies. For more details on individual ethics matters please see:
https://www.springer.com/gp/authors-editors/journal-author/journal-author-helpdesk/publishing-ethics/14214

More information about this series at http://www.springer.com/series/61

Ruiyun Qi · Gang Tao · Bin Jiang

Fuzzy System Identification and Adaptive Control

Springer

Ruiyun Qi
College of Automation Engineering
Nanjing University of Aeronautics
and Astronautics
Nanjing, Jiangsu, China

Gang Tao
School of Engineering and Applied Science
University of Virginia
Charlottesville, VI, USA

Bin Jiang
College of Automation Engineering
Nanjing University of Aeronautics
and Astronautics
Nanjing, Jiangsu, China

ISSN 0178-5354 ISSN 2197-7119 (electronic)
Communications and Control Engineering
ISBN 978-3-030-19881-7 ISBN 978-3-030-19882-4 (eBook)
https://doi.org/10.1007/978-3-030-19882-4

MATLAB® and Simulink® are registered trademarks of The MathWorks, Inc., 1 Apple Hill Drive, Natick, MA 01760-2098, USA, http://www.mathworks.com.

Mathematics Subject Classification (2010): 93C40, 93C42, 93B52, 70Q05, 93C83

This Springer imprint is published by the registered company Springer Nature Switzerland AG
The registered company address is: Gewerbestrasse 11, 6330 Cham, Switzerland

To my family and the memory of Mietek A. Brdys

Ruiyun Qi

To our families

Gang Tao, Bin Jiang

Preface

Control of uncertain nonlinear dynamic systems has been a challenging topic in the last few decades. For nonlinear systems with parametric uncertainties, adaptive nonlinear control combined with backstepping, feedback linearization, sliding model control techniques has been extensively studied and applied to solve many practical problems successfully. For nonlinear systems with functional uncertainties, adaptive approximation-based control designs, using neural networks, fuzzy systems, and traditional adaptive approximators, have attracted much research interest in the last two decades and evolved to create powerful approaches which can help achieve stability and tracking results for a wide class of uncertain nonlinear systems.

In adaptive approximation-based control designs, neural networks, fuzzy systems, or other traditional function approximators such as polynomials, splines, and wavelets are employed to approximate unknown nonlinear functions. From a mathematical perspective, those function approximators can be used in a similar way when their role is to approximate some static nonlinear functions in dynamic nonlinear systems. Actually, most adaptive approximation-based control designs have used neural networks or fuzzy systems in this way: for some classes of nonlinear systems (strict-feedback, pure feedback, feedback linearizable, etc.), suitable nonlinear control techniques, such as backstepping and feedback linearization, are applied to develop nonlinear controllers where the unknown nonlinear functions are approximated by neural networks or fuzzy systems. The parameters or weights of neural networks/fuzzy systems are adaptively adjusted to compensate the nonlinear effects to achieve desirable control performance.

When not only some nonlinear functions but also the structure of a nonlinear dynamic system are unknown, dynamic neural networks and dynamic fuzzy systems provide effective tools to represent the whole unknown dynamic nonlinear system. Based on some available qualitative (i.e., experience, expert knowledge) and quantitative (i.e., input–output data) information on the nonlinear system, the initial structure and parameters of the dynamic neural networks or fuzzy system can be determined through some offline identification methods. When those models are used in online control, their structures and parameters can be further adjusted through evolutionary and adaptive algorithms.

Among various dynamic approximation models, the fuzzy dynamic system, usually called Takagi–Sugeno (T–S) fuzzy system, has been widely studied. Different from conventional Mamdani fuzzy systems, the rule consequents of a T–S fuzzy system are represented by local linear models rather than fuzzy sets. The overall T–S fuzzy system is a nonlinear dynamic model by smoothly blending all the local linear models using fuzzy membership functions. Compared with dynamic neural networks, T–S fuzzy modeling is intuitively easier to understand since each of its local models represents the local behavior of a nonlinear dynamic system in certain operating region defined by the fuzzy sets in the rule premise. From the control synthesis and analysis perspective, T–S fuzzy systems are very suitable for systematic control designs and stability analysis.

From the 1990s to 2000s, extensive studies were carried out on T–S fuzzy systems, leading to many fundamental theoretical results on identification, stability analysis, and control synthesis of T–S fuzzy systems. Since there exist uncertain parameters in real-life nonlinear systems, when T–S fuzzy systems are used to represent real-life nonlinear systems, they have uncertain parameters as well, which motivated the adaptive control designs for T–S fuzzy systems. Since T–S fuzzy systems can be viewed as a class of nonlinear dynamic systems, adaptive control of T–S fuzzy systems is not only an important branch of T–S fuzzy-model-based control synthesis and analysis but also provides solutions for adaptive control of a more general class of uncertain nonlinear systems.

This book aims at providing a systematic and unified framework for identification and adaptive control of fuzzy systems, especially T–S fuzzy systems. It is well-known that adaptive control algorithms for both linear and nonlinear systems are related closely to the system structures, which determines the design conditions, the controller structure, the estimation error model, the parameter adaptive laws, etc. In literature, T–S fuzzy systems appear in different forms: state-space form, input–output form, continuous-time form, discrete-time form, single-input–single-output (SISO) form, multi-input–multi-output (MIMO) form. On the other hand, from the perspective of control objective, it can be state tracking or output tracking; from the perspective of available information for control designs, one can choose state feedback control or output feedback control. Different system forms, control objectives, and available information lead to different adaptive control design methodologies for T–S fuzzy systems, which will be carefully considered and rigorously addressed within a sound analytical framework in this book.

This book has been written at the first-year graduate level with prerequisite knowledge of linear algebra, dynamic systems, and feedback control. This book can be used as a textbook for graduate students in a course of fuzzy modeling and control, and also a technical reference for graduate students, scholars, engineers in the fields of mechanical engineering, electrical engineering, aerospace engineering, computer science, applied mathematics, and others who have some basic knowledge of linear feedback control.

Organization. With 10 chapters, this book systematically presents identification, adaptive state tracking control and output tracking control algorithms for T–S fuzzy systems of different forms and with unknown parameters.

Chapter 1 introduces some background knowledge. The basic concepts of fuzzy sets, fuzzy logic, and fuzzy inference system are presented. Some typical fuzzy systems including Mamdani fuzzy system and T–S fuzzy system are briefly introduced. A literature review on fuzzy system identification and T–S fuzzy model-based adaptive nonlinear control is also given.

Chapter 2 aims at giving readers a clear and concise description on some important features and properties of T–S fuzzy systems, which relate closely to the problems of T–S fuzzy system identification and control designs, including different system architectures, approximation capability, stability conditions, stabilization control, and tracking control.

Chapter 3 reviews basic concepts and design methodologies of adaptive control. Since T–S fuzzy systems are locally linear models and semi-globally nonlinear models, their adaptive control designs can be carried out using both adaptive linear control and adaptive nonlinear control approaches. The goal of this chapter is to help readers who are not familiar with adaptive control to understand the basic ideas, classical design methods, and key issues of adaptive linear control and adaptive nonlinear control.

Chapter 4 presents the identification methods for T–S fuzzy systems. Both linearization based and data-based methods are provided. In the data-based methods, both offline identification and online identification approaches are studied, where fuzzy clustering methods are used to identify the model structure and linear and nonlinear parameter learning methods are adopted to estimate model parameters.

Chapter 5 deals with adaptive state tracking for continuous-time T–S fuzzy systems in state-space forms. This chapter presents an extensive study on how state tracking can be achieved for T–S fuzzy systems via various controller structures and reference model structures. Actually, the control design for T–S fuzzy systems can be very flexible due to their structure features. It is revealed in this chapter that by choosing suitable controller structure and reference model, the necessary plant-model matching conditions for state tracking control can be relaxed. State tracking control problems for both single-input and multi-input T–S fuzzy systems are handled with detailed design and analysis. The issue on how to guarantee the input matrix with estimated parameters nonsingular in the adaptive controller for multi-input T–S fuzzy systems is also addressed.

Chapter 6 develops a solution framework for adaptive state feedback output tracking control of general discrete-time state-space T–S fuzzy systems. A normal form is derived for a global T–S fuzzy model, which has an explicit relative degree structure and a proper causality property. Such a model is an approximation of a discrete-time nonlinear system with a general relative degree, based on which an adaptive state feedback control scheme can be developed with desired stability and output tracking performance.

Chapter 7 considers adaptive output feedback control for discrete-time SISO T–S fuzzy systems described by input–output models. A key advantage of using an input–output approach over a state-space approach is that desired system tracking performance can be characterized and achieved based on much relaxed design conditions for adaptive control, in the presence of system parameter uncertainties. Two nonlinear prediction models for T–S fuzzy systems with multiple input–output delays are developed and for each of them, adaptive control schemes with complete stability and convergence analysis are presented.

Chapter 8 extends the results in Chap. 7 to adaptive control of MIMO discrete-time T–S fuzzy systems with general delay matrices. The results of this chapter include the derivation of a global prediction fuzzy system model with a general delay structure, parametrization and parameter estimation, development of an adaptive control scheme, stability and tracking analysis of such an adaptive control system, and illustration of new features and concepts of adaptive control for MIMO fuzzy systems.

Chapter 9 considers discrete-time input–output multiple-delay T–S fuzzy systems with unknown membership function parameters and consequent parameters, and derived a solution for it. A fuzzy system with such uncertain parameters leads to a nonlinearly parametrized estimation error model, for which the gradient algorithm can be applied to derive an adaptive law to adaptively update the estimates of both the membership function parameters and rule consequent parameters. The properties of the parameter adaptive laws and closed-loop stability have been studied with some key design conditions specified and established for the Gaussian membership functions.

Chapter 10 presents two adaptive actuator fault compensation scheme for nonlinear systems represented by T–S fuzzy systems with redundant actuators.

Acknowledgements This book cannot be completed without the inspiration and knowledge from many contributors in the fields of fuzzy systems and adaptive control on whose work our book is based. In particular, we want to mention: Lotfi A. Zadeh, Tomohiro Takagi, Michio Sugeno, Li-Xin Wang, Gang Feng, Hao Ying, Graham Goodwin, Ioan Landau, Petros A. Ioannou, Marios M. Polycarpou, Jay A. Farrell, Jing Sun, Miroslav Krstić, Hua O. Wang, Kazuo Tanaka, Ronald Yager, B. Babuska, Shaocheng Tong, Han-Xiong Li, Kevin M. Passino, and Plamen P. Angelov.

This first author would like to thank Narasimhan Sundarajan, who introduced her to the field of fuzzy systems back in 2002, and Mietek A. Brdys who supervised her Ph.D. work in adaptive fuzzy control from 2003 to 2007.

We would like to thank our former Ph.D. students Chang Tan, Xuelian Yao, Liyan Wen, Yanjun Zhang, Zhifeng Gao, Qikun Shen and master students Yuhai Huang, Limeng Zhu, Xiapeng Zhou, and Lin Jiang for their collaboration in the development and implementation of various fuzzy identification and adaptive control techniques, and Yicheng Chen, Le Wang, Yue Liang, Zhiyu Peng, and Xinlei Dong for their assistance in carrying out some of the simulation studies contained in this book and proofreading the manuscript.

The authors are also grateful to Oliver Jackson, Editor; Balaganesh Sukumar, Project Coordinator; Meertinus Faber, Project Coordinator; Yasmin Brookes, Editorial Assistant; Divya Prabha Karthikesan at Springer for their kind help and generous support during the long project period.

The research work in this book was supported by the National Natural Science Foundation of China (No. 61873127, 61773201, 60904042), the National Aeronautical Science Foundation of China (No. 2017ZA52013) and the Six Talent Peaks Project in Jiangsu Province (No. HKHT-010), and the Priority Academic Program Development of Jiangsu Higher Education Institutions. The first and the third authors are also with the Key Laboratory of Navigation, Control and Health-Management Technologies of Advanced Aerocraft (Nanjing Univ. of Aeron. & Astron.), Ministry of Industry and Information Technology of China.

Nanjing, China Ruiyun Qi
Charlottesville, VI, USA Gang Tao
Nanjing, China Bin Jiang

Contents

Chapter 1
Introduction

This book presents fuzzy system identification and fuzzy system based adaptive control methodologies that employ fuzzy systems as dynamic approximation models of nonlinear systems. Fuzzy system identification can be carried out using offline input/output (I/O) data collection or in an online mode. Dynamic fuzzy systems are treated as the design models of nonlinear systems, whose structure and parameters serve as a foundation for adaptive control designs. This book aims at providing a systematic and unified framework for identification and adaptive control of fuzzy system, especially Takagi–Sugeno (T–S) fuzzy systems.

This introductory chapter presents some basic knowledge related closely to other chapters in this book, such as basics of fuzzy sets, fuzzy inference system, and typical fuzzy systems. We also provide an overview of fuzzy system identification and fuzzy system based adaptive control and make it clear to readers what this book is about.

1.1 Basic Concepts of Fuzzy Systems

In this section, a self-contained yet succinct introduction to the basic concepts in developing a fuzzy system is presented. We mainly focus on the basic knowledge necessary for understanding the concepts and techniques in other chapters of this book. The main references for this section are Lilly (2010), Farrell and Polycarpou (2006), Passino and Yurkovich (1998), and Wang (1994).

1.1.1 Fuzzy Sets

The foundation of fuzzy logic system is fuzzy set. The concept of fuzzy set was first introduced by Prof. Lotfi A. Zadeh in his paper published in *Information and Control* (Zadeh 1965). Fuzzy set is a generalization of the classical set which has clear and

R. Qi et al., *Fuzzy System Identification and Adaptive Control*, Communications and Control Engineering,
https://doi.org/10.1007/978-3-030-19882-4_1

crisp boundaries. For a classical set, an element can either belong to or not belong to it, that is, the truth value of the belongingness is either 1 or 0. In contrast with classical sets, a fuzzy set does not have crisp boundaries and an element can belong to it with a membership degree between 0 and 1.

Definition 1.1 (Zadeh 1965) A fuzzy set is a class of objects with a continuum of grades of membership. Such a set is characterized by a membership function which assigns to each object a grade of membership ranging between 0 and 1.

Mathematically, consider a real number x in a universe of discourse $\mathscr{X} \subseteq R$. Let A denote a fuzzy set defined on $\mathscr{X}$. A membership function $\mu_A(x)$ of x in A is a function that maps $\mathscr{X}$ into a membership value $\in [0, 1]$, which characterizes how much x belongs to A. Usually, a fuzzy set A is described as

$$A = \{(x, \mu_A(x)) \,|x \in \mathscr{X}\}. \tag{1.1}$$

The universe of discourse can be either discrete or continuous, as shown in the following two examples.

Example 1.1 Consider a university of discourse $\mathscr{X} = \{1, 2, 3, 4, 5\}$. Let A denote a fuzzy set "about 3". Then, the fuzzy set A can be represented by

$$A = \{(1, 0.1), (2, 0.5), (3, 1), (4, 0.5), (5, 0.1)\}, \tag{1.2}$$

in which each pair consists of a member and its degree of membership.

Example 1.2 Let M denote a fuzzy set "fast" defined on the universe of discourse $\mathscr{X} = [0, 100](\text{m/s})$ for the variable x (which represents speed). The membership function of M for $x \in \mathscr{X}$ can be defined as

$$\mu_M(x) = \begin{cases} 0, & x < 30 \\ \frac{x}{30} - 1, & 30 \le x < 60 \\ 1, & x \ge 60, \end{cases} \tag{1.3}$$

which is shown in Fig. 1.1.

Fuzzy singleton. A special fuzzy set which contains only a single point x with $\mu(x) = 1$ is called a fuzzy singleton.

Fuzzy singleton plays an important role for fuzzification in Mamdani fuzzy systems and Takagi–Sugeno (T–S) fuzzy systems.

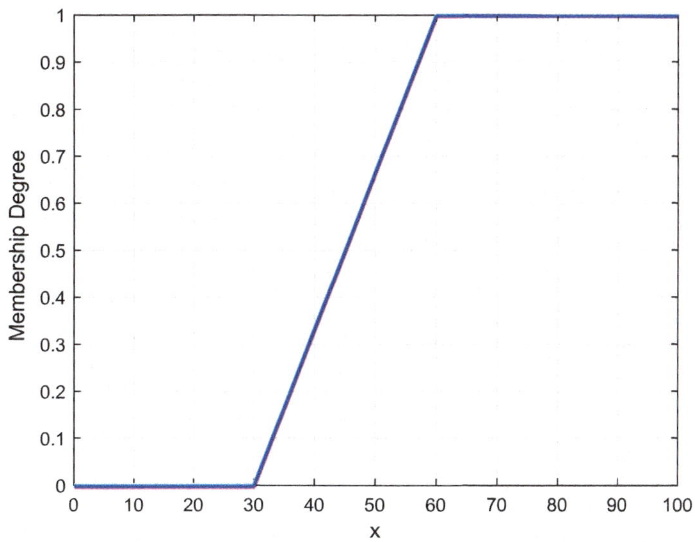

Fig. 1.1 Membership function of fuzzy set "fast"

1.1.2 Fuzzy Logic Operations

For classical sets, there are some logic operations such as AND, OR, and NOT. For fuzzy sets, typically, most fuzzy logic applications make use of the following operations:

- fuzzy intersection (AND)
- fuzzy union (OR)
- fuzzy complement (NOT)

Consider two fuzzy sets A and B defined on the universe of discourse $\mathscr{X}$, whose membership functions are $\mu_A(x)$ and $\mu_B(x)$, respectively, for $x \in \mathscr{X}$.

Fuzzy intersection (AND). The intersection of two fuzzy sets A and B, that is, "A AND B" is represented as "$A \cap B$". The membership function for the fuzzy set $A \cap B$ is denoted by $\mu_{A\cap B}(x)$. $\mu_{A\cap B}(x)$ is specified in general by a binary mapping T, aggregating $\mu_A(x)$ and $\mu_B(x)$ as follows:

$$\mu_{A\cap B}(x) = \mu_A(x) \star \mu_B(x) = T(\mu_A(x), \mu_B(x)), \tag{1.4}$$

where $T(\cdot)$ is called a *T-norm* operator.

Two typical *T-norm* operators for fuzzy intersection are *minimum*:

$$T_{\min}(\mu_A(x), \mu_B(x)) = \min(\mu_A(x), \mu_B(x)) \tag{1.5}$$

and *algebraic product*:

$$T_{\text{ap}}(\mu_A(x), \mu_B(x)) = \mu_A(x)\mu_B(x). \tag{1.6}$$

Fuzzy union (OR). The union of two fuzzy sets A and B, that is, "A OR B" is represented as "$A \cup B$". The membership function for the fuzzy set $A \cup B$ is denoted by $\mu_{A\cup B}(x)$. $\mu_{A\cup B}(x)$ is specified in general by a binary mapping S, aggregating $\mu_A(x)$ and $\mu_B(x)$ as follows:

$$\mu_{A\cup B}(x) = \mu_A(x) \oplus \mu_B(x) = S(\mu_A(x), \mu_B(x)), \tag{1.7}$$

where $S(\cdot)$ is called a *S-norm* operator.

Two typical *S-norm* operators for fuzzy union are *maximum*:

$$S_{\max}(\mu_A(x), \mu_B(x)) = \max(\mu_A(x), \mu_B(x)) \tag{1.8}$$

and *algebraic sum*:

$$S_{\text{as}}(\mu_A(x), \mu_B(x)) = \mu_A(x) + \mu_B(x) - \mu_A(x)\mu_B(x). \tag{1.9}$$

Fuzzy complement (NOT). The complement of a fuzzy set A, that is, "NOT A" is represented by "$\bar{A}$". The membership function for the fuzzy set $\bar{A}$ is defined as

$$\mu_{\bar{A}}(x) = 1 - \mu_A(x). \tag{1.10}$$

The fuzzy intersection and union operations introduced above are performed on fuzzy sets with the same variable and defined on the same universe of discourse. When we want to perform some operations on fuzzy sets defined on different universes of discourse, we need to introduce another important concept called fuzzy Cartesian product.

Definition 1.2 (*Cartesian Product*) A Cartesian product is a mathematical operation that returns a set (or product set or simply product) from multiple sets. That is, for sets $\mathscr{X}$ and $\mathscr{Y}$, the Cartesian product $\mathscr{X} \times \mathscr{Y}$ is the set of all ordered pairs (x, y) where $x \in \mathscr{X}$ and $y \in \mathscr{Y}$:

$$\mathscr{X} \times \mathscr{Y} = \{(x, y)|x \in \mathscr{X},\ y \in \mathscr{Y}\}. \tag{1.11}$$

Definition 1.3 (*Fuzzy Relation*) A fuzzy relation R between the university of discourses $\mathscr{X}$ and $\mathscr{Y}$ is a fuzzy set defined on $\mathscr{X} \times \mathscr{Y}$ with ordered pairs (x, y) whose membership function is $\mu_R(x, y) \in [0, 1]$:

$$R = \{(x, y),\ \mu_R(x, y) | x \in \mathscr{X},\ y \in \mathscr{Y}\}. \tag{1.12}$$

Definition 1.4 (*Fuzzy Cartesian Product*) The Cartesian product between two fuzzy sets A for $x \in \mathscr{X}$ and B for $y \in \mathscr{Y}$ results in a fuzzy relation defined on $\mathscr{X} \times \mathscr{Y}$:

$$Q = A \times B = \{(x, y),\ \mu_Q(x, y) | x \in \mathscr{X},\ y \in \mathscr{Y}\}, \tag{1.13}$$

where the membership function

$$\mu_Q(x, y) = \mu_{A \times B}(x, y) = \mu_A(x) \star \mu_B(y), \tag{1.14}$$

with the $\star$ operation taking the T-norm operator minimum or algebraic product.

Definition 1.5 (*Linguistic Variable*) A linguistic variable is characterized by a quintuple

$$(x,\ \mathscr{X},\ \Sigma,\ G,\ M), \tag{1.15}$$

where

x the name of variable
$\mathscr{X}$ the universe of discourse of x
Σ a term set, i.e., the set of linguistic values of x
G a way for generating linguistic values of x
M a way for making each linguistic value associate with its meaning.

Example 1.3 If *outdoor temperature* denoted by T is interpreted as a *linguistic variable* defined on the *universe of discourse* $\mathscr{X} = [-20, 45]$(°C), then its *term set* could be

$$\Sigma = \{Very\ cold,\ Cold,\ Cool,\ Warm,\ Hot,\ Very\ hot\} \tag{1.16}$$

with *Very cold*, *Cold*, *Cool*, *Warm*, *Hot*, and *Very hot* being *linguistic values* of T. The meaning of each linguistic value may be interpreted as follows:

"*Very cold*" means "temperature below about −5 °C",
"*Cold*" means "temperature around 0 °C",
"*Cool*" means "temperature around 10 °C",
"*Warm*" means "temperature around 20 °C",
"*Hot*" means "temperature around 30 °C",
"*Very hot*" means "temperature above about 35 °C".

Each linguistic value can be characterized by a fuzzy set. For example, fuzzy sets A_1, A_2, A_3, A_4, A_5, and A_6 represent linguistic values *Very cold*, *Cold*, *Cool*, *Warm*, *Hot*, and *Very hot*, respectively, whose membership functions are shown in Fig. 1.2.

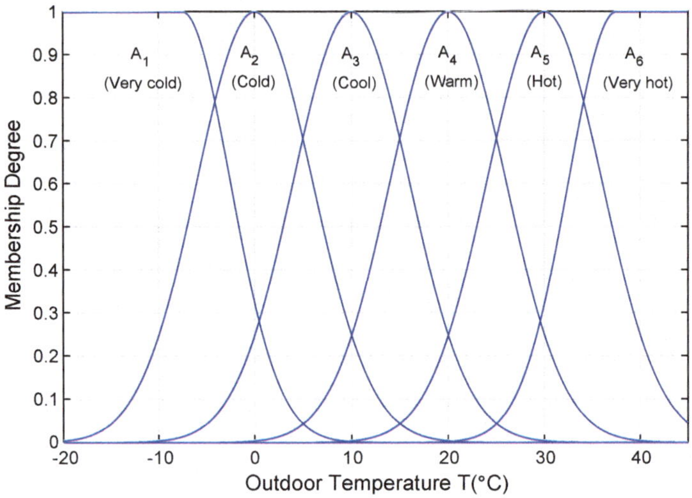

Fig. 1.2 Membership functions defined for *outdoor temperature*

1.1.3 Fuzzy Inference System

Fuzzy inference system is an important application of fuzzy sets and fuzzy logic. Figure 1.3 depicts a typical fuzzy inference system which consists of four essential components: *fuzzy rule base*, *fuzzification*, *fuzzy inference*, and *defuzzification*.

Fuzzy rule base. Fuzzy rule base is the foundation of the whole fuzzy inference system, which consists of a group of "IF-THEN" rules.

For instance, a fuzzy rule base with four rules has the following form:

$$\begin{aligned}
R^1: &\quad \text{IF } \bar{x}_1 \text{ is } \bar{A}_1 \text{ and } \bar{x}_2 \text{ is } \bar{B}_1, \text{ THEN } \bar{y} \text{ is } \bar{C}_1\\
R^2: &\quad \text{IF } \bar{x}_1 \text{ is } \bar{A}_1 \text{ and } \bar{x}_2 \text{ is } \bar{B}_2, \text{ THEN } \bar{y} \text{ is } \bar{C}_2\\
R^3: &\quad \text{IF } \bar{x}_1 \text{ is } \bar{A}_2 \text{ and } \bar{x}_2 \text{ is } \bar{B}_1, \text{ THEN } \bar{y} \text{ is } \bar{C}_3\\
R^4: &\quad \text{IF } \bar{x}_1 \text{ is } \bar{A}_2 \text{ and } \bar{x}_2 \text{ is } \bar{B}_2, \text{ THEN } \bar{y} \text{ is } \bar{C}_4,
\end{aligned} \tag{1.17}$$

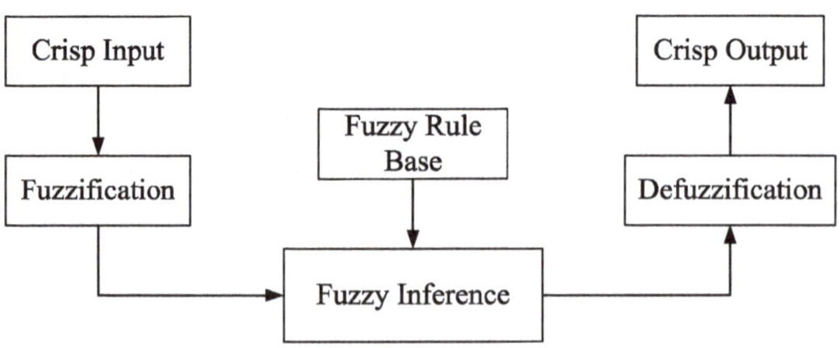

Fig. 1.3 A typical fuzzy inference system

where $\bar{x}_1$, $\bar{x}_2$, and $\bar{y}$ are *linguistic variables* on the universe of discourses $\mathscr{X}_1$, $\mathscr{X}_2$, and $\mathscr{Y}$, respectively; $\bar{A}_1$ and $\bar{A}_2$ are two *linguistic values* for $\bar{x}_1$, $\bar{B}_1$ and $\bar{B}_2$ are two linguistic values for $\bar{x}_2$, and $\bar{C}_1$, $\bar{C}_2$, $\bar{C}_3$, and $\bar{C}_4$ are four linguistic values for $\bar{y}$.

The first part of the statement, "$\bar{x}_1$ is $\bar{A}_1$ and $\bar{x}_2$ is $\bar{B}_1$" is called the *premise* of the rule, and the second part of the statement, "$\bar{y}$ is $\bar{C}_1$" is called the *consequent* of the rule.

Define two fuzzy sets A_1 and A_2 to characterize the two linguistic values $\bar{A}_1$ and $\bar{A}_2$, two fuzzy sets B_1, B_2 for $\bar{B}_1$ and $\bar{B}_2$, and four fuzzy sets C_i for $\bar{C}_i$, $i = 1, 2, 3, 4$.

Fuzzification. Since $\bar{x}_1$ is a linguistic variable, it can only take linguistic values (fuzzy sets), and if the input to the fuzzy inference system is a crisp value x_1, it needs to be fuzzified first, that is, to make the crisp value into a fuzzy set. A most popular fuzzification method is to assign a *fuzzy singleton* to the crisp value x_1, that is, $\{(x_1, \mu(x_1) = 1)\}$. Throughout this book, we use fuzzy singleton in fuzzification in a fuzzy system or fuzzy controller.

Fuzzy inference. Each premise in (1.17) defines a fuzzy relation, e.g.,

$$P_1 = A_1 \times B_1 \tag{1.18}$$

on $\mathscr{X}_1 \times \mathscr{X}_2$, whose membership function is determined once the *T-norm* and *S-norm* representations of the "and" and "or" operations are selected. Therefore, the confidence or firing strength of the rule R^1 can be calculated by

$$\mu_{P_1}(\bar{x}_1, \bar{x}_2) = \mu_{A_1 \cap B_1}(\bar{x}_1, \bar{x}_2) = \mu_{A_1}(\bar{x}_1) \star \mu_{B_1}(\bar{x}_2), \tag{1.19}$$

where $\star$ can be implemented as *minimum* or *algebraic product*. Using fuzzy singleton in fuzzification and choosing $\star$ operation as *minimum*, the firing strength for two crisp values x_1 and x_2 is computed as

$$\mu_{P_1}(x_1, x_2) = \mu_{A_1 \cap B_1}(x_1, x_2) = \min(\mu_{A_1}(x_1), \mu_{B_1}(x_2)). \tag{1.20}$$

Now, we have completed the interpreting work of the premise of the rule R^1. Interpreting the premises of the other rules can be similarly done, which gives $\mu_{P_2}(x_1, x_2)$, $\mu_{P_3}(x_1, x_2)$ and $\mu_{P_4}(x_1, x_2)$.

Next, we consider how to interpret the whole rule R^1. The fuzzy IF-THEN rule of the form

$$R^1: \quad \text{IF } \bar{x}_1 \text{ is } \bar{A}_1 \text{ and } \bar{x}_2 \text{ is } \bar{B}_1, \text{ THEN } \bar{y} \text{ is } \bar{C}_1 \tag{1.21}$$

for $x_1 \in \mathscr{X}_1$, $x_2 \in \mathscr{X}_2$ and $y \in \mathscr{Y}$ can be interpreted as a fuzzy relation (implication) represented by

$$\begin{aligned} I_1 &= A_1 \times B_1 \longrightarrow C_1 \\ &= P_1 \longrightarrow C_1. \end{aligned} \tag{1.22}$$

Definition 1.6 (*Fuzzy Implication*) Let A and B be two fuzzy sets for two variables u and v on the universe of discourses $\mathscr{U}$ and $\mathscr{V}$, respectively. A fuzzy implication is defined as a fuzzy relation on the Cartesian product $\mathscr{U} \times \mathscr{V}$ denoted by

$$I = A \rightarrow B, \tag{1.23}$$

and the membership function $\mu_I(u, v)$ can be calculated from $\mu_A(u)$ and $\mu_B(v)$.

The membership function corresponding to the fuzzy relation (implication) I can be derived by various fuzzy implication methods such as

Larsen implication: $\mu_I(u, v) = \mu_A(u)\mu_B(v)$
Kleene-Dienes implication: $\mu_I(u, v) = \max(1 - \mu_A(u), \mu_B(v))$
Mamdani implication: $\mu_I(u, v) = \min(\mu_A(u), \mu_B(v))$.

If the *T-norm* operator for (1.19) is selected as *minimum* and the fuzzy implication method for (1.22) is selected as Mamdani implication, then the membership function for the fuzzy implication $I_1 = P_1 \rightarrow C_1$ can be calculated as

$$\begin{aligned} \mu_{I_1}(x_1, x_2, y) &= \min(\mu_{P_1}(x_1, x_2), \mu_{C_1}(y)) \\ &= \min(\mu_{A_1}(x_1), \mu_{B_1}(x_2), \mu_{C_1}(y)). \end{aligned} \tag{1.24}$$

Similarly, we can obtain the membership functions for $\mu_{I_2}(x_1, x_2, y)$, $\mu_{I_3}(x_1, x_2, y)$, and $\mu_{I_4}(x_1, x_2, y)$.

Till now, each crisp input value has been converted into a fuzzy singleton and each fuzzy rule has been converted into a fuzzy implication. There still remain two questions. How can the fuzzy set in $\mathscr{Y}$ inferred from a single rule be determined? How can the fuzzy set in $\mathscr{Y}$ inferred from a group of rules be determined?

Definition 1.7 (*Compositional Rule of Inference*) Consider a fuzzy rule

$$\text{IF } x \text{ is } A, \text{ THEN } y \text{ is } B, \tag{1.25}$$

where A and B are two fuzzy sets defined on $\mathscr{X}$ and $\mathscr{Y}$, respectively. Given a fact "x is A'", the consequence is "y is B'", where A' and B' are two fuzzy sets on $\mathscr{X}$ and $\mathscr{Y}$, respectively. B' is determined as a composition of the fact and the fuzzy implication operator

$$B' = A' \circ (A \rightarrow B) \tag{1.26}$$

and the membership function of B' is determined by

$$\mu_{B'}(y) = \sup_{x \in \mathscr{X}} T(\mu_{A'}(x), \mu_{A \rightarrow B}(x, y)), \quad y \in \mathscr{Y}, \tag{1.27}$$

where "sup" denotes the least upper bound.

Based on the *compositional rule of inference*, given two specific crisp input values x_1^* and x_2^* which are fuzzified as fuzzy sets A_1' and B_1', the inferred fuzzy set C_1' in $\mathscr{Y}$ from the rule R^1 is determined by

$$C_1' = (A_1' \times B_2') \circ I_1 \tag{1.28}$$

and the membership function of C_1' is determined by

$$\mu_{C_1'}(y) = \sup_{(x_1,x_2)\in\mathscr{X}_1\times\mathscr{X}_2} T(\mu_{A_1'\times B_1'}(x_1,x_2), \mu_{I_1}(x_1,x_2,y)), \quad y \in \mathscr{Y}. \tag{1.29}$$

If A_1' and B_1' are selected as fuzzy singletons and *minimum* is used as the *T-norm* operator in (1.27), the membership function of C_1' can be determined by

$$\mu_{C_1'}(y) = \min(\mu_{A_1}(x_1^*), \mu_{B_1}(x_2^*), \mu_{C_1}(y)), \quad y \in \mathscr{Y}, \tag{1.30}$$

which is also the way to obtain the output fuzzy set in a Mamdani fuzzy system.

Now we have answered the first question: how can the fuzzy set in $\mathscr{Y}$ inferred from a single rule be determined? The output fuzzy set $C_i' = \{(y, \mu_{C_i'}(y))|y \in \mathscr{Y}\}$ corresponding to each individual rule is determined according to (1.30). The overall output of the fuzzy inference system can be obtained by either

$$C' = C_1' \cup C_2' \cup \cdots \cup C_N' \tag{1.31}$$

or

$$C' = C_1' \cap C_2' \cap \cdots \cap C_N', \tag{1.32}$$

where N is the number of rules.

When each rule is treated as an independent conditional statement that can work for all possible operational situations, the overall output fuzzy set is obtained by (1.31). When the rule base represents a strongly coupled set of conditional statements that all of them should be taken into account for a given operating situation, the overall output fuzzy set is obtained by (1.32).

In a Mamdani fuzzy system, the overall output fuzzy set is inferred by (1.31). If *maximum* is selected as the fuzzy union operator, the membership function of C' is calculated by

$$\mu_{C'}(y) = \max_{i=1}^{N}(\mu_{C_i'}(y)), \quad y \in \mathscr{Y}. \tag{1.33}$$

Now we have finished the procedure of fuzzy inference and obtained the overall output fuzzy set $C' = \{(y, \mu_{C'}(y))|y \in \mathscr{Y}\}$. However, the output fuzzy set cannot be directly used in applications. It is necessary to convert the fuzzy sets into crisp values for further processing.

Defuzzification. Defuzzification means the fuzzy-to-crisp conversion, which is a process to select a representative element (unique single-valued quantity) from the fuzzy output inferred from a fuzzy rule-based system. In literature, there are many defuzzification approaches such as *maximum defuzzificaton, center of gravity (COG) defuzzification, center-average (CA) defuzzification*, etc.

1. Maximum defuzzification

$$\begin{aligned} \Omega(y) &= \{y \in \mathscr{Y} \mid \mu_{C'}(y) = \sup_{y \in \mathscr{Y}} (\mu_{C'}(y))\} \\ y^* &= g(\Omega(y)), \end{aligned} \tag{1.34}$$

where "sup" denotes the least upper bound, $\Omega(y)$ is a set containing all the values of y that achieve the maximum membership function value of C' in $\mathscr{Y}$, and the function g represents a way to determine the unique crisp output y^* from $\Omega(y)$. For instance, g could be used to select the maximum, mean, or minimum value of y in $\Omega(y)$.

2. Center of gravity (COG) defuzzification

The center of gravity method is also called the center of area method. The crisp output y^* is chosen as the center of area for the membership function of the overall output fuzzy set C':

$$y^* = \frac{\int_{\mathscr{Y}} \mu_{C'}(y) y dy}{\int_{\mathscr{Y}} \mu_{C'}(y) dy}. \tag{1.35}$$

3. Center-average (CA) defuzzification

Define the center of area of the ith output fuzzy set C_i' as a point y_i^* satisfying

$$\int_{-\infty}^{y_i^*} \mu_{C_i'}(y) dy = \int_{y_i^*}^{\infty} \mu_{C_i'}(y) dy. \tag{1.36}$$

The crisp output of the overall fuzzy system can be calculated as

$$y^* = \frac{\sum_{i=1}^{N} y_i^* \sup_{y \in \mathscr{Y}} \{\mu_{C_i'}(y)\}}{\sum_{i=1}^{N} \sup_{y \in \mathscr{Y}} \{\mu_{C_i'}(y)\}}. \tag{1.37}$$

1.2 Typical Fuzzy Systems

In the area of fuzzy system identification and fuzzy logic control, there are two types of fuzzy systems that are most popular in many identification and control applications. One is Mamdani fuzzy system and the other is Takagi–Sugeno (T–S) fuzzy system. The main difference between them lies in the consequent of the rule. The consequent of a Mamdani fuzzy rule is a fuzzy set while the consequent of a T–S fuzzy rule is a crisp function.

1.2.1 Mamdani Fuzzy Systems

With $x_1 \in \mathscr{X}_1$, $x_2 \in \mathscr{X}_2, \ldots,$, $x_L \in \mathscr{X}_L$ and $y \in \mathscr{Y}$, a Mamdani fuzzy system is a nonlinear mapping from $\mathscr{X} = \mathscr{X}_1 \times \ldots \times \mathscr{X}_L$ to $\mathscr{Y}$, which admits the following form:

$$R^i: \quad \text{IF } \bar{x}_1 \text{ is } \bar{A}_1^j \text{ and } \bar{x}_2 \text{ is } \bar{A}_2^k \text{ and } \ldots \text{ and } \bar{x}_L \text{ is } \bar{A}_L^l, \text{ THEN } \bar{y} \text{ is } \bar{C}_i, \tag{1.38}$$

where $i \in [1, \ldots, N]$, $\bar{A}_1^j$ represents the jth linguistic value of $\bar{x}_1$, $\bar{A}_2^k$ represents the kth linguistic value of $\bar{x}_2$, and so on.

For the simplicity of notation, in the rest of this book, we will dispense with the bar over the linguistic variables and values, resulting in the fuzzy rule described by

$$R^i: \quad \text{IF } x_1 \text{ is } A_1^j \text{ and } x_2 \text{ is } A_2^k \text{ and } \ldots \text{ and } x_L \text{ is } A_L^l, \text{ THEN } y \text{ is } C_i, \tag{1.39}$$

where x_1 is deemed as a linguistic variable by default, A_1^j represents both the jth linguistic value $\bar{A}_1^j$ and the fuzzy set associated with $\bar{A}_1^j$, and so on.

Using *singleton fuzzification* and *minimum* for the "and" operation, the firing strength of each rule denoted by $\lambda_i(x)$ is calculated by

$$\lambda_i(x) = \min(\mu_{A_1^j}(x_1), \mu_{A_2^k}(x_2), \ldots, \mu_{A_L^l}(x_L)). \tag{1.40}$$

Using the *Mamdani implication* method, the output fuzzy set C_i' of the ith rule is characterized by the membership function

$$\mu_{C_i'}(y) = \min(\lambda_i(x), \mu_{C_i}(y)), \quad y \in \mathscr{Y}. \tag{1.41}$$

Interpreting the relation among different rules as "or" and using *maximum* for the "or" operation, the overall output fuzzy set is obtained as

$$C' = \bigcup_{i=1}^{N} C_i' \tag{1.42}$$

whose membership function $\mu_{C'}(y)$ is computed by

$$\mu_{C'}(y) = \max_{i=1}^{N}(\mu_{C_i'}(y)), \quad y \in \mathscr{Y}. \tag{1.43}$$

Using the *center of gravity (COG) defuzzification* method, the overall crisp output is computed by

$$y^* = \frac{\int_{\mathscr{Y}} \mu_{C'}(y) y dy}{\int_{\mathscr{Y}} \mu_{C'}(y) dy}. \tag{1.44}$$

1.2.2 Takagi–Sugeno (T–S) Fuzzy Systems

Takagi–Sugeno (T–S) fuzzy systems are one important class of fuzzy systems, which were proposed by Takagi and Sugeno (1985). Compared with Mamdani fuzzy systems, T–S fuzzy systems are more suitable systematic control design and synthesis and their parametric form makes it convenient to employ various parameter estimation algorithms. In this section, we present a brief introduction to T–S fuzzy systems, including their rule structure and inference method.

The main difference between T–S fuzzy systems and Mamdani fuzzy systems is in the form of the consequent parts of their rules. The consequents of T–S fuzzy rules are usually real-valued functions instead of fuzzy sets. T–S fuzzy systems have the following general form:

$$\begin{aligned} R^i :\ & \text{IF } x_1 \text{ is } F_1^i \text{ and } x_2 \text{ is } F_2^i \text{ and } \ \dots \ \text{ and } x_L \text{ is } F_L^i \\ & \text{THEN } y = g_i(x_1, x_2, \dots, x_L), \end{aligned} \tag{1.45}$$

where R^i denotes the ith fuzzy rule, $x_1, x_2, \dots, x_L$ are premise (input) variables on the universe of discourses $\mathscr{X}_1$, $\mathscr{X}_2$, ..., $\mathscr{X}_L$, respectively; $g_i(\cdot)$ is a crisp function, and F_j^i denotes a fuzzy set associated with which there is a membership function $\mu_{F_j^i}(x_j)$ $i = 1, 2, \dots, N$, $j = 1, \dots, L$, N is the number of fuzzy rules, and L is the number of premise variables.

The consequent part $y = g_i(\cdot)$ of T–S fuzzy system can represent either a static mapping or a dynamic relation, which leads to two kinds of T–S fuzzy systems: static T–S fuzzy system and dynamic T–S fuzzy system, which will be presented in details in Chap. 2.

Using *singleton fuzzification* and *product* for the “and” operation, the *firing strength* of each rule is calculated by

$$\lambda_i(x) = \prod_{j=1}^{L} \mu_{F_j^i}(x_j), \tag{1.46}$$

where $x = [x_1, x_2, \dots, x_L]^T$.

The overall output of the fuzzy model (1.45) is inferred by taking the *weighted average* of all local models:

$$y = \frac{\sum_{i=1}^{N} \lambda_i(x) g_i(x)}{\sum_{i=1}^{N} \lambda_i(x)}, \tag{1.47}$$

which can be further formulated into

$$y = \sum_{i=1}^{N} \mu_i(x) g_i(x), \tag{1.48}$$

where $\mu_i(x)$ is the *normalized firing strength*

$$\mu_i(x) = \frac{\lambda_i(x)}{\sum_{i=1}^{N} \lambda_i(x)}. \tag{1.49}$$

A more extensive introduction to T–S fuzzy systems including static and dynamic T–S fuzzy systems, their universal approximation property, stability conditions, stabilization control, and state and output tracking control designs will be presented in Chap. 2.

1.3 Fuzzy System Identification

In the early approaches, most of fuzzy identification problems were dealt with by preselecting structures and then adjusting membership functions by *trial-and-error*. A landmark approach on synthesizing an algorithm that tackles structure and parameter identification of T–S fuzzy models was made by Takagi and Sugeno (1985). Since then, a number of algorithms for structure and/or parameter determination from I/O data have been proposed. A successive identification for synthesizing a T–S fuzzy model was proposed by Sugeno and Kang (1988), which determined the model structure by combining least-square estimation (LSE) with an unbiasedness criterion and adjusted parameters using a weighted recursive least-square estimation (wRLSE). In the early 1990s, the idea of tuning the parameters of fuzzy models using I/O data attracted much research interest. Inspired by the award winning paper on the identification of neural networks using static or dynamic back propagation gradient descent (BP/GD) learning algorithm (Narendra and Parthasarathy 1990), a BP/GD algorithm for training fuzzy models was developed based on the basic concept that the BP algorithm can be applied to any feedforward networks (Wang and Mendel 1992). Jang (1993) proposed a fuzzy-neuro method for tuning parameters of T–S fuzzy models. Yager and Filev proposed a unified structure and parameter identification approach to identify fuzzy models (Yager and Filex 1993) and the mountain clustering method for fuzzy rule generation (Yager and Filex 1994). Following Yager and Filev's work, Chiu (1994) proposed a modified mountain clustering (MMC) method for fuzzy model identification. Jang (1993) proposed an adaptive network-based fuzzy inference system (ANFIS). Sun (1994) extended Jang's ANFIS algorithm to identify structure using the concept of binary box tree. In Barada and Singh (1998), a method to generate optimal adaptive T–S fuzzy models from I/O data was described. This method achieved structure determination by a combination of MMC algorithm, RLSE and GMDH. Parameter adjustment is achieved by training the initial T–S fuzzy model using ANFIS. An algorithm to improve the interpretability of T–S fuzzy models by combing global learning and local learning was proposed in Yen et al. (1998). A novel framework for fuzzy modeling with multivariate membership functions and model-based control design was described in Abonyi et al. (2001).

More recent research shows more interest in supervised fuzzy clustering, probabilistic fuzzy clustering, and online identification of fuzzy models. Supervised fuzzy clustering uses the class label of each data point to identify the optimal set of clusters that describe the data. A supervised fuzzy clustering method for the identification of fuzzy classifiers was proposed in Abonyi and Szeifert (2003). In Liu and Huang (2003), an evolution semi-supervised fuzzy clustering algorithm was described for learning classifier from labeled and unlabeled data. Some drawbacks exist in classical fuzzy clustering algorithms such as fuzzy c-means (FCM) (Bezdek 1976, 1981). When applying standard FCM into fuzzy modeling with Gaussian distributions, the application of Euclidean distance usually equalizes the effect of mean and standard deviation and leads to some rather medium results. A modified FCM algorithm was proposed in Nefti and Oussalah (2004) which used a probabilistic distance structure in standard FCM. In Abonyi et al. (2002), a modified Gath-Geva (MGG) fuzzy clustering, which interpreted the Gath-Geva (GG) clustering algorithm (Gath and Geva 1989) in the probabilistic framework, was proposed to construct interpretable T–S fuzzy models based on the expectation–maximization (EM) identification of Gaussian mixture of models.

Most of the above methods suppose that all the data is available at the start of training process and are appropriate for offline applications only. This form of learning sounds easy to guarantee its success in reaching an optimal solution based on its learning objective function. Furthermore, it has the flexibility in recalling the stored training data to improve the quality of learning. However, the offline learning algorithms often take a long time and need a large amount of memory. The whole set of rules have to be generated from the scratch when new data are coming. Besides, they cannot be applied to an environment where data are acquired online, such as closed-loop control and adaptive systems. Online learning methods, on the other hand, consider training data one by one. Clusters are built up incrementally. Kasabov et al. (2002) proposed an evolving, distance-based connectionist clustering method to partition the input space for the purpose of creating fuzzy inference rules. Lee and Ouyang (2003) designed a self-constructing rule generation algorithm (SCRG) to extract fuzzy rules based on similarity measures.

In Angelov (2002), evolving Rule-based (eR) models used the informative potential of the new data (accumulated spatial proximity information) as trigger to update the rule base, which ensured greater generality of the structure changes. In Angelov and Filev (2004), the concept of eR was further developed with respect to online identification of T–S fuzzy models. Recursive procedures for calculation of the information potential of the new data and of the consequent parameters were introduced, which are vitally important for real-time applications. This method was extended to identify MIMO T–S fuzzy models in Angelov et al. (2004).

In recent decade, more researches have been done on evolving T–S fuzzy models (Kalhor et al. 2013; Lin et al. 2013; Baruah and Angelov 2014; Rastegar et al. 2017). Several approaches based on clustering technique have been proposed for optimal structure selection of the T–S fuzzy models, such as a unified min-max approach to fuzzy clustering, estimation, and identification which attempts to minimize worst-case effect of data uncertainties and modeling errors on estimation performance (Kumar and Stoll 2006), an online local learning with generic fuzzy input T–S fuzzy framework for nonlinear system estimation (Quah and Quek 2006), a structure identification algorithm using dual kernel-based learning machines (Li and Yang 2008), a fuzzy c-regression model clustering algorithm (Li et al. 2009), a hyperplane-shaped clustering (HPSC) model based on the gravitational search algorithm was proposed in Li et al. (2012), and a supervised, hierarchical clustering algorithm which is effective in solving the problem of global model accuracy, together with the interpretability of local models as valid linearizations of the modeled nonlinear system (Hartmann et al. 2011). Besides, Hou et al. (2007) proposed a method using fuzzy average with fuzzy cluster distribution to identify the key variables for complex systems with high-dimensional input space. Li et al. (2017) designed a hyperplane-shaped fuzzy membership function to match HPSC for T–S fuzzy model identification. Tasi and Chen (2018) proposed an identification method for the T–S fuzzy model based on the Xie-Beni index and an improved particle swarm optimization algorithm.

The parameter identification of fuzzy systems can be carried out by either series–parallel configuration or parallel configuration. Banakar and Azeem (2011) made a comparative study on them. Liu et al. (2014) developed a novel cost function which is the combination of these terms constituted of decomposing least-square support vector machine (LS-SVM) and error terms for the identification of T–S fuzzy system.

Most fuzzy identification algorithms consider multiple-input single-output (MISO) fuzzy systems. Multi-input multi-output (MIMO) fuzzy systems are usually treated as a group of MISO systems. Luo et al. (2014) took into account the block structure information in the MIMO T–S fuzzy system and cast the problem of multidimensional output fuzzy model identification as a joint structure sparse optimization problem, where the consequent parameters are estimated with a common block structured sparsity pattern over all dimensions of the output variable.

1.4 Fuzzy System Based Adaptive Control

Control design and analysis for nonlinear systems is always a challenging task, especially when there exist uncertainties in a nonlinear system. Adaptive control provides effective methodologies to deal with parametric uncertainties in linear and nonlinear systems (Goodwin and Sin 1984; Astrom and Wittenmark 1995; Krstić et al. 1995; Ioannou and Sun 1996; Tao 2003; Landau and Lozano et. al. 2011).

Approximation-based adaptive control provides an effective way for control designs for nonlinear systems with functional and parametric uncertainties (Farrell and Polycarpou 2006). Fuzzy systems, as a class of universal approximators (Buckley 1993; Kosko 1994; Zeng and Singh 1994, 1995; Castro 1995; Castro and Delgado 1996; Ying 1998b; Cao et al. 1997a; Robatti 1998; Ying 1999a; Zeng et al. 2000; Rastegar et al. 2017), can represent static nonlinear functions or dynamic nonlinear systems. The former employs fuzzy systems as static nonlinear mapping from input space to output space while the latter treats fuzzy systems, especially dynamic T–S fuzzy systems, as dynamic models of nonlinear systems.

In this section, we give an overview of both static fuzzy system approximator based adaptive control and dynamic T–S fuzzy system based adaptive control.

1.4.1 Fuzzy Systems as Static Function Approximators

In order to deal with the uncertainties of nonlinear systems, in the fuzzy control system literature, a considerable amount of adaptive approximation-based control schemes have been suggested (Wang 1994; Spooner et al. 1997; Chai and Tong 1999; Ordóñez and Passino 1999; Wang and Lin 1999; Chen and Zhang 2000; Jagannathan et al. 2000; Zhang and Bien 2000; Han et al. 2001; Yu and Sun 2001; Gao and Er 2003; Golea et al. 2002a, b; Li and Lee 2003; Li and Tong 2003, 2016; Nounou and Passino 2004; Chiu 2005; Cheng and Chien 2006; Lin and Xu 2006; Phan and Gale 2007; Wang et al. 2007, 2010; Choi 2008; Moustakidis et al. 2008; Zou et al. 2008; Shi 2008; Chiang et al. 2009; Chen et al. 2009, 2010, 2012; Boulkrounea et al. 2010; Hojati and Gazor 2010; Li et al. 2010; Nekoukar and Erfanian 2011; Pan et al. 2011; Theodoridis et al. 2011; Gao et al. 2012; Lin and Li 2012; Tong et al. 2012; Lin et al. 2013; Qi et al. 2013; Shi 2014; Liu and Tong 2014, 2015; Chwa 2015; Long and Zhao 2016; Wu and Yang 2016; Lai et al. 2017; Wiktorowicz 2017; Wu et al. 2017; Zhang et al. 2017).

In the studied approaches, depending on the structure features of the nonlinear systems (e.g., feedback linearizable, strict feedback, pure feedback), different nonlinear controllers (e.g., feedback linearization control, backstepping control, sliding mode control, etc) can be designed based on nonlinear control theory (Nijmeijer and Van der Schaft 1990; Isidori 1995; Khalil 2002). Since there exist unknown nonlinear functions in those nonlinear systems, the nonlinear controllers designed based on those nonlinear functions cannot be implemented in reality. Under such circumstances, fuzzy systems are employed as static function approximators to approximate the unknown nonlinear functions such that nonlinear controllers can be implemented using the approximated values. Depending on whether the fuzzy systems are used to approximate the nonlinear functions in the nonlinear systems or the whole nonlinear controllers, the adaptive fuzzy system approximation-based control can be classified as indirect adaptive fuzzy control and direct adaptive fuzzy control.

The role of fuzzy systems as static function approximators in adaptive approximation-based control schemes is similar to that of other function approximators, such as polynomials, B-Splines, neural networks, etc (Farrell and Polycarpou 2006).

1.4.2 Fuzzy Systems as Dynamic Systems

For dynamic fuzzy system based control designs, the dynamic fuzzy system is treated as a dynamic nonlinear system rather than a static nonlinear mapping function. As a class of nonlinear systems, dynamic T–S fuzzy systems have their own structure features, e.g., each subsystem in a rule is a linear system while the overall T–S fuzzy system is a nonlinear system due to the participation of membership functions. For dynamic T–S fuzzy systems, it is convenient to apply mature linear control theory to design local linear controllers that can stabilize each local linear system. However, how to prove the overall nonlinear controller constructed from local linear controller can make the overall T–S fuzzy system stable is a problem. In the past two decades, massive research work has been done on the stability of T–S fuzzy systems (Tanaka and Sugeno 1992; Cuesta et al. 1999; Johansson et al. 1999; Chadli et al. 2000; Chou and Chen 2001; Liu and Zhang 2003; Pang and Guu 2003; Feng 2004; Sun and Wang 2006; Kalhor et al. 2013; Liu et al. 2014). Besides, stabilization control of T–S fuzzy systems (Tanaka et al. 1996; Wang et al. 1996; Feng et al. 1997; Cao et al. 1995, 1997a, b, 1999; Li and Li 2004; Xiu and Ren 2005; Zhou et al. 2005; Ting 2006) also attracts much attention. In Feng (2006) and Feng (2010), a comprehensive overview is given on stability conditions and stabilization control of dynamic T–S fuzzy systems. In the Lyapunov-based stability analysis and stabilization control designs of T–S fuzzy systems, the selection of Lyapunov functions plays a critical role in relaxing the stability conditions, which usually requires finding solutions to a group of linear matrix inequalities (LMIs). In Chap. 2, we will present some results on stability conditions and stabilization control of T–S fuzzy systems based on single quadratic Lyapunov function, piecewise quadratic Lyapunov functions, and fuzzy quadratic Lyapunov function.

For output tracking control of dynamic T–S fuzzy systems, it is important to derive the relative degree, zero dynamics and define the minimum phase property of T–S fuzzy systems such that feedback linearization control techniques can be applied. The relative degrees of discrete-time T–S fuzzy systems and continuous-time T–S fuzzy systems are derived in Qi et al. (2013b) and Zhang et al. (2017), respectively, where the corresponding zero dynamics and minimum phase property are also defined.

When there exist uncertain parameters in T–S fuzzy systems, T–S fuzzy system based control designs can be combined with adaptive control methodologies to handle parametric uncertainties (Cho et al. 2002; Feng 2002; Feng et al. 2002; Park and Cho 2004; Lin and Xu 2006; Qi and Brdys 2005, 2008; Hyun et al. 2010; Wang et al. 2010; Khanesar et al. 2011; Qi et al. 2011, 2012a, b, 2013b, 2014; Huang et al. 2014; Zhang et al. 2017).

1.5 What This Book Is About

This book is about identification and adaptive control of T–S fuzzy systems.

Identification of T–S fuzzy systems. The identification algorithms introduced in Chap. 4 can be applied to identify either static T–S fuzzy systems or dynamic discrete-time T–S fuzzy systems. Both offline and online identification algorithms will be presented which enable one to obtain the system structure (premise variables and number of rules) and parameters (membership function parameters and consequent parameters) completely from input–output (I/O) data. The T–S fuzzy system identified offline can serve as an initial model for control design whose structure and parameters can be further tuned using online data in closed-loop control.

It is not our intention to give a comprehensive overview on the identification methods for fuzzy systems. We want to present this topic in a tutorial way that is concise and easy to follow for graduate students to construct a T–S fuzzy system from I/O data of a nonlinear plant, which can be used in T–S fuzzy system based adaptive control design.

Adaptive control of T–S fuzzy systems. T–S fuzzy systems are widely used in adaptive approximation-based control, which serve as static nonlinear function approximators. The underlying technical issues of such approaches are similar to other adaptive approximation-based control approaches, such as adaptive neural control approaches.

In this book, we specifically consider adaptive control designs of dynamic T–S fuzzy systems. Dynamic T–S fuzzy systems can be deemed as a class of dynamic nonlinear systems with a special structure. It is well known that the control design for nonlinear systems is related closely to system structure. Dynamic T–S fuzzy systems have several forms: continuous-time state-space form, discrete-time state-space form and discrete-time input–output form and they may be single-input single-output (SISO) systems or multi-input multi-output (MIMO) systems. Different forms of T–S fuzzy systems own different structure characteristics which should by analyzed carefully such that suitable control schemes can be designed.

This book provides a systematic study on how adaptive control schemes can be designed for different forms of dynamic T–S fuzzy systems, covering both continuous-time and discrete-time T–S fuzzy system, both state-space form and input–output form T–S fuzzy systems, both SISO and MIMO T–S fuzzy systems, to achieve both state tracking and output tracking. Some important issues such as relative degree and zero dynamics of T–S fuzzy systems, minimum phase property of T–S fuzzy systems are discussed and defined. Parametrization models of T–S fuzzy systems are developed based on which parameter adaptive laws are derived. Closed-loop stability and tracking performance analysis are provided. The effects of modeling errors and disturbances are also addressed and parameter adaptive laws are modified to guarantee the robustness of parameter adaptation.

This book builds a framework for adaptive control of uncertain nonlinear systems based on dynamic T–S fuzzy systems with rigorous theoretical formulation and detailed design techniques, which is a stimulating text for researchers and graduate students in the area of nonlinear system control and a valuable reference for practical control engineers.

References

Abonyi J, Szeifert F (2003) Supervised fuzzy clustering for the identification of fuzzy classifiers. Pattern Recognit Lett 24:2195–2207

Abonyi J, Babuska R, Szeifert F (2001) Fuzzy modelling with multivariable membership functions: grey-box identification and control design. IEEE Trans Syst Man Cybern 31:755–767

Abonyi J, Babuska R, Szeifert F (2002) Modified Gath-Geva fuzzy clustering for identification of Takagi–Sugeno fuzzy models. IEEE Trans Syst Man Cybern 32(5):755–767

Angelov PP (2002) Evolving rule-based models: a tool for design of flexible adaptive systems. Springer-Verlag, Heidelberg

Angelov PP, Filev DP (2004) An approach to online identification of Takagi–Sugeno fuzzy models. IEEE Trans Syst Man Cybern Part B: Cybern 34(1):484–498

Angelov PP, Xydeas C, Filev D (2004) On-line identification of MIMO evolving Takgi–Sugeno fuzzy models. In: Proceedings of IEEE fuzzy system, Budapest, Hungary, pp 55–60

Astrom K, Wittenmark B (1995) Adaptive control, 2nd edn. Addison-wesley, Reading

Banakar A, Azeem MF (2011) Parameter identification of TSK neuro-fuzzy models. IEEE Trans Syst Man Cybern Part C: Appl Rev 28(3):297–313

Barada S, Singh H (1998) Generating optimal adaptive fuzzy-neural models of dynamical systems with applications to control. IEEE Trans Syst Man Cybern - Part C: Appl Rev 28(3):297–313

Baruah RD, Angelov PP (2014) Dynamically evolving clustering and its application to structure identification of evolving fuzzy models. IEEE Trans Cybern 44(9):1619–1631

Bezdek J (1976) A physical interpretation of fuzzy isodata. IEEE Trans Syst Man Cybern 387–389

Bezdek J (1981) Pattern recognition with fuzzy objective function. Plenum Press, New York

Boulkrounea A, Tadjineb M, M'Saadc M, Farzac M (2010) Fuzzy adaptive controller for MIMO nonlinear systems with known and unknown control direction. Fuzzy Sets Syst 161:797–820

Buckley JJ (1993) Sugeno type controllers are universal controllers. Fuzzy Sets Syst 53:299–303

Cao SG, Rees NW, Feng G (1995) Analysis and design of fuzzy control systems using dynamic fuzzy global models. Fuzzy Sets Syst 75:47–62

Cao SG, Rees NW, Feng G (1997a) Analysis and design for a class of complex control systems - Part I: fuzzy modeling and identification. Automatica 33:1017–1028

Cao SG, Rees NW, Feng G (1997b) Analysis and design for a class of complex control systems, Part II: fuzzy controller design. Automatica 33:1029–1039

Cao SG, Rees NW, Feng G (1999) Analysis and design of fuzzy control systems using dynamic fuzzy-state space models. IEEE Trans Fuzzy Syst 7(2):192–200

Castro JL (1995) Fuzzy logic controllers are universal approximators. IEEE Trans Syst Man Cybern 25:629–635

Castro JL, Delgado M (1996) Fuzzy systems with defuzzification are universal approximators. IEEE Trans Syst Man Cybern 26:149–152

Chadli M, Maquin D, Ragot J (2000) Relaxed stability conditions for Takagi–Sugeno fuzzy systems. In: Proceedings of IEEE international conference on systems, man, and cybernetics, Nashville, TN, pp 3514–3519

Chai T, Tong S (1999) Fuzzy direct adaptive control for a class of nonlinear systems. Fuzzy Sets Syst 103(3):379–387

Cheng C-C, Chien S-H (2006) Adaptive sliding mode controller design based on T–S fuzzy system models. Automatica 42:1005–1010
Chen W, Zhang Z (2000) Globally stable adaptive backstepping fuzzy control for output-feedback systems with unknown high-frequency gain sign. Fuzzy Sets Syst 161(6):821–836
Chen B, Liu X, Liu K, Lin C (2009) Direct adaptive fuzzy control of nonlinear strict-feedback systems. Automatica 45(6):1530–1535
Chen WL, Jiao R, Li J (2010) Adaptive backstepping fuzzy control for nonlinearly parameterized systems with periodic disturbances. IEEE Trans Fuzzy Syst 18(4):674–685
Chen B, Liu X, Ge SS, Lin C (2012) Adaptive fuzzy control of a class of nonlinear systems by fuzzy approximation approach. IEEE Trans Fuzzy Syst 20(6):1012–1021
Chiang T-S, Chiu C-S, Liu P (2009) Adaptive TS-FNN control for a class of uncertain multi-time-delay systems: the exponentially stable sliding mode-based approach. Int J Adapt Control Signal Process 23(4):378–399
Chiu SL (1994) Fuzzy model identification based on cluster estimation. J Intell Fuzzy Syst 2:267–278
Chiu C-S (2005) Robust adaptive control of uncertain MIMO non-linear systems - feedforward Takagi–Sugeno fuzzy approximation based approach. IEE Proc Control Theory Appl 152(2):157–164
Cho Y-W, Park C-W, Park M (2002) An indirect model reference adaptive fuzzy control for SISO Takagi–Sugeno model. Fuzzy Sets Syst 131:197–215
Choi HH (2008) Adaptive controller design for uncertain fuzzy systems using variable structure control approach. Automatica 45(11):2646–2650
Chou J-H, Chen S-H (2001) Stability analysis of the discrete Takagi–Sugeno fuzzy model with time-varying consequent uncertainties. Fuzzy Sets Syst 118:271–279
Chwa D (2015) Fuzzy adaptive output feedback tracking control of VTOL aircraft with uncertain input coupling and input-dependent disturbances. IEEE Trans Fuzzy Syst 23(5):1505–1518
Cuesta F, Gordillo F et al (1999) Stability analysis of nonlinear multivariable Takagi–Sugeno fuzzy control systems. IEEE Trans Fuzzy Syst 7(5):508–520
Farrell JA, Polycarpou MM (2006) Adaptive approximation based control: unifying neural, fuzzy and traditional adaptive approximation approaches. Wiley, New York
Feng G (2002) An approach to adaptive control of fuzzy dynamic systems. IEEE Trans Fuzzy Syst 10(2):268–275
Feng G (2004) Stability analysis of discrete-time fuzzy dynamic systems based on piecewise Lyapunov functions. IEEE Trans Fuzzy Syst 12(1):22–28
Feng G (2006) A survey on analysis and design of model-based fuzzy control systems. IEEE Trans Fuzzy Syst 14(5):676–697
Feng G (2010) Analysis and synthesis of fuzzy control systems: a model-based approach. CRC Press, Boca Raton
Feng G, Cao SG et al (1997) Design of fuzzy control systems with guaranteed stability. Fuzzy Sets Syst 85:1–10
Feng G, Cao SG, Rees NW (2002) Stable adaptive control of fuzzy dynamic systems. Fuzzy Sets Syst 131(2):217–224
Gao Y, Er MJ (2003) Online adaptive fuzzy neural identification and control of a class of MIMO nonlinear systems. IEEE Trans Fuzzy Syst 11(4):462–477
Gao Q, Feng G, Wang Y, Qiu J (2012) Universal fuzzy controllers based on generalized T–S fuzzy models. Fuzzy Sets Syst 201:55–70
Gath I, Geva AB (1989) Unsupervised optimal fuzzy clustering. IEEE Trans Pattern Mach Intell 7:773–781
Golea N, Golea A, Benmahammed K (2002a) Fuzzy model reference adaptive control. IEEE Trans Fuzzy Syst 10(4):436–444
Golea N, Golea A, Benmahammed K (2002b) Stable indirect fuzzy adaptive control. Fuzzy Sets Syst 137:353–366

Goodwin GC, Sin KS (1984) Adaptive filtering prediction and control. Prentice-Hall, Englewood Cliffs
Han H, Su C-Y, Stepanenko Y (2001) Adaptive control of a class of nonlinear systems with nonlinearly parameterized fuzzy approximators. IEEE Trans Fuzzy Syst 9(2):315–323
Hartmann B, Bänfer O et al (2011) Supervised hierarchical clustering in fuzzy model identification. IEEE Trans Fuzzy Syst 9(6):1163–1176
Hojati M, Gazor S (2010) Hybrid adaptive fuzzy identification and control of nonlinear systems. IEEE Trans Fuzzy Syst 10:198–210
Hou Y, Zurada JM et al (2007) Identification of key variables using fuzzy average with fuzzy cluster distribution. IEEE Trans Fuzzy Syst 15(4):673–685
Huang Y, Qi R, Tao G (2014) An adaptive state tracking control scheme for T–S fuzzy models in non-canonical form and with uncertain parameters. J Frankl Inst 351(7):3610–3632
Hyun C-H, Park C-W, Kim S (2010) Takagi–Sugeno fuzzy model based indirect adaptive fuzzy observer and controller design. Inf Sci 180(11):2314–2327
Ioannou PA, Sun J (1996) Robust adaptive control. Prentice-Hall, Englewood Cliffs
Isidori A (1995) Nonlinear control systems, 3rd edn. Springer-Verlag, London
Jagannathan S, Vandegrift MW, Lewis FL (2000) Adaptive fuzzy ontrol of discrete-time dynamical systems. Automatica 36(2):229–241
Jang JS (1993) ANFIS: adaptive-network-based fuzzy inference system. IEEE Trans Syst Man Cybern 23(3):665–685
Johansson M, Rantzer A, Årzé K-E (1999) Piecewise quadratic stability of fuzzy systems. IEEE Trans Fuzzy Syst 7(6):713–722
Kalhor A, Araabi BN, Lucas C (2013) Piecewise quadratic stability of fuzzy systems. Appl Soft Comput 13:939–946
Kasabov NK, Song Q (2002) DENFIS: dynamic evolving neural-fuzzy inference system and its application for time-series prediction. IEEE Trans Fuzzy Syst 10:144–154
Khalil HK (2002) Nonlinear systems. Prentice-Hall, Upper Saddle River
Khanesar MA, Kaynak O, Teshnehlab M (2011) Direct model reference Takagi–Sugeno fuzzy control of SISO nonlinear systems. IEEE Trans Fuzzy Syst 19(5):914–924
Kosko B (1994) Fuzzy systems as universal approximators. IEEE Trans Comput 43:1329–1333
Krstić M, Kanellakopoulos I, Kokotović P (1995) Nonlinear and adaptive control design. Wiley, New York
Kumar M, Stoll R, Stoll N (2006) A min-max approach to fuzzy clustering, estimation, and identification. IEEE Trans Fuzzy Syst 14(2):248–262
Lai G, Liu Z, Zhang Y et al (2017) Adaptive inversion-based fuzzy compensation control of uncertain pure-feedback systems with asymmetric actuator backlash. IEEE Trans Fuzzy Syst 25(1):141–155
Landau ID, Lozano R, M'Saad M, Karimi A (2011) Adaptive control algorithms, analysis and applications. Springer-Verlag, London
Lee S-J, Ouyang C-S (2003) A neuralfuzzy system modelling with self-construcing rule generation and hybrid SVD based leanring. IEEE Trans Fuzzy Syst 11:341–353
Li C, Lee C-Y (2003) Self-organizing neuro-fuzzy system for control of unknown plants. IEEE Trans Fuzzy Syst 11(1):135–150
Li N, Li S-Y (2004) Stability analysis and design of T–S fuzzy control system with simplified linear rule consequent. IEEE Trans Syst Man Cybern B 34(1):788–795
Li H-X, Tong S (2003) A hybrid adaptive fuzzy control for a class of nonlinear MIMO systems. IEEE Trans Fuzzy Syst 11(1):24–34
Li Y, Tong S (2016) Hybrid adaptive fuzzy control for uncertain MIMO nonlinear systems with unknown dead-zones. Inf Sci 328(20):97–114
Li W, Yang Y (2008) A new approach to TS fuzzy modeling using dual Kernel-based learning machines. Neurocomputing 71:3660–3665
Li C, Zhou J, Xiang X, Li Q, An X (2009) T–S fuzzy model identification based on a novel fuzzy c-regression model clustering algorithm. Inf Sci 165:247–263

Li T-S, Tong S-C, Feng G (2010) A novel robust adaptive-fuzzy-tracking control for a class of nonlinear multi-input/multi-output systems. IEEE Trans Fuzzy Syst 18(1):150–160
Li C, Zhou J, Fu B, Kou P, Xiao J (2012) T–S fuzzy model identification with a gravitational search-based hyperplane clustering algorithm. IEEE Trans Fuzzy Syst 20(2):305–317
Li C, Zhou J et al (2017) T–S fuzzy model identification based on a novel hyperplane-shaped membership function. IEEE Trans Fuzzy Syst 25(5):1364–1370
Lilly JH (2010) Fuzzy control and identification. Wiley, Hoboken
Lin C-M, Li H-Y (2012) TSK fuzzy CMAC-based robust adaptive backstepping control for uncertain nonlinear systems. IEEE Trans Fuzzy Syst 20(6):1147–1154
Lin CJ, Xu Y-J (2006) The design of TSK-type fuzzy controllers using a new hybrid learning approach. Int J Adapt Control Signal Process 20(1):1–25
Lin Y-Y, Chang J-Y, Lin C-T (2013) Identification and prediction of dynamic systems using an interactively recurrent self-evolving fuzzy neural network. IEEE Trans Neural Netw Learn Syst 24(2):310–321
Liu H, Huang S-T (2003) Evolutionary semi-supervised fuzzy clusering. Pattern Recognit Lett 24:3105–3113
Liu Y-J, Tong S (2014) Adaptive fuzzy control for a class of nonlinear discrete-time systems with backlash. IEEE Trans Fuzzy Syst 22(5):1359–1365
Liu Y-J, Tong S (2015) Adaptive fuzzy control for a class of unknown nonlinear dynamical systems. Fuzzy Sets Syst 263(15):49–70
Liu X, Zhang Q (2003) Approaches to quadratic stability conditions and H1 control design for T–S fuzzy systems. IEEE Trans Fuzzy Syst 11(6):830–839
Liu X, Fang H, Chen Z (2014) Approaches to quadratic stability conditions and H1 control design for T–S fuzzy systems. IET Control Theory Appl 8(5):338–347
Long L, Zhao J (2016) Adaptive fuzzy output-feedback dynamic surface control of MIMO switched nonlinear systems with unknown gain signs. Fuzzy Sets Syst 302:27–51
Luo M, Sun F, Liu H (2014) Joint block structure sparse representation for multi-input-multi-output (MIMO) T–S fuzzy system identification. IEEE Trans Fuzzy Syst 22(6):1387–1400
Moustakidis SP, Rovithakis GA, Theocharis JB (2008) An adaptive neuro-fuzzy tracking control for multi-input nonlinear dynamic systems. Automatica 44(5):1418–1425
Narendra KS, Parthasarathy K (1990) Identification and control of dynamical systems using neural networks. IEEE Trans Neural Netw 1:4–27
Nefti S, Oussalah M (2004) Probabilistic-fuzzy clustering algorithm. In: Proceedings of IEEE international conference on systems, man, and cybernetics, vol 5. Hague, Netherlands, pp 4786-4791. 10–13 Oct 2004
Nekoukar V, Erfanian A (2011) Adaptive fuzzy terminal sliding mode control for a class of MIMO uncertain nonlinear systems. Fuzzy Sets Syst 179(1):34–49
Nijmeijer H, Van der Schaft A (1990) Nonlinear dynamical control systems. Spring-Verlag, New York
Nounou HN, Passino KM (2004) Stable auto-tuning of adaptive fuzzy/neural controllers for non-linear discrete-time systems. IEEE Trans Fuzzy Syst 12(1):70–83
Ordóñez R, Passino KM (1999) Stable multi-input multi-output adaptive fuzzy/neural control. IEEE Trans Fuzzy Syst 7(3):345–353
Pan Y, Er MJ, Huang D, Wang Q (2011) Adaptive fuzzy control with guaranteed convergence of optimal approximation error. IEEE Trans Fuzzy Syst 19(5):807–818
Pang C-T, Guu S-M (2003) Sufficient conditions for the stability of linear Takagi–Sugeno free fuzzy systems. IEEE Trans Fuzzy Syst 11(5):695–700
Park C-W, Cho Y-W (2004) T–S model based indirect adaptive fuzzy control using online parameter estimation. IEEE Trans Syst Man Cybern - Part B: Cybern 34(6):2293–2302
Passino KM, Yurkovich S (1998) Fuzzy control. Addison Wesley Longman Inc.,
Phan PA, Gale T (2007) Two-mode adaptive fuzzy control with approximation error estimator. IEEE Trans Fuzzy Syst 15(5):943–955

Qi R, Brdys MA (2005) Adaptive fuzzy modelling and control for discrete-time nonlinear uncertain systems. In: Proceedings of American control conference 2005, Portland, US, pp 1108–1113
Qi R, Brdys M (2008) Stable indirect adaptive control based on discrete-time T–S fuzzy model. Fuzzy Sets Syst 159(8):900–925
Qi R, Tao G, Tan C (2011) An adaptive control scheme for discrete-time T–S fuzzy systems with unknown membership parameters. In: Proceedings of the 8th Asia control conference, Taiwan, pp 1164–1169
Qi R, Tao G, Jiang B, Tan C (2012a) Adaptive control schemes for discrete-time T–S fuzzy systems with unknown parameters and actuator failures. IEEE Trans Fuzzy Syst 20:471–486
Qi R, Tao G, Tan C, Yao X (2012b) Adaptive prediction and control of discrete-time T–S fuzzy systems. Int J Adapt Control Signal Process 26(7):560–575
Qi R, Zhu L, Jiang B (2013a) Fault-tolerant reconfigurable control for MIMO systems using online fuzzy identification. Int J Innov Comput Inf Control 9(10):3915–3928
Qi R, Tao G, Tan C, Yao X (2013b) Adaptive control of discrete-time state-space T–S fuzzy systems with general relative degree. Fuzzy Sets Syst 217:22–40
Qi R, Tao G, Jiang B (2014) Adaptive control of MIMO time-varying systems with indicator function based parametrization. Automatica 50(5):1369–1380
Quah KH, Quek C (2006) FITSK: online local learning with generic fuzzy input Takagi–Sugeno–Kang fuzzy framework for nonlinear system estimation. IEEE Trans Syst Man Cybern 36(1):166–178
Rastegar S, Araújo R, Mendes J (2017) Fuzzy piecewise multilinear and piecewise linear systems as universal approximators in Sobolev norms. Appl Math Model 45:606–620
Rovatti R (1998) Fuzzy piecewise multilinear and piecewise linear systems as universal approximators in Sobolev norms. IEEE Trans Fuzzy Syst 6:235–249
Shi W (2008) Indirect adaptive fuzzy control for a class of nonlinear discrete-time systems. J Syst Eng Electron 19(6):1203–1207
Shi W (2014) Adaptive fuzzy control for MIMO nonlinear systems with nonsymmetric control gain matrix and unknown control direction. IEEE Trans Fuzzy Syst 22(5):1288–1300
Spooner TJ, Ordonez R, Passino KM(1997) Indirect adaptive fuzzy control for a class of discrete-time systems. In: Proceedings of ACC, Albuquerque, NM, pp 3311–3315
Sugeno M, Kang GT (1988) Structure identification of fuzzy model. Fuzzy Sets Syst 28:15–33
Sun CT (1994) Rule-base structure identification in an adaptive-network-based fuzzy inference system. IEEE Trans Fuzzy Syst 2:64–73
Sun C-H, Wang W-J (2006) An improved stability criterion for T–S fuzzy discrete systems via vertex expression. IEEE Trans Syst Man Cybern B 36(3):672–678
Takagi T, Sugeno M (1985) Fuzzy identification of systems and its applications to modeling and control. IEEE Trans Syst Man Cybern 15(1):116–132
Tanaka K, Sugeno M (1992) Stability analysis and design of fuzzy control systems. Fuzzy Sets Syst 45:135–156
Tanaka K, Ikeda T, Wang HO (1996) Robust stabilization of a class of uncertain nonlinear systems via fuzzy control: quadratic stabilizability, H^{∞} control theory, and linear matrix inequalities. IEEE Trans Fuzzy Syst 4:1–13
Tao G (2003) Adaptive control design and analysis. Wiley, New York
Theodoridis DC, Boutalis YS, Christodoulou MA (2011) Indirect adaptive control of unknown multi variable nonlinear systems with parametric and dynamic uncertainties using a new neuro-fuzzy system description. Int J Neural Syst 20(2):129–148
Ting C-S (2006) Stability analysis and design of Takagi–Sugeno fuzzy systems. Inf Sci 176:2817–2845
Tong S, Li M, Shi P (2012) Observer-based adaptive fuzzy backstepping output feedback control of uncertain MIMO pure-feedback nonlinear systems. IEEE Trans Fuzzy Syst 20(4):771–785
Tsai S-H, Chen Y-W (2018) A novel identification method for Takagi–Sugeno fuzzy model.pdf. Fuzzy Sets Syst 338:117–135

Wang L-X (1994) Adaptive fuzzy systems and control-design and stability analysis. Prentice-Hall, Englewood Cliffs

Wang WJ, Lin HR (1999) Fuzzy control design for the trajectory tracking on uncertain nonlinear systems. IEEE Trans Fuzzy Syst 7(1):53–62

Wang LX, Mendel JM (1992) Fuzzy basis functions, universal approximation, and orthogonal least-square learning. IEEE Trans Neural Netw 3:807–814

Wang HO, Tanaka K, Griffin MF (1996) An approach to fuzzy control of nonlinear systems: stability and design issues. IEEE Trans Fuzzy Syst 4:14–23

Wang M, Chen B, Dai S-L (2007) Direct adaptive fuzzy tracking control for a class of perturbed strict-feedback nonlinear systems. Fuzzy Sets Syst 158(24):2655–2670

Wang W-Y, Chien Y-H, Leu Y-G, Lee T-T (2010) Adaptive T–S fuzzy-neural modeling and control for general MIMO unknown nonaffine nonlinear systems using projection update laws. Automatica 46:852–863

Wiktorowicz K (2017) Design of state feedback adaptive fuzzy controllers for second-order systems using a frequency stability criterion. IEEE Trans Fuzzy Syst 25(3):499–510

Wu L-B, Yang G-H (2016) Adaptive fuzzy tracking control for a class of uncertain nonaffine nonlinear systems with dead-zone inputs. Fuzzy Sets Syst 290:1–21

Wu T-S, Karkoub M, Wang H et al (2017) Robust tracking control of MIMO underactuated nonlinear systems with dead-zone band and delayed uncertainty using an adaptive fuzzy control. IEEE Trans Fuzzy Syst 25(4):905–918

Xiu Z-H, Ren G (2005) Stability analysis and systematic design of Takagi–Sugeno fuzzy control systems. Fuzzy Sets Syst 151:119–138

Yager RR, Filev DP (1993) Unified structure and parameter identification of fuzzy models. IEEE Trans Syst Man Cybern 23:1198–1205

Yager RR, Filev DP (1994) Approximate clustering via the mountain method. IEEE Trans Syst Man Cybern 24:1279–1284

Yen J, Wang L, Gillespie CW (1998) Improving the interpretability of TSK fuzzy models by combining global learning and local learning. IEEE Trans Fuzzy Syst 6:530–537

Ying H (1998b) Sufficient conditions on uniform approximation of multivariate functions by general Takagi–Sugeno fuzzy systems with linear rule consequent. IEEE Trans Syst Man Cybern 28:515–520

Ying H (1999) Comparison of necessary conditions for typical Takagi–Sugeno and Mamdani fuzzy systems as universal approximators. IEEE Trans Syst Man Cybern 29:508–514

Yu W-S, Sun C-J (2001) Fuzzy model based adaptive control for a class of nonlinear systems. IEEE Trans Fuzzy Syst 9:41

Zadeh LA (1965) Fuzzy sets. Inf Control 8(3):338–353

Zeng XJ, Singh MG (1994) Approximation theory of fuzzy systems - SISO case. IEEE Trans Fuzzy Syst 2:162–176

Zeng XJ, Singh MG (1995) Approximation theory of fuzzy systems - MIMO case. IEEE Trans Fuzzy Syst 3(2):219–235

Zeng K, Zhang N-Y, Xu W-L (2000) A comparative study on sufficient conditions for Takagi–Sugeno fuzzy systems as universal approximators. IEEE Trans Fuzzy Syst 8(6):773–780

Zhang H, Bien Z (2000) Adaptive fuzzy control of MIMO nonlinear systems. Fuzzy Sets Syst 115(2):191–204

Zhang H, Cui Y, Wang Y (2017) Hybrid fuzzy adaptive fault-tolerant control for a class of uncertain nonlinear systems with unmeasured states. IEEE Trans Fuzzy Syst 25(5):1041–1050

Zhang YJ, Tao G, Chen M (2017) Parametrization and adaptive control of multivariable non-canonical T–S fuzzy systems. IEEE Trans Fuzzy Syst 25(1):156–171

Zhou SS, Feng G, Lam J, Xu SY (2005) Robust H-infinity control for discrete fuzzy systems via basis-dependent Lyapunov functions. Inf Sci 174(3–4):197–217

Zou A-M, Hou Z-G, Tan M (2008) Adaptive control of a class of nonlinear pure-feedback systems using fuzzy backstepping approach. IEEE Trans Fuzzy Syst 16(4):886–897

Chapter 2
T–S Fuzzy Systems

This chapter presents an overview of ideas and techniques of Takagi–Sugeno (T–S) fuzzy systems. In Chap. 1, we have introduced the basic concepts of fuzzy sets, fuzzy logic, and fuzzy inference mechanism. This chapter aims to present a necessarily selective review on T–S fuzzy systems including their architectures, important properties, and applicability in the field of nonlinear system identification and control with particular emphasis on the issues which are directly related to the main topics addressed in the following chapters.

For readers who want to know more on fuzzy sets, fuzzy logic, fuzzy inference system, and fuzzy control, here are some good reference books. An introductory-level exposure to fuzzy set, fuzzy logic, fuzzy modeling, and control can be gained by reading the book Lilly (2010), which is very suitable for beginners. A pragmatic engineering approach to the design, analysis, performance evaluation, and implementation of fuzzy control systems can be found in the book (Passino and Yurkovich 1998). A unique reference devoted to the systematic analysis and synthesis of model-based fuzzy control systems is offered by Feng (2010).

2.1 Static T–S Fuzzy Systems

In Sect. 1.2.2, we have presented a brief introduction to general T–S fuzzy systems in the form (1.45), that is,

$$\begin{aligned} R^i: \ & \text{IF } \xi_1 \text{ is } F_1^i \text{ and } \xi_2 \text{ is } F_2^i \text{ and } \ \ldots \ \text{ and } \xi_L \text{ is } F_L^i \\ & \text{THEN } y = g_i(\xi_1, \xi_2, \ldots, \xi_L), \end{aligned} \tag{2.1}$$

where R^i denotes the ith fuzzy rule, $\xi_1, \xi_2, \ldots, \xi_L$ are premise (input) variables on the universe of discourses $\mathscr{X}_1$, $\mathscr{X}_2$, …, $\mathscr{X}_L$, respectively, y is an output variable on

R. Qi et al., *Fuzzy System Identification and Adaptive Control*, Communications and Control Engineering,
https://doi.org/10.1007/978-3-030-19882-4_2

the university of discourse $\mathscr{Y}$, g_i is a crisp function, and F_j^i denotes a fuzzy set for ξ_j on $\mathscr{X}_j$, $i = 1, 2, \ldots, N, j = 1, \ldots, L$.

Remark 2.1 In order to distinguish premise variables and state variables in a fuzzy system, from now on, we use $\xi_1, \xi_2, \ldots$ to denote premise variables and $x_1, x_2, \ldots$ to denote state variables.

In a static T–S fuzzy system, the function g_i is a static mapping from $\mathscr{X}_1 \times \mathscr{X}_2 \times \cdots \times \mathscr{X}_L$ to $\mathscr{Y}$, which usually takes the following form:

$$\begin{aligned} R^i :\ & \text{IF } \xi_1 \text{ is } F_1^i \text{ and } \xi_2 \text{ is } F_2^i \text{ and } \ \ldots \ \text{ and } \xi_L \text{ is } F_L^i \\ & \text{THEN } y = \theta_{i,0} + \theta_{i,1}\xi_1 + \cdots + \theta_{i,L}\xi_L, \end{aligned} \tag{2.2}$$

where $\theta_{i,0}, \theta_{i,1}, \ldots, \theta_{i,L}$ are coefficients of the ith function. If $\theta_{i,0} = 0$, the consequent part represents a linear mapping and if $\theta_{i,0} \neq 0$, the mapping is called "affine".

Using *singleton fuzzification*, *product inference*, and *weighted average*, the fuzzy model (2.2) can be transformed into the following global model:

$$y = \sum_{i=1}^{N} \mu_i(\xi)(\theta_{i,0} + \theta_{i,1}\xi_1 + \cdots + \theta_{i,L}\xi_L), \tag{2.3}$$

where $\xi = [\xi_1, \xi_2, \ldots, \xi_L]^T$, $\mu_i(\xi)$ is the *normalized firing strength* of the ith rule:

$$\mu_i(\xi) = \frac{\lambda_i(\xi)}{\sum_{i=1}^{N} \lambda_i(\xi)} \tag{2.4}$$

which is calculated from the *firing strength* $\lambda_i(\xi)$:

$$\lambda_i(\xi) = \prod_{j=1}^{L} F_j^i(\xi_j). \tag{2.5}$$

Remark 2.2 To maintain notational simplicity, for the rest of this book, we use F_j^i to represent a fuzzy set and $F_j^i(\xi_j)$ to represent its membership function.

The membership functions $F_j^i(\xi_j)$ can be chosen as typical fuzzy membership functions such as triangular, trapezoidal, Gaussian functions, etc. For example, a Gaussian membership function has the following form:

$$F_j^i(\xi_j) = \exp\left[-\frac{(\xi_j - c_j^i)^2}{2\sigma_j^{i\,2}}\right], \tag{2.6}$$

where c_j^i and σ_j^i are the center and radius of Gaussian function, respectively.

From the definition of μ_i in (2.4), it can be concluded that μ_i, $i = 1, 2, \ldots, N$ satisfy the following properties:

$$\mu_i \geq 0, \quad \sum_{i=1}^{N} \mu_i = 1. \tag{2.7}$$

It can be observed that the global T–S fuzzy system (2.3) essentially performs a nonlinear interpolation between the linear/affine mappings. $\mu_i(\xi)$ plays a role to determine the weight of each linear/affine mapping in the overall nonlinear mapping.

Example 2.1 Consider two variables $\xi_1 \in [-1, 1]$ and $\xi_2 \in [-2, 2]$. For each variable, two fuzzy sets are defined on their universe of discourses, that is, $[-1, 1]$ and $[-2, 2]$ respectively. Specifically, the fuzzy sets for ξ_1 are F_1^1 and F_1^2 and for ξ_2 are F_2^1 and F_2^2. Then, a static T–S fuzzy system with four rules ($N = 4$) can be formed:

$$\begin{aligned}
R^1 &: \text{ IF } \xi_1 \text{ is } F_1^1 \text{ and } \xi_2 \text{ is } F_2^1, \text{ THEN } y = 1 + 2\xi_1 + \xi_2, \\
R^2 &: \text{ IF } \xi_1 \text{ is } F_1^1 \text{ and } \xi_2 \text{ is } F_2^2, \text{ THEN } y = 2 + \xi_1 - 0.5\xi_2, \\
R^3 &: \text{ IF } \xi_1 \text{ is } F_1^2 \text{ and } \xi_2 \text{ is } F_2^1, \text{ THEN } y = 3 - 2\xi_1 + \xi_2, \\
R^4 &: \text{ IF } \xi_1 \text{ is } F_1^2 \text{ and } \xi_2 \text{ is } F_2^2, \text{ THEN } y = 2 + 0.5\xi_1 - \xi_2.,
\end{aligned} \tag{2.8}$$

where the fuzzy sets F_1^1, F_1^2, F_2^1, and F_2^2 are characterized by the following Gaussian membership functions:

$$F_1^1(\xi_1) = \begin{cases} \exp\left[-\frac{1}{2}\left(\frac{\xi_1+1}{0.8}\right)^2\right], & \text{if } \xi_1 \geq -1 \\ 1, & \text{otherwise} \end{cases} \tag{2.9}$$

$$F_1^2(\xi_1) = \begin{cases} \exp\left[-\frac{1}{2}\left(\frac{\xi_1-1}{0.8}\right)^2\right], & \text{if } \xi_1 \leq 1 \\ 1, & \text{otherwise} \end{cases} \tag{2.10}$$

$$F_2^1(\xi_2) = \begin{cases} \exp\left[-\frac{1}{2}\left(\frac{\xi_2+2}{1.5}\right)^2\right], & \text{if } \xi_2 \geq -2 \\ 1, & \text{otherwise} \end{cases} \tag{2.11}$$

$$F_2^2(\xi_2) = \begin{cases} \exp\left[-\frac{1}{2}\left(\frac{\xi_2-2}{1.5}\right)^2\right], & \text{if } \xi_2 \leq 2 \\ 1, & \text{otherwise} \end{cases} \tag{2.12}$$

which are shown in Figs. 2.1 and 2.2.

The firing strengths of the four rules are calculated by

$$\begin{aligned}
\lambda_1(\xi) &= F_1^1(\xi_1)F_2^1(\xi_2), \quad \lambda_2(\xi) = F_1^1(\xi_1)F_2^2(\xi_2), \\
\lambda_3(\xi) &= F_1^2(\xi_1)F_2^1(\xi_2), \quad \lambda_4(\xi) = F_1^2(\xi_1)F_2^2(\xi_2).
\end{aligned} \tag{2.13}$$

The normalized firing strength μ_i can be obtained using

$$\mu_i(\xi) = \frac{\lambda_i(\xi)}{\sum_{i=1}^{N} \lambda_i(\xi)}, \quad i = 1, 2, 3, 4. \tag{2.14}$$

The overall output of the static T–S fuzzy system is obtained as

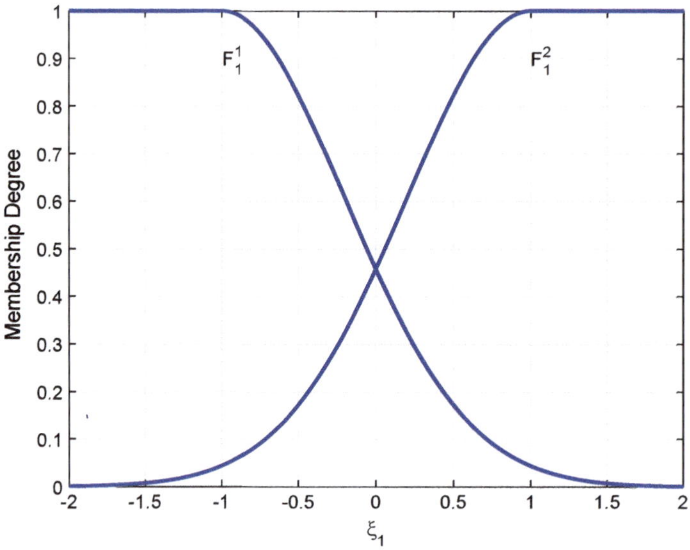

Fig. 2.1 Membership functions of F_1^1 and F_1^2

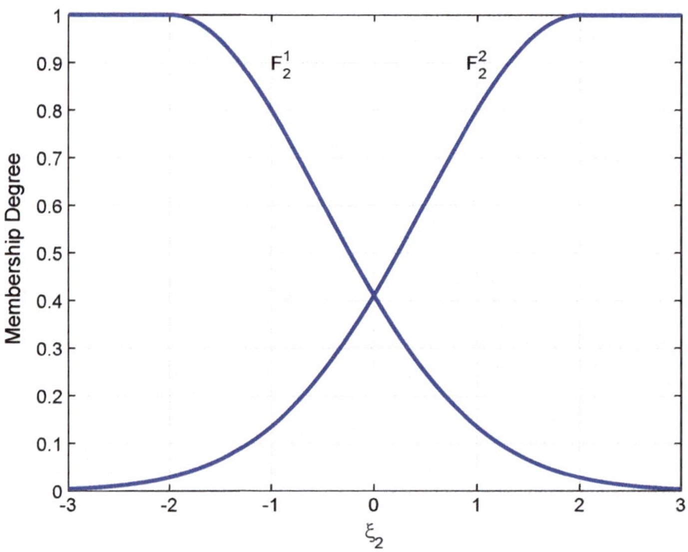

Fig. 2.2 Membership functions of F_2^1 and F_2^2

$$
\begin{aligned}
y = \;& \mu_1(\xi)(1 + 2\xi_1 + \xi_2) + \mu_2(\xi)(2 + \xi_1 - 0.5\xi_2) \\
& + \mu_3(\xi)(3 - 2\xi_1 + \xi_2) + \mu_4(\xi)(2 + 0.5\xi_1 - \xi_2).
\end{aligned} \tag{2.15}
$$

Given a particular crisp input (ξ_1^*, ξ_2^*), the corresponding overall crisp output can be calculated through (2.9)–(2.15).

The static T–S fuzzy system with the form (2.2) represents a static nonlinear mapping from input space to output space, which is usually used for function approximation. Its universal approximation capability has been proven and will be presented in Sect. 2.3.

In the following section, we will introduce the dynamic T–S fuzzy system which is more suitable for systematic control system analysis and synthesis.

2.2 Dynamic T–S Fuzzy Systems

If the consequents of the fuzzy rule are designed as linear dynamic systems, the T–S fuzzy system can represent a dynamic system, which is also called dynamic fuzzy system (Cao et al. 1995; Feng 2010).

2.2.1 *Continuous-Time State-Space Form*

Consider a continuous-time nonlinear dynamic system

$$\begin{aligned} \dot{x}(t) &= f(x(t), u(t)) \\ y(t) &= h(x(t)), \end{aligned} \tag{2.16}$$

where $x(t) \in R^n$ is the state vector, $u(t) \in R^m$ is the input, $y(t) \in R^l$ is the output, $f : R^n \times R^m \rightarrow R^n$ is an n-dimensional nonlinear vector function, and $h : R^n \rightarrow R^l$ is an l-dimensional nonlinear vector function. Let x_{0i} and u_{0i} be a set (constant) operation points of interest, $i = 1, 2, \ldots, N$, at some representative (and properly separated) points in the (x, u) space, and set $x = x_{0i} + \delta x_i$ and $u = u_{0i} + \delta u_i$, $i = 1, 2, \ldots, N$. Using Taylor series expansion of $f(x, u)$, we obtain

$$\dot{x}(t) \approx f(x_{0i}, u_{0i}) + A_i \delta x_i(t) + B_i \delta u_i(t), \tag{2.17}$$

where $A_i = \left.\frac{\partial f}{\partial x}\right|_{(x_{0i}, u_{0i})}$, $B_i = \left.\frac{\partial f}{\partial u}\right|_{(x_{0i}, u_{0i})}$, and the high-order terms are neglected.

If (x_{0i}, u_{0i}), $i = 1, 2, \ldots, N$, are the equilibrium points of the system, i.e., $f(x_{0i}, u_{0i}) = 0$, we have

$$\dot{x}(t) \approx A_i \delta x_i(t) + B_i \delta u_i(t). \tag{2.18}$$

For the output equation $y = h(x)$, letting $y = y_{0i} + \delta y_i$ with $y_{0i} = h(x_{0i}, u_{0i})$, we have

$$\delta y_i(t) \approx C_i \delta x_i(t),\ \ i = 1, 2, \ldots, N, \tag{2.19}$$

where the Jacobian matrices C_i are $C_i = \frac{\partial h}{\partial x}\Big|_{(x_{0i}, u_{0i})}$.

For simplicity of presentation and development, we will consider the generic form linear models:

$$\dot{x}(t) = A_i x(t) + B_i u(t),\ \ y(t) = C_i x(t) \tag{2.20}$$

at each operation point, for $i = 1, 2, \ldots, N$.

Remark 2.3 If (x_{0i}, u_{0i}), $i = 1, 2, \ldots, N$, are not equilibrium points, $f(x_{0i}, u_{0i}) \neq 0$ in (2.17). The effect of (x_{0i}, u_{0i}) and the neglected high-order terms will appear as an additional term in (2.20), which can be treated with an additional compensation signal in an adaptive control design.

Based on the derivation (2.17)–(2.20), the nonlinear system (2.16) can be approximated by the dynamic T–S fuzzy system with the following rule:

$$\begin{aligned} R^i :\ & \text{IF } \xi_1(t) \text{ is } F_1^i \text{ and } \xi_2(t) \text{ is } F_2^i \text{ and } \ \ldots\ \text{ and } \xi_L(t) \text{ is } F_L^i \\ & \text{THEN } \dot{x}(t) = A_i x(t) + B_i u(t), \\ & \quad\quad\ \ y(t) = C_i x(t), \end{aligned} \tag{2.21}$$

where $\xi_1, \xi_2, \ldots, \xi_L$ are some measurable system signals which are used to determine the operating area of the ith subsystem, $x \in R^n$ is the state vector, $u \in R^m$ is the input vector, $y \in R^l$ is the output vector, and $A_i \in R^{n\times n}$, $B_i \in R^{n\times m}$, and $C_i \in R^{l\times n}$ are the matrices of the ith subsystem, $F_1^i, F_2^i, \ldots, F_L^i$ are fuzzy sets.

Using *singleton fuzzification*, *product inference*, and *weighted average* introduced in Chap. 1, we obtain the following global dynamic T–S fuzzy system:

$$\dot{x}(t) = \sum_{i=1}^{N} \mu_i(\xi)(A_i x(t) + B_i u(t)), \tag{2.22}$$

$$y(t) = \sum_{i=1}^{N} \mu_i(\xi) C_i x(t), \tag{2.23}$$

where $\xi = [\xi_1, \xi_2, \ldots, \xi_L]^T$, $\mu_i(\xi)$ is the normalized firing strength given by (2.4) of the ith rule.

Example 2.2 Suppose $\xi_1 = x_1$. A dynamic T–S fuzzy system with two rules is given by

$$R^1: \quad \text{IF } x_1(t) \text{ is } F_1^1, \quad \text{THEN } \dot{x}(t) = A_1 x(t) + B_1 u(t), \ y(t) = C_1 x(t)$$
$$R^2: \quad \text{IF } x_1(t) \text{ is } F_1^2, \quad \text{THEN } \dot{x}(t) = A_2 x(t) + B_2 u(t), \ y(t) = C_2 x(t),$$

where $x(t) = [x_1(t), x_2(t)]^T$ and

$$\begin{aligned} A_1 &= \begin{bmatrix} 1 & -2 \\ -1 & 3 \end{bmatrix}, \ A_2 = \begin{bmatrix} 2 & 1 \\ 1 & -3 \end{bmatrix}, \ B_1 = \begin{bmatrix} 1 \\ 2 \end{bmatrix}, \ B_2 = \begin{bmatrix} 2 \\ 1 \end{bmatrix} \\ C_1 &= C_2 = [1, \ 1] \end{aligned} \tag{2.24}$$

and F_1^1 and F_1^2 are two fuzzy sets defined on $x_1 \in [-1, 1]$ with the corresponding membership functions:

$$F_1^1(x_1) = \begin{cases} \exp\left[-\frac{1}{2}\left(\frac{x_1+1}{0.8}\right)^2\right], & \text{if } x_1 \geq -1 \\ 1, & \text{otherwise} \end{cases} \tag{2.25}$$

$$F_1^2(x_1) = \begin{cases} \exp\left[-\frac{1}{2}\left(\frac{x_1-1}{0.8}\right)^2\right], & \text{if } x_1 \leq 1 \\ 1, & \text{otherwise.} \end{cases} \tag{2.26}$$

Since there is only one premise variable in this example, we have $\lambda_1(x_1) = F_1^1(x_1)$ and $\lambda_2(x_1) = F_1^2(x_1)$, and the normalized firing strength

$$\mu_1(x_1) = \frac{F_1^1(x_1)}{F_1^1(x_1) + F_1^2(x_1)}, \quad \mu_2(x_1) = \frac{F_1^2(x_1)}{F_1^1(x_1) + F_1^2(x_1)}. \tag{2.27}$$

The overall dynamic T–S fuzzy system is derived as

$$\begin{aligned} \dot{x}(t) &= \mu_1(x_1)(A_1 x(t) + B_1 u(t)) + \mu_2(x_1)(A_2 x(t) + B_2 u(t)) \\ y(t) &= \mu_1(x_1) C_1 x(t) + \mu_2(x_1) C_2 x(t). \end{aligned} \tag{2.28}$$

It can be seen that the overall T–S fuzzy system (2.28) performs nonlinear interpolation between two linear systems. When the value of $x_1(t)$ changes, $\mu_1(x_1)$ and $\mu_2(x_1)$ also change, which determine the contribution of the two linear systems in the overall nonlinear system.

In practice, we usually can build a group of linear models of a nonlinear plant that are accurate enough in certain local regions. When the nonlinear plant works in different local regions, we can pick different linear models to represent it. The current local operating region can be characterized by some key signals. In Example 2.2, the state variable x_1 can be viewed as the key signal, whose value determines the current operating region. If we want to obtain the global nonlinear model of the nonlinear plant, the T–S fuzzy system provides a very intuitive representation of a nonlinear plant as a nonlinear interpolation between a group of linear models.

2.2.2 Discrete-Time State-Space Form

Consider a discrete-time nonlinear dynamic system

$$
\begin{aligned}
x(t+1) &= f(x(t), u(t)) \\
y(t) &= h(x(t)),
\end{aligned} \tag{2.29}
$$

with t being the discrete-time sequence, where $x(t) \in R^n$ is the state vector, $u(t) \in R^m$ is the input, $y(t) \in R^l$ is the output, $f : R^n \times R^m \rightarrow R^n$ is an n-dimensional nonlinear vector function, and $g : R^n \rightarrow R^l$ is an l-dimensional vector function.

Consider a set of operation points of interest: (x_{0i}, u_{0i}), and set $\delta x_i = x - x_{0i}$ and $\delta u_i = u - u_{0i}$, $i = 1, 2, \ldots, N$. Using Taylor series expansion of $f(x, u)$, we obtain

$$
\delta x_i(t+1) \approx f(x_{0i}, u_{0i}) + A_i \delta x_i(t) + B_i \delta u_i(t), \tag{2.30}
$$

where $A_i = \frac{\partial f}{\partial x}\Big|_{(x_{0i}, u_{0i})}$, $B_i = \frac{\partial f}{\partial u}\Big|_{(x_{0i}, u_{0i})}$, and the high-order terms are neglected.

For the output equation $y = h(x)$, letting $\delta y_i = y - y_{0i}$ with $y_{0i} = h(x_{0i}, u_{0i})$, we have

$$
\delta y_i(t) \approx C_i \delta x_i(t), \ \ i = 1, 2, \ldots, N, \tag{2.31}
$$

where the Jacobian matrices C_i are $C_i = \frac{\partial h}{\partial x}\Big|_{(x_{0i}, u_{0i})}$.

For simplicity of presentation and development, we will consider the generic form linear models:

$$
x(t+1) = A_i x(t) + B_i u(t), \ \ y(t) = C_i x(t) \tag{2.32}
$$

at each operation point, for $i = 1, 2, \ldots, N$. We note that the offset term $f(x_{0i}, u_{0i})$ in (2.30) and the effect of x_{0i}, u_{0i}, not included in (2.32), can be treated with an additional compensation signal in an adaptive control design.

Based on the derivation (2.30)–(2.32), the discrete-time nonlinear system (2.29) can be approximated by the the dynamic T–S fuzzy system with the following rule:

$$
\begin{aligned}
R^i :\ & \text{IF } \xi_1(t) \text{ is } F_1^i \text{ and } \xi_2(t) \text{ is } F_2^i \text{ and } \ \ldots \ \text{ and } \xi_L(t) \text{ is } F_L^i \\
& \text{THEN } x(t+1) = A_i x(t) + B_i u(t), \\
& \qquad\quad\ \ y(t) = C_i x(t),
\end{aligned} \tag{2.33}
$$

where $\xi_1, \xi_2, \ldots, \xi_L$ are some measurable system signals which are used to determine the operating area of the $i$the subsystem, $x \in R^n$ is the state vector, $u \in R^m$ is the input vector, and $y \in R^l$ is the output vector, $A_i \in R^{n\times n}$, $B_i \in R^{n\times m}$, and $C_i \in R^{l\times n}$ are matrices of the ith subsystem. $F_1^i, F_2^i, \ldots, F_L^i$ are fuzzy sets defined as before.

Using the standard technique of *singleton fuzzification*, *product inference*, and *weighted average*, we obtain the following global dynamic T–S fuzzy system

$$x(t+1) = \sum_{i=1}^{N} \mu_i(\xi)(A_i x(t) + B_i u(t)), \tag{2.34}$$

$$y(t) = \sum_{i=1}^{N} \mu_i(\xi) C_i x(t), \tag{2.35}$$

where $\xi = [\xi_1, \xi_2, \ldots, \xi_L]^T$, $\mu_i(\xi)$ is the normalized firing strength (2.4) of the ith rule.

2.2.3 *Discrete-Time Input–Output Form*

If the consequent of each rule is a linear single-input single-output (SISO) dynamic system in discrete-time input–output form, the dynamic T–S fuzzy system has the following form:

$$\begin{aligned} R^i: \ & \text{IF } \xi_1(t) \text{ is } F_1^i \text{ and } \xi_2(t) \text{ is } F_2^i \text{ and } \ \ldots \ \text{ and } \xi_L(t) \text{ is } F_L^i \\ & \text{THEN } y(t) + a_{i,1}y(t-1) + \cdots + a_{i,n}y(t-n) = b_{i,0}u(t-d) \\ & \quad +b_{i,1}u(t-d-1) + \cdots + b_{i,\,n-d}u(t-n), \ \ b_{i,0} \neq 0, \end{aligned} \tag{2.36}$$

where $\xi_1, \xi_2, \ldots, \xi_L$ are some measurable system signals which are used to determine the operating area of the $i$the subsystem, and $u \in R$ and $y \in R$ are the input and output variables, respectively. $a_{i,1}, \ldots, a_{i,n}, b_{i,0}, \ldots, b_{i,\,n-d}$ are coefficients of the ith subsystem, where $d \geq 1$ denotes the input–output delay.

For the case $d = 1$, the consequent of (2.36) becomes

$$\begin{aligned} y(t) + a_{i,1}y(t-1) + \cdots + a_{i,n}y(t-n) = \ & b_{i,0}u(t-1) + b_{i,1}u(t-2) + \cdots \\ & +b_{i,\,n-d}u(t-n), \end{aligned} \tag{2.37}$$

which leads to the following one step prediction model:

$$\begin{aligned} y(t+1) + a_{i,1}y(t) + \cdots + a_{i,n}y(t-n+1) = \ & b_{i,0}u(t) + b_{i,1}u(t-1) + \cdots \\ & +b_{i,\,n-d}u(t-n+1). \end{aligned} \tag{2.38}$$

Using the standard technique of *singleton fuzzification*, *product inference*, and *weighted average*, we obtain the following global dynamic T–S fuzzy system:

$$\begin{aligned} y(t+1) = & \sum_{i=1}^{N} \mu_i(\xi) \big(-a_{i,1}y(t) - \cdots - a_{i,n}y(t-n+1) \\ & +b_{i,0}u(t) + b_{i,1}u(t-1) + \cdots + b_{i,\,n-d}u(t-n+1))\big), \end{aligned} \tag{2.39}$$

where $\xi = [\xi_1, \xi_2, \ldots, \xi_L]^T$ and $\mu_i(\xi)$ is the normalized firing strength (2.4) of the ith rule.

If $\sum_{i=1}^{N} \mu_i(\xi) b_{i,0} \neq 0$, it means the relative degree of the fuzzy system (2.39) is one. Based on (2.39), it is easy to select a control signal $u(t)$ to achieve the output tracking of a bounded reference signal $y_m(t+1)$ by $y(t+1)$, e.g., solving $u(t)$ from the following equation:

$$
\begin{aligned}
y_m(t+1) = \sum_{i=1}^{N} \mu_i(\xi) \big(& -a_{i,1} y(t) - \cdots - a_{i,n} y(t-n+1) \\
& + b_{i,0} u(t) + b_{i,1} u(t-1) + \cdots + b_{i,n-d} u(t-n+1) \big) . \qquad (2.40)
\end{aligned}
$$

For the general case $d > 1$, how to derive a global fuzzy prediction model is crucial for the control design, which will be elaborated in Chap. 7.

In practice, when a collection of input–output (I/O) data can be obtained from a nonlinear plant, the T–S fuzzy model (2.36) can be identified from the I/O data, including the premise variables ξ_j, the number of rules N, the parameters of the membership functions, and the parameters of the consequents. In Chap. 4, we will present how to identify a T–S fuzzy system from I/O data systematically.

2.3 Universal Approximation Property

When T–S fuzzy systems are used to represent nonlinear plants, a fundamental question is “Are T–S fuzzy systems universal approximators?” This question means whether T–S fuzzy systems are capable of approximating any real continuous functions on a compact set to arbitrary degree of accuracy. The universal approximation property of T–S fuzzy systems is the basis for identification and control designs based on T–S fuzzy systems. The studies on the universal approximation property of fuzzy systems have been extensively conducted in the 1990s, covering both Mamdani fuzzy systems and T–S fuzzy systems—the most popular two classes of fuzzy systems. Important results on proving Mamdani fuzzy systems are universal approximators can be found in Wang and Mendel (1992), Wang (1994, 1998), Zeng and Singh (1994, 1995), Ying (1994), Castro (1995), Castro and Delgado (1996), Robatti (1998).

Compared with Mamdani fuzzy systems, T–S fuzzy systems are more suitable for nonlinear identification and control synthesis. Buckley (1993) proved that two-input single-output T–S fuzzy systems with linear defuzzifier instead of the commonly used centroid defuzzifier are universal approximators. Ying (1998a) proved that the general single-input single-output (SISO) T-S fuzzy systems are universal approximators, which use any continuous input fuzzy sets, T–S fuzzy rules with affine consequents, and a generalized defuzzifier containing the widely used centroid defuzzifier as a special case. Then, the result was extended to general multi-input single-output (MISO) T–S fuzzy systems in Ying (1998b).

For general MISO, T–S fuzzy systems with L premise (input) variables, that is,

$$\begin{aligned} R^i: \ & \text{IF } \xi_1 \text{ is } F_1^i \text{ and } \xi_2 \text{ is } F_2^i \text{ and } \ \ldots \ \text{ and } \xi_L \text{ is } F_L^i \\ & \text{THEN } y = \theta_{i,0} + \theta_{i,1}\xi_1 + \cdots + \theta_{i,L}\xi_L, \end{aligned} \tag{2.41}$$

the following universal approximation theorem has been proven.

Theorem 2.1 (Ying 1998b) (Universal Approximation Theorem) *The general MISO T–S fuzzy systems with linear rule consequent can uniformly approximate any multivariate continuous function on a compact domain to any degree of accuracy.*

It should be noted that the "linear" in Theorem 2.1 actually means both affine and linear. The proof of Theorem 2.1 consists of two steps. First, it is proven that the general MISO T–S fuzzy models can approximate any multivariate polynomial to any degree of accuracy. Second, it is proven that the general T–S fuzzy models can uniformly approximate any multivariate continuous function with arbitrary precision by utilizing the fact that any multivariate continuous function can always be approximated by a multivariate polynomial arbitrarily well with the Weierstrass approximation theorem (Bronshtein and Semendyayev 1985).

Furthermore, Ying (1998b) also derived a formula of sufficient conditions for the fuzzy approximation that can compute the minimal upper bound on the number of input fuzzy sets and rules needed for any given continuous function and specified approximation error bound.

An important application of Theorem 2.1 is to use a T–S fuzzy system to approximate an unknown function of a nonlinear system on a compact set. Consider a general class of nonlinear discrete-time system described by the following state space model:

$$x(t+1) = f(x(t), u(t)), \tag{2.42}$$

where $x(t) \in R^n$ and $u(t) \in R^m$ are state and input vectors, respectively, and the function $f(x, u)$ is unknown by satisfying the following assumption.

Assumption 2.1 There exists an equilibrium $x_0 = 0 \in R^n$ such that $f(0, 0) = 0$ and $f \in C^2$, that is, the function $f(x, u)$ has a second-order continuous derivative with respect to x and u.

Let X and U be two compact sets in R^n and R^m, respectively. For $x \in X$ and $u \in U$, a discrete-time T–S fuzzy system (2.34)

$$\begin{aligned} x(t+1) &= \sum_{i=1}^{N} \mu_i(\xi)(A_i x(t) + B_i u(t)) \\ &= \hat{f}(x, u) \end{aligned} \tag{2.43}$$

is employed to approximate $f(x, u)$ on the compact set $X \times U \subset R^n \times R^m$. It has been proven that the T–S fuzzy system (2.43) is universal approximator in the sense that it can approximate any smooth nonlinear function $f(x)$ on a compact set $X \times U$.

Let $\sum_n$ be the set of all nonlinear functions satisfying Assumption 2.1. The universal approximation capability of the T–S fuzzy system (2.43) is given by the following theorem.

Theorem 2.2 (Cao et al. 1997a) *For any given $f(x, u)$ in $\sum_n$ on the compact set $X \times U \subset R^n \times R^m$ and arbitrary $\varepsilon > 0$, there exists a fuzzy system $\hat{f}(x, u)$ such that*

$$\sup_{x \in X, u \in U} \|f(x, u) - \hat{f}(x, u)\| < \varepsilon. \tag{2.44}$$

Remark 2.4 In Theorem 2.2, x and u of the functions $\hat{f}$ and f are assumed to be the same and the approximation accuracy can only be guaranteed under this condition. In practice, when a T–S fuzzy system is employed to represent a dynamic plant, the state of the T–S fuzzy system, $\hat{x}$, is usually different from that of the plant, x. Hence, the approximation error between $\hat{f}(\hat{x}, u)$ and $f(x, u)$ may grow with the evolution of $\hat{x}$ and x, which should be carefully considered.

2.4 Stability and Stabilization Control

Since the consequent of each rule in a T–S fuzzy system is usually a linear system, it is natural to apply mature linear control theory to design local linear controllers for each subsystem. Two important questions are: how can the overall controller for the overall T–S fuzzy system be calculated from the local linear controllers? can the overall controller make the closed-loop T–S fuzzy system stable? In this section, we will present some basic yet important results on the stability and stabilization control of T–S fuzzy systems.

2.4.1 *Stability of T–S Fuzzy Systems*

By setting $u(t) = 0$, the continuous-time T–S fuzzy system (2.22) becomes an autonomous system:

$$\dot{x}(t) = \sum_{i=1}^{N} \mu_i(\xi) A_i x(t), \tag{2.45}$$

and the discrete-time T–S fuzzy system (2.34) becomes the following autonomous system:

$$x(t+1) = \sum_{i=1}^{N} \mu_i(\xi) A_i x(t). \tag{2.46}$$

Single quadratic Lyapunov function. Based on (2.46) and using a *single quadratic Lyapunov function*

$$V(x) = x^T P x, \tag{2.47}$$

Tanaka and Sugeno (1992) proposed an important criterion for checking the stability of the T–S fuzzy systems in the early 90s. The stability criterion for (2.46) is formulated in the following theorem.

Theorem 2.3 (Tanaka and Sugeno 1992) *The equilibrium state of the fuzzy system (2.46) (namely, $x = 0$) is globally asymptotically stable if there exists a common positive definite matrix P such that*

$$A_i^T P A_i - P < 0, \quad \text{for all} \quad i = 1, 2, \ldots, N. \tag{2.48}$$

Using the same single quadratic Lyapunov function $V(x) = x^T P x$, the stability criterion for continuous-time T–S fuzzy systems (2.45) can also be derived.

Theorem 2.4 *The equilibrium state of the fuzzy system (2.45) (namely, $x = 0$) is globally asymptotically stable if there exists a common positive definite matrix P such that*

$$A_i^T P + P A_i < 0, \quad \text{for all} \quad i = 1, 2, \ldots, N. \tag{2.49}$$

The stability criteria in both Theorems 2.3 and 2.4 require a common symmetric positive definite matrix P to satisfy all the inequalities (2.48) or (2.49). However, both theorems do not provide a systematic way to find such a common matrix P. A systematic way of searching the common matrix P was suggested in the paper (Joh et al. 1998) assuming that all the A_i, $i = 1, 2, \ldots, N$, are Schur and pairwise commutative. However, the assumption on A_i is not easy to be satisfied for general T–S fuzzy systems. Based on this approach, Li and Li (2004) provided a systematic way to find P for T–S fuzzy systems with simplified linear consequent parts of rules without the assumption on A_i. The subsystem matrix A_i is decomposed into a pairwise part $\bar{A}_i$ and an additional part ΔA_i. Hence, the iteration method in (Joh et al. 1998) can be applied to find a common matrix P for pairwise commutative matrices $\bar{A}_i$, $i = 1, 2, \ldots, N$. The stability of the global system is guaranteed if ΔA_i satisfies certain conditions.

Theorem 2.3 is for discrete-time T–S fuzzy models with linear consequent part of rules. It is extended to T–S fuzzy models with affine consequent parts of rules in

(Kim and Kim 2001):

$$x(t+1) = \sum_{i=1}^{N} \mu_i(\xi)(A_i x(t) + \nu_i), \tag{2.50}$$

where ν_i is the affine term in the ith rule.

Finding the common matrix P in the above approaches can be stated numerically as a convex optimization problem involving linear matrix inequalities (LMIs) (Boyd et al. 1994). Such problems can be solved efficiently using publicly available software such as LMI Toolbox (Gahinet et al. 1995).

Example 2.3 Consider the T–S fuzzy system

$$\begin{aligned} R^1: &\quad \text{IF } x_1(t) \text{ is} F_1^1, \quad \text{THEN } x(t+1) = A_1 x(t) \\ R^2: &\quad \text{IF } x_1(t) \text{ is } F_1^2, \quad \text{THEN } x(t+1) = A_2 x(t), \end{aligned}$$

where $x(t) = [x_1(t), x_2(t)]^T$ and

$$A_1 = \begin{bmatrix} -0.6 & 0.3 \\ 0.1 & -0.5 \end{bmatrix}, \quad A_2 = \begin{bmatrix} -0.5 & 0.2 \\ 0.3 & -0.5 \end{bmatrix}. \tag{2.51}$$

The membership functions of the two fuzzy sets F_1^1 and F_1^2 are illustrated in Fig. 2.3. Since the eigenvalues of A_1 are -0.7303 and -0.3697, of A_2 are -0.2551 and -0.7449, the two linear subsystems are stable. However, the stability of the linear subsystems cannot ensure the stability of the overall T–S fuzzy system:

$$x(t+1) = \mu_1(x_1) A_1 x(t) + \mu_2(x_1) A_2 x(t). \tag{2.52}$$

To verify the stability of the overall T–S fuzzy system (2.52) using Theorem 2.3, we need to find a common positive matrix P satisfying the following two inequalities:

$$\begin{aligned} A_1^T P A_1 - P &< 0 \\ A_2^T P A_2 - P &< 0. \end{aligned} \tag{2.53}$$

With the assistant of MATLAB LMI toolbox, we find the solution

$$P = \begin{bmatrix} 88.0594 & -10.9344 \\ -10.9344 & 87.8949 \end{bmatrix}. \tag{2.54}$$

Then, we can conclude that the overall T–S fuzzy system is stable based on Theorem 2.3.

It should be noted that since Theorems 2.3 and 2.4 only provide sufficient conditions for the stability of T–S fuzzy systems, we cannot conclude the T–S fuzzy system is unstable even we cannot find a common positive matrix P to solve the LMIs (2.48)

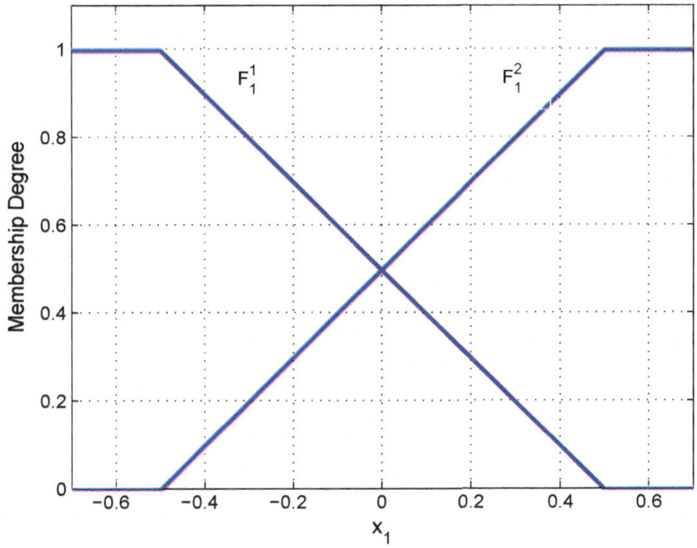

Fig. 2.3 Membership functions of F_1^1 and F_1^2

or (2.49). Actually, the standard LMIs conditions for quadratic stability are often found to be conservative when applied to fuzzy systems. One drawback is the fact that the state is dependent on the membership functions is disregarded. Another major restriction is that Lyapunov function should be quadratic. Many exponentially stable systems do not admit a globally quadratic Lyapunov function.

Piecewise quadratic Lyapunov functions. To reduce these restrictions mentioned above, Johansson et al. (1999) proposed stability conditions for continuous-time T–S fuzzy systems based on *piecewise quadratic Lyapunov functions*:

$$V(x) = x^T P_i x, \quad \xi \in S_i, \quad i = 1, 2, \ldots, N, \tag{2.55}$$

where $P_i, i = 1, 2, \ldots, N$ are positive definite matrices and S_i defines a region in the space of the premise variable ξ:

$$S_i = \{\xi \,|\, \mu_i(\xi) > \mu_j(\xi), \ \ j = 1, 2, \ldots, N, \ \ j \neq i\}, \ \ i = 1, 2, \ldots, N. \tag{2.56}$$

The stability results for continuous-time T–S fuzzy systems were extended to discrete-time T–S fuzzy systems by Feng (2004). Johansson et al. (1999) and Feng (2004) also considered T–S fuzzy systems with affine subsystems:

$$\dot{x}(t) = \sum_{i=1}^{N} \mu_i(\xi)(A_i x(t) + a_i)$$

$$x(t+1) = \sum_{i=1}^{N} \mu_i(\xi)(A_i x(t) + a_i), \tag{2.57}$$

where $a_i, i = 1, 2, \ldots, N$ are additional offset terms, and developed stability conditions for them based on piecewise quadratic Lyapunov functions.

Fuzzy quadratic Lyapunov function. In single quadratic Lyapunov function based approaches and piecewise Lyapunov functions based approaches, the membership functions are not used in the Lyapunov functions. One way to include the membership functions is to build the Lyapunov function by following the way similar to building the overall T–S fuzzy fuzzy system. Chadli et al. (2000) proposed the following fuzzy rules:

$$\begin{aligned} R^i: \ & \text{IF } \xi_1(t) \text{ is } F_1^i \text{ and } \xi_2(t) \text{ is } F_2^i \text{ and } \ \ldots \ \text{ and } \xi_L(t) \text{ is } F_L^i \\ & \text{THEN } V(x(t)) = x(t)^T P_i x(t), \end{aligned} \tag{2.58}$$

where $P_i, i = 1, 2, \ldots, N$ are positive definite matrices.

Using the same fuzzy inference method as we obtain the overall T–S fuzzy system, a *fuzzy quadratic Lyapunov function* is constructed as

$$V(x) = \sum_{i=1}^{N} \mu_i(\xi) x^T P_i x \tag{2.59}$$

based on which the following stability conditions are derived for continuous-time T–S fuzzy systems (2.45).

Theorem 2.5 (Chadli et al. 2000) *Suppose there exist positive definite matrices Q and P_i such that the following LMIs hold:*

$$A_i^T P_i + P_i A_i < -Q, \quad i = 1, 2, \ldots, N, \tag{2.60}$$

$$\frac{A_i^T P_j + P_j A_i + A_j^T P_i + P_i A_j}{2} < -Q, \quad i, j = 1, 2, \ldots, N, \quad i \neq j \tag{2.61}$$

for $\mu_i(\xi)\mu_j(\xi) \neq 0$. Then, the global model (2.45) is globally exponentially stable if the following constraint on the variation of the state is satisfied:

$$\|\dot{x}(t)\| \leq \frac{\lambda_M(Q)}{\gamma \sum_{i=1}^{N} \lambda_M(P_i)}, \tag{2.62}$$

where $\lambda_M(\cdot)$ represents the largest eigenvalues of the corresponding matrix, γ is defined as $\gamma = \max_{i=1}^{N} \|\frac{\partial \mu_i(\xi)}{\partial x}\|$.

For the discrete-time T–S fuzzy system (2.46), based on the fuzzy quadratic Lyapunov function, Zhou et al. (2005) developed the following stability conditions.

Theorem 2.6 (Zhou et al. 2005) *The T–S fuzzy system (2.46) is globally exponentially stable if there exists a set of positive definite matrices* P_i *such that the following LMIs are satisfied:*

$$A_i^T P_j A_i - P_j < 0, \quad i, j = 1, 2, \ldots, N. \tag{2.63}$$

Compared with the stability conditions in Theorem 2.3, 2.6 does not require a common positive definite matrix P but a group of positive definite matrices P_i, which relaxes the conditions.

T–S fuzzy systems with time-varying linear subsystems. All the above approaches assume that all the subsystems of T–S fuzzy system are linear time-invariant (LTI). However, if the real nonlinear plant is time-varying, when a T–S fuzzy system is employed to approximate the time-varying nonlinear plant, its subsystems will be time-varying as well (Jin 2003), that is, A_i, $i = 1, 2, \ldots, N$ in (2.46) are not deterministic but varying with t. The T–S fuzzy system (2.46) becomes

$$x(t+1) = \sum_{i=1}^{N} \mu_i(\xi) A_t^i x(t), \tag{2.64}$$

where A_t^i is the system matrix of the ith local T–S model at time step t. It is important to investigate the stability properties of the T–S fuzzy system in the form (2.64). Jin (2003) derived stability conditions of (2.64) by employing a single quadratic Lyapunov function.

Theorem 2.7 (Jin 2003) *Fuzzy system (2.64) is exponentially stable at* $x = 0$, *if*

1. *There exist two constants a and b and a symmetric matrix* P_t *for all* A_t^i *such that*

$$0 < aI < P_t < bI \quad \text{holds for all } t \tag{2.65}$$

2. *There exists a common matrix* H_t *such that*

$$(A_t^i)^T P_{t+1} A_t^i - P_t = -H_t H_t^T \tag{2.66}$$

3. *There exist a constant C and a function* $S_n(t)$, *for a certain n and for all t such that*

$$S_n(t) = \sum_{i=0}^{p-1} \Phi(t+i, t)^T H(t+i) H(t+i)^T \Phi(t+i, t) \geq CI > 0, \tag{2.67}$$

where $\Phi(\cdot)$ *is the transfer matrix of system* A_t^i.

Theorem 2.7 can be viewed as an extension of Theorem 2.3 to time-varying T–S fuzzy systems. At each time step t, a common positive definite matrix solution of Lyapunov functions needs to be found. However, it is computationally incontinent to use the conditions in Theorem 2.7 to check the stability of T–S fuzzy models in practice. Chou and Chen (2001) proposed stability criteria that do not need solving any Lyapunov equation. In their approach, the time-varying system matrix of the ith rule A_t^i is split into a constant nominal system matrix A^i and a time-varying matrix ΔA_t^i representing the time-varying uncertainty. With an assumption on the boundedness of these uncertainty matrices ΔA_t^i, $i = 1, 2, \ldots, N$, stability criteria are derived, which enable one to analyze the stability of the discrete T–S fuzzy model whose rules do not have a common positive definite matrix solution of Lyapunov equations.

2.4.2 Stabilization Control of T–S Fuzzy Systems

Since each subsystem of a T–S fuzzy system is linear, it is natural to design a linear controller for each subsystem first, and then build the overall controller. Such an idea is called parallel distribution compensation (PDC) (Tanaka and Sano 1994; Wang et al. 1996).

Parallel distributed compensation (PDC). For the T–S fuzzy system (2.21), its PDC controller shares the same fuzzy sets with it in the IF parts of the rules:

$$\begin{aligned} R^i:\ &\text{IF } \xi_1(t) \text{ is } F_1^i \text{ and } \xi_2(t) \text{ is } F_2^i \text{ and } \ldots \text{ and } \xi_L(t) \text{ is } F_L^i \\ &\text{THEN } u(t) = K_i x(t), \quad i = 1, 2, \ldots, N, \end{aligned} \tag{2.68}$$

where $K_i, i = 1, 2, \ldots, N$ are feedback gain matrices. K_i can be chosen using mature state feedback control design methods, such as pole placement, linear quadratic regulator (LQR), etc, as long as the subsystem is controllable.

(1) Fuzzily blended controller

Following the similar way to derive the overall T–S fuzzy system (2.34), the overall T–S fuzzy controller is obtained as

$$u(t) = \sum_{i=1}^{N} \mu_i(\xi) K_i x(t) \tag{2.69}$$

which is a nonlinear controller in general.

Although the local state feedback gain K_i designed based on linear state feedback control theory can guarantee the stability of the local linear subsystem, there is no guarantee that the overall T–S fuzzy controller (2.69) can stabilize the overall T–S fuzzy system (2.34). In the following study, we will derive the closed-loop T–S fuzzy system and analyze under which conditions it is stable.

By substituting (2.69) into (2.34), we obtain the closed-loop T–S fuzzy system:

$$x(t+1) = \sum_{i=1}^{N}\sum_{j=1}^{N} \mu_i(\xi)\mu_j(\xi)(A_i + B_iK_j)x(t). \tag{2.70}$$

which can be further formulated as

$$x(t+1) = \sum_{i=1}^{N} \mu_i(\xi)\mu_i(\xi)(A_i + B_iK_i)x(t) + 2\sum_{i<j}^{N} \mu_i(\xi)\mu_j(\xi)G_{ij}x(t), \tag{2.71}$$

where

$$G_{ij} = \frac{(A_i + B_iK_j) + (A_j + B_jK_i)}{2}. \tag{2.72}$$

Using a *single quadratic Lyapunov function* $V(x) = x^T Px$, the following sufficient condition can be derived for the closed-loop T–S fuzzy system (2.70).

Theorem 2.8 (Wang et al. 1996) *The equilibrium of a fuzzy control system (2.70) is asymptotically stable in the large if there exists a common positive definite matrix P such that the following two conditions are satisfied:*

$$(A_i + B_iK_i)^T P(A_i + B_iK_i) - P < 0, \tag{2.73}$$

$$G_{ij}^T PG_{ij} - P < 0, \quad i < j, \tag{2.74}$$

for all $i, j = 1, 2, \ldots, N$ *except the pairs* (i, j) *such that* $\mu_i(\xi)\mu_j(\xi) = 0$.

Remark 2.5 The PDC controller (2.68)–(2.69) is also applicable for continuous-time T–S fuzzy system. With a single quadratic Lyapunov function $V(x) = x^T Px$, the stability condition for the closed-loop continuous-time T–S fuzzy system becomes (Tanaka et al. 1998)

$$(A_i + B_iK_i)^T P + P(A_i + B_iK_i) < 0, \tag{2.75}$$

$$G_{ij}^T P + PG_{ij} < 0, \quad i < j \tag{2.76}$$

for all $i, j = 1, 2, \ldots, N$ except the pairs (i, j) such that $\mu_i(\xi)\mu_j(\xi) = 0$.

The conditions (2.73)–(2.74) and (2.75)–(2.76) require finding a common definite matrix P. The control design task is formulated into solving a group of LMIs. If the solution P can be found, it means the stabilization constraints are met and the local state feedback gains K_i can be obtained simultaneously. However, there is no guarantee that such a solution exists. With the increase of the number of fuzzy rules, leading to a massive number of LMIs, it may become infeasible to find a common solution P. Many research efforts have been made around how to relax the stabilization control design conditions.

(2) Switched controller

In this approach, the overall controller switches among different local linear controllers according to certain conditions, rather than fuzzily blending all the local linear controllers together.

The switching controller is defined as

$$u(t) = K_i x(t), \quad \xi \in S_i, \quad i = 1, 2, \ldots, N, \tag{2.77}$$

where S_i defines a region in the space of the premise variable ξ:

$$S_i = \{\xi \,|\, \mu_i(\xi) > \mu_j(\xi), \;\; j = 1, 2, \ldots, N, \;\; j \neq i\}, \; i = 1, 2, \ldots, N. \tag{2.78}$$

By substituting (2.78) into (2.34), we obtain the closed-loop T–S fuzzy system:

$$x(t+1) = A_{ci} x(t), \quad \xi \in S_i, \tag{2.79}$$

where $A_{ci} = A_i + \Delta A_i(\mu) + (B_i + \Delta B_i(\mu))K_i$, with $\Delta A_i(\mu)$ and $\Delta B_i(\mu)$ defined as

$$\begin{aligned} \Delta A_i(\mu) &= \sum_{j=1, j\neq i}^{N} \mu_j(\xi)(A_j - A_i), \\ \Delta B_i(\mu) &= \sum_{j=1, j\neq i}^{N} \mu_j(\xi)(B_j - B_i). \end{aligned} \tag{2.80}$$

Based on piecewise quadratic Lyapunov functions

$$V(x) = x^T X_i^{-1} x, \quad \xi \in S_i, \quad i = 1, 2, \ldots, N, \tag{2.81}$$

the stability condition of the closed-loop system (2.79) is derived by Feng (2010), which requires finding a set of positive definite matrices X_i, $i = 1, 2, \ldots, N$ to satisfy a group of LMIs.

2.5 Tracking Control of T–S Fuzzy Systems

In this section, we consider the tracking control designs for the T–S fuzzy systems (2.22) and (2.34). Generally, there are two kinds of tracking control: state tracking control and output tracking control. The state tracking control objective is to design a control law that generates the control input $u(t)$ such that $u(t)$ and $x(t)$ remain bounded and $x(t)$ tracks a bounded reference $x_m(t)$. The output tracking control objective is to design a controller $u(t)$ to make $u(t)$ and $x(t)$ bounded and the output $y(t)$ track a bounded reference $y_m(t)$.

2.5.1 State Tracking Control

In state tracking control, the reference $x_m(t)$ is usually generated by a stable reference model, which produces desired state trajectories. The reference model can be a linear model or a T–S fuzzy model.

Linear reference model. We start by considering a linear reference model:

$$\dot{x}_m(t) = A_m x_m(t) + B_m r(t), \tag{2.82}$$

where $x_m \in R^n$ is the state vector, $r \in R^l$ is the bounded input vector, $A_m \in R^{n\times n}$ is a Hurwitz matrix, and $B_i \in R^{n\times l}$.

Based on the idea of PDC, the local controller for each subsystem can be designed as

$$\begin{aligned} R^i:\ & \text{IF } \xi_1(t) \text{ is } F_1^i \text{ and } \xi_2(t) \text{ is } F_2^i \text{ and } \ \dots \ \text{ and } \xi_L(t) \text{ is } F_L^i \\ & \text{THEN } u(t) = K_1^i x(t) + K_2^i r(t), \quad i = 1, 2, \dots, N, \end{aligned} \tag{2.83}$$

where $K_1^i \in R^{m\times n}$ and $K_2^i \in R^{m\times l}$.

By fuzzily blending all the local controllers, the overall fuzzy controller is obtained as

$$u(t) = \sum_{i=1}^{N} \mu_i(\xi) K_1^i x(t) + \sum_{i=1}^{N} \mu_i(\xi) K_2^i r(t). \tag{2.84}$$

Substituting (2.84) into (2.22) yields the closed-loop fuzzy system:

$$\dot{x}(t) = \sum_{i=1}^{N}\sum_{j=1}^{N} \mu_i(\xi)\mu_j(\xi)[(A_i + B_i K_1^j)x(t) + B_i K_2^j r(t)]. \tag{2.85}$$

Since μ_i satisfies (2.7), if the matrices K_1^j and K_2^j meet the following matching conditions:

$$A_i + B_i K_1^j = A_m, \quad B_i K_2^j = B_m, \quad i, j = 1, 2, \dots, N, \tag{2.86}$$

the fuzzy system (2.85) becomes

$$\dot{x}(t) = A_m x(t) + B_m r(t), \tag{2.87}$$

which has the same dynamics as the reference model (2.82). After some transient process, the state $x(t)$ will approach the reference $x_m(t)$ and the state tracking objective is achieved.

From the matching conditions (2.86) and with $j = 1, 2, \ldots, N$, we have

$$\begin{aligned} A_i + B_i K_1^1 &= A_m, \quad B_i K_2^1 = B_m, \\ A_i + B_i K_1^2 &= A_m, \quad B_i K_2^2 = B_m, \\ &\vdots \\ A_i + B_i K_1^N &= A_m, \quad B_i K_2^N = B_m, \end{aligned} \tag{2.88}$$

which lead to

$$\begin{aligned} B_i K_1^1 = B_i K_1^2 = \ldots = B_i K_1^N &= A_m - A_i, \\ B_i K_2^1 = B_i K_2^2 = \ldots = B_i K_2^N &= B_m, \end{aligned} \tag{2.89}$$

for all $i = 1, 2, \ldots, N$.

It can be observed from (2.89) that except for some matrices A_i, B_i, A_m, and B_m with very special structures, it is almost impossible to design K_1^j and K_2^j to meet the matching conditions (2.86).

Another way to design a state tracking controller is to use a fuzzy reference model instead of a linear reference model.

Fuzzy reference model. A fuzzy reference model has the following rules:

$$\begin{aligned} R^i: \ & \text{IF } \xi_1(t) \text{ is } F_1^i \text{ and } \xi_2(t) \text{ is } F_2^i \text{ and } \ \ldots \ \text{ and } \xi_L(t) \text{ is } F_L^i \\ & \text{THEN } \dot{x}_m(t) = A_{mi} x_m(t) + B_{mi} r(t), \quad i = 1, 2, \ldots, N, \end{aligned} \tag{2.90}$$

where $x_m \in R^n$ is the state vector, $r \in R^l$ is the bounded input vector, $A_{mi} \in R^{n \times n}$, and $B_{mi} \in R^{n \times l}$. The fuzzy reference model has the same premise variables and membership functions as the T–S fuzzy system (2.21). The overall fuzzy reference model is inferred as

$$\dot{x}_m(t) = \sum_{i=1}^{N} \mu_i(\xi)(A_{mi} x_m(t) + B_{mi} r(t)). \tag{2.91}$$

To ensure the stability of the fuzzy reference model (2.91), the matrices A_{mi} should be chosen to satisfy the following conditions:

$$A_{mi}^T P + P A_{mi} < 0, \quad \text{for all} \quad i = 1, 2, \ldots, N, \tag{2.92}$$

which is a sufficient condition for the stability of (2.91) based on Theorem 2.4.

Based on (2.85) and (2.91), we can derive the new matching conditions for the fuzzy system (2.22) to track the states of the fuzzy reference model (2.91):

$$A_i + B_i K_1^j = A_{mi}, \quad B_i K_2^j = B_{mi}, \quad i, j = 1, 2, \ldots, N. \tag{2.93}$$

Comparing (2.93) with (2.86), we can see A_{mi} and B_{mi} on the right side of the matching conditions (2.93) instead of A_m and B_m, which gives us more design freedom. However, it is still not easy to meet the matching conditions (2.93) since the coupling between the jth controller and the ith subsystem still exists.

In summary, for general T–S fuzzy systems, it is difficult to achieve state tracking since it is almost impossible to select the controller parameters to meet the stringent matching conditions. For certain T–S fuzzy systems with special structures, such as A_i and B_i in Canonical form or the number of inputs equals the number of states, i.e., $m = n$, the state tracking problem can be solved successfully under some feasible design conditions, which will be discussed in details in Chap. 5.

2.5.2 *Output Tracking Control*

Consider the output tracking control for the continuous-time T–S fuzzy system (2.22)–(2.23):

$$\dot{x} = \sum_{i=1}^{N} \mu_i(\xi)(A_i x + B_i u) \tag{2.94}$$

$$y = \sum_{i=1}^{N} \mu_i(\xi) C_i x, \tag{2.95}$$

where it is assumed $y \in R$ and $u \in R$ here.

Since the premise variables $\xi_j, j = 1, \ldots, L$ are usually selected from system state variables, we can define

$$f(x) = \sum_{i=1}^{N} \mu_i(\xi) A_i x, \quad g(x) = \sum_{i=1}^{N} \mu_i(\xi) B_i \tag{2.96}$$

$$h(x) = \sum_{i=1}^{N} \mu_i(\xi) C_i x. \tag{2.97}$$

The fuzzy system (2.94)–(2.95) can be written into the following form:

$$\dot{x} = f(x) + g(x)u \tag{2.98}$$

$$y = h(x). \tag{2.99}$$

In the following study, we assume the membership functions are designed as smooth functions, e.g., Gaussian functions, such that the f, g, and h are smooth vector functions.

Feedback linearization. One of the powerful methods for nonlinear tracking control design is *feedback linearization*, which employs a change of coordinates and feedback control to transform a nonlinear system into a system whose dynamics are linear. Feedback linearization techniques have been extensively studied in the nonlinear control literature (Isidori 1995; Khalil 2002; Marino and Tomei 1995; Nijmeijer and Van der Schaft 1990). In this subsection, we only briefly review some basic techniques of feedback linearization that are related to the output tracking control of T–S fuzzy systems.

Our goal is to design a state feedback control law such that the output y tracks a given reference y_m. To achieve this goal, a direct relationship between the output y (or its derivative) and the input u is desirable. From (2.98) to (2.99), the derivative of y is given by

$$\dot{y} = \frac{\partial h}{\partial x}(x)f(x) + \frac{\partial h}{\partial x}(x)g(x)u. \tag{2.100}$$

If we have $\frac{\partial h}{\partial x}(x_0)g(x_0) \neq 0$, then the system (2.98)–(2.99) is said to have *relative degree* one at x_0. Since the functions f, g, h are smooth, $\frac{\partial h}{\partial x}(x_0)g(x_0) \neq 0$ implies there exists a neighborhood B_0 of x_0 on which $\frac{\partial h}{\partial x}(x)g(x) \neq 0$. The system with relative degree one means its output and input are separated by one integrator only.

If $\frac{\partial h}{\partial x}(x)g(x) = 0$, then we compute the second derivative of y:

$$\ddot{y} = \frac{\partial}{\partial x}\left(\frac{\partial h}{\partial x}(x)f(x)\right)f(x) + \frac{\partial}{\partial x}\left(\frac{\partial h}{\partial x}(x)f(x)\right)g(x)u. \tag{2.101}$$

If $\frac{\partial}{\partial x}\left(\frac{\partial h}{\partial x}(x)f(x)\right)g(x) \neq 0$ for any $x \in B_0$, then the system (2.98)–(2.99) is said to have *relative degree* two on B_0.

If $\frac{\partial}{\partial x}\left(\frac{\partial h}{\partial x}(x)f(x)\right)g(x) = 0$, we keep on computing the derivatives of y until u appears explicitly. The following definition on relative degree ρ is introduced (Krstić et al. 1995).

Definition 2.1 The system (2.98)–(2.99) is said to have relative degree ρ if there exists a neighborhood B_0 of x_0 on which

$$\frac{\partial \psi_1}{\partial x}(x)g(x) = \frac{\partial \psi_2}{\partial x}(x)g(x) = \cdots = \frac{\partial \psi_{\rho-1}}{\partial x}(x)g(x) = 0, \tag{2.102}$$

$$\frac{\partial \psi_\rho}{\partial x}(x)g(x) \neq 0, \tag{2.103}$$

where

$$\psi_1(x) = h(x), \quad \psi_i(x) = \frac{\partial \psi_{i-1}}{\partial x}(x)f(x), \quad i = 2, \ldots, \rho. \tag{2.104}$$

If (2.102) and (2.103) are valid for all $x \in R^n$, then the relative degree of (2.98)–(2.99) is said to be globally defined.

According to Definition 2.1, the ρth derivative of y can be written as

$$y^{(\rho)} = \frac{\partial \psi_\rho}{\partial x}(x)f(x) + \frac{\partial \psi_\rho}{\partial x}(x)g(x)u, \tag{2.105}$$

where $\frac{\partial \psi_\rho}{\partial x}(x)g(x) \neq 0$ on B_0. By choosing a state feedback control law

$$u = \frac{1}{\frac{\partial \psi_\rho}{\partial x}(x)g(x)}\left(-\frac{\partial \psi_\rho}{\partial x}(x)f(x) + v\right), \tag{2.106}$$

the nonlinearities in (2.105) can be canceled, leading to the following linear input–output mapping:

$$y^{(\rho)} = v, \tag{2.107}$$

where v is an auxiliary signal to be designed. It is easy to design v to achieve output tracking.

It should be noted that the linearized input–output dynamics (2.107) is equivalent to a chain of ρ integrators. Since the original nonlinear system (2.98)–(2.99) has dimension n, there are another $n - \rho$ states. Using differential geometric tools, it can be shown that for a system with relative degree ρ, there always exist $n - \rho$ functions $\psi_{\rho+1}(x), \ldots, \psi_n(x)$ satisfying

$$\frac{\partial \psi_i(x)}{\partial x}(x)g(x) = 0, \quad i = \rho + 1, \ldots, n, \tag{2.108}$$

such that the change of coordinates

$$\begin{aligned}
&\zeta_1 = y = \psi_1(x), \ \zeta_2 = \dot{y} = \psi_2(x), \ \ldots, \ \zeta_\rho = y^{(\rho-1)} = \psi_\rho(x) \\
&\eta_1 = \psi_{\rho+1}(x), \ \ldots, \ \eta_{n-\rho} = \psi_n(x)
\end{aligned} \tag{2.109}$$

is locally invertible.

Define $\zeta = [\zeta_1, \zeta_2, \ldots, \zeta_\rho]^T$, $\eta = [\eta_1, \eta_2, \ldots, \eta_{n-\rho}]^T$ and

$$\begin{aligned}
\phi_1(\zeta, \eta) &= \frac{\partial \psi_{\rho+1}(x)}{\partial x}(x)f(x), \\
&\vdots \\
\phi_{n-\rho}(\zeta, \eta) &= \frac{\partial \psi_n(x)}{\partial x}(x)f(x).
\end{aligned} \tag{2.110}$$

Then, the nonlinear system (2.98)–(2.99) can be transformed into

$$\begin{cases} \dot{\zeta}_1 = \zeta_2 \\ \quad \vdots \\ \dot{\zeta}_{\rho-1} = \zeta_\rho \\ \dot{\zeta}_\rho = \frac{\partial \psi_\rho}{\partial x}(x) f(x) + \frac{\partial \psi_\rho}{\partial x}(x) g(x) u, \end{cases} \tag{2.111}$$

$$\begin{cases} \dot{\eta}_1 = \phi_1(\zeta, \eta) \\ \quad \vdots \\ \dot{\eta}_{n-\rho} = \phi_{n-\rho}(\zeta, \eta). \end{cases} \tag{2.112}$$

The first ρ Eq. (2.111) related to ζ represent dynamics that are input–output linearizable. The last $n - \rho$ Eq. (2.112) related to η are called the *internal dynamics* of the system, which cannot be directly manipulated by u. When $\zeta = 0$, the internal dynamics become so-called *zero dynamics* (in literature, the internal dynamics are sometimes called zero dynamics even when $\zeta \neq 0$). If the zero dynamics are stable, the nonlinear system is said to be *minimum phase*. For a minimum phase system, it is only necessary to design a control law for the dynamics of ζ. For bounded ζ, η would be bounded as well.

To apply feedback linearization techniques to design an output tracking controllers for a T–S fuzzy system, it is important to transform the T–S fuzzy system into an input–output feedback linearizable canonical form, i.e., (2.111)–(2.112). Since the membership functions and the forms of A_i and B_i in each subsystem are different for different T–S fuzzy systems, it is difficult to determine the relative degree and the internal dynamics for general T–S fuzzy systems.

However, for a specified T–S fuzzy system, it is possible to derive its relative degree and internal dynamics. We will show how to obtain them by the following example.

Example 2.4 Consider the following T–S fuzzy system with two rules:

$$R^1 : \text{IF } x_1(t) \text{ is } F_1^1, \quad \text{THEN } \begin{cases} \dot{x}(t) = A_1 x(t) + B_1 u(t) \\ y(t) = C_1 x(t) \end{cases}$$

$$R^2 : \text{IF } x_1(t) \text{ is } F_1^2, \quad \text{THEN } \begin{cases} \dot{x}(t) = A_2 x(t) + B_2 u(t) \\ y(t) = C_2 x(t), \end{cases}$$

where the membership functions for F_1^1 and F_1^2 are chosen as

$$F_1^1(x_1) = \exp\left[-\frac{(x_1 - c_1^1)^2}{2(\sigma_1^1)^2}\right], \quad F_1^2(x_1) = \exp\left[-\frac{(x_1 - c_1^2)^2}{2(\sigma_1^2)^2}\right].$$

To simplify the problem, let us consider the case $C_1 = C_2 = C$. Then, we have the overall T–S fuzzy system

$$
\begin{aligned}
\dot{x} &= \mu_1(x_1)(A_1x + B_1u) + \mu_2(x_1)(A_2x + B_2u) \\
y &= Cx
\end{aligned} \tag{2.113}
$$

with $\mu_1(x_1) = F_1^1(x_1)$ and $\mu_2(x_1) = F_1^2(x_1)$.

By taking the first derivative of y, we obtain

$$\dot{y} = C\dot{x} = (\mu_1 CA_1 + \mu_2 CA_2)x + (\mu_1 CB_1 + \mu_2 CB_2)u.$$

If $\mu_1 CB_1 + \mu_2 CB_2 \neq 0$, the relative degree is 1. If $\mu_1 CB_1 + \mu_2 CB_2 = 0$, we continue to compute the second derivative of y:

$$
\begin{aligned}
\ddot{y} &= \frac{\partial \mu_1}{\partial x}\dot{x}CA_1x + \mu_1 CA_1\dot{x} + \frac{\partial \mu_2}{\partial x}\dot{x}CA_2x + \mu_2 CA_2\dot{x} \\
&= \underbrace{(CA_1x\frac{\partial \mu_1}{\partial x} + CA_2x\frac{\partial \mu_2}{\partial x} + \mu_1 CA_1 + \mu_2 CA_2)}_{F(x)}\dot{x}.
\end{aligned} \tag{2.114}
$$

Substituting (2.113) into (2.114) yields

$$\ddot{y} = F(x)(\mu_1 A_1 + \mu_2 A_2)x + F(x)(\mu_1 B_1 + \mu_2 B_2)u. \tag{2.115}$$

If $F(x)(\mu_1 B_1 + \mu_2 B_2) \neq 0$, the relative degree is 2. If $F(x)(\mu_1 B_1 + \mu_2 B_2) = 0$, we continue to compute the third derivative of y. Since the membership functions are usually nonlinear functions of states, the calculation of the high-order derivatives of y can be very complicated.

For general continuous-time T–S fuzzy systems, the topic on how to define their relative degrees was addressed by Zhang et al. (2017).

For general discrete-time T–S fuzzy systems described by

$$
\begin{aligned}
x(t+1) &= \sum_{i=1}^{N} \mu_i(\xi)(A_i x(t) + B_i u(t)) \\
y(t) &= \sum_{i=1}^{N} \mu_i(\xi) C_i x(t),
\end{aligned} \tag{2.116}
$$

a normal form for them will be derived in Chap. 6, which has an explicit relative degree structure and a specific input–output signal causality relationship desired for a feedback control design.

2.6 Summary

T–S fuzzy systems have some special features that make them become a powerful tool for nonlinear control system designs. T–S fuzzy systems have been proven to be universal approximators that enable them to serve as function approximators in approximation-based nonlinear control designs. When dynamic T–S fuzzy systems are used to represent dynamic nonlinear systems, the control design is based on the structure and parameters of T–S fuzzy systems.

In this chapter, we have studied three forms of dynamic T–S fuzzy systems: continuous-time state-space form, discrete-time state-space form, and discrete-time input–output form. Based on the three forms of T–S fuzzy systems, we have reviewed some important results on their stability conditions, stabilization control designs and tracking control designs, and some key design problems.

With different Lyapunov functions, i.e., single quadratic Lyapunov function, piecewise quadratic Lyapunov functions, and fuzzy quadratic Lyapunov function, different stability conditions of T–S fuzzy systems can be derived, which require finding either a common positive definite matrix or several positive definite matrices by solving a group of LMIs. It should be noted that all those conditions are only sufficient conditions for the stability of T–S fuzzy systems.

For the control designs for T–S fuzzy systems, generally there are two approaches. One approach is to design a local controller for each local linear subsystem first and then obtain the global controller for the overall fuzzy system with certain aggregating or switching mechanism. This approach is usually called parallel distributed compensation (PDC). The other approach is to design the global controller directly based on the overall T–S fuzzy systems. The PDC control design can use mature linear control theory when designing the local linear controllers. The difficulty lies in how to prove that the global controller can ensure the stability of the overall closed-loop systems. The parameters of the local controllers can also be obtained by solving a group of LMIs. However, there is no guarantee that the solutions can be found all the time. Research efforts are still being made on how to relax those design conditions.

For the tracking control design based on the global T–S fuzzy system, the T–S fuzzy systems are treated as a class of nonlinear systems with certain structure and parameters. If we want to employ popular nonlinear control design techniques such as feedback linearization, important questions include how to determine the relative degree of a T–S fuzzy system and how to define its internal dynamics and its minimum phase property. This problem has been discussed in this chapter and will be further addressed in Chap. 6.

All the designs in this chapter are assumed that the parameters of T–S fuzzy systems are known. However, in reality, there exist parameter uncertainties in nonlinear systems. Correspondingly, the T–S fuzzy systems used to represent such nonlinear systems will also have parameter uncertainties. As is well known, adaptive control is effective in dealing with parametric uncertainties. In Chap. 3, we will first give a tutorial introduction on the basic ideas and design techniques of adaptive control. Then in Chaps. 5–10, we will present a systematic study on the adaptive control designs and analysis of T–S fuzzy systems.

References

Boyd S, El Ghaoui L, Feron E, Balakrishnan V (1994) Linear matrix inequalities in systems and control theory. SIAM, PA, Philadelphia

Bronshtein IN, Semendyayev KA (1985) Handbook of mathematics. Van Nostrand Reinhold, New York

Buckley JJ (1993) Sugeno type controllers are universal controllers. Fuzzy Sets Syst 53:299–303

Cao SG, Rees NW, Feng G (1995) Analysis and design of fuzzy control systems using dynamic fuzzy global models. Fuzzy Sets Syst 75:47–62

Cao SG, Rees NW, Feng G (1997a) Analysis and design for a class of complex control systems - Part I: fuzzy modeling and identification. Automatica 33:1017–1028

Castro JL (1995) Fuzzy logic controllers are universal approximators. IEEE Trans Syst Man Cybern 25:629–635

Castro JL, Delgado M (1996) Fuzzy systems with defuzzification are universal approximators. IEEE Trans Syst Man Cybern 26:149–152

Chadli M, Maquin D, Ragot J (2000) Relaxed stability conditions for Takagi–Sugeno fuzzy systems. In: Proceedings of IEEE international conference on systems, man, and cybernetics, Nashville, TN, pp 3514–3519

Chou J-H, Chen S-H (2001) Stability analysis of the discrete Takagi–Sugeno fuzzy model with time-varying consequent uncertainties. Fuzzy Sets Syst 118:271–279

Feng G (2004) Stability analysis of discrete-time fuzzy dynamic systems based on piecewise Lyapunov functions. IEEE Trans Fuzzy Syst 12(1):22–28

Feng G (2010) Analysis and synthesis of fuzzy control systems: a model-based approach. CRC Press, Boca Raton

Gahinet P, Nemirovski A, Laub AJ, Chilali M (1995) LMI control toolbox for use with Matlab. The Mathworks, MA, Natick

Isidori A (1995) Nonlinear control systems, 3rd edn. Springer-Verlag, London

Jin Y (2003) Advanced fuzzy systems design and applications. Physica-Verlag, Heidelberg

Joh J, Chen YH, Langari R (1998) On the stability issues of linear Takagi-Sugeno fuzzy models. IEEE Trans Fuzzy Syst 6:402–410

Johansson M, Rantzer A, Årzé K-E (1999) Piecewise quadratic stability of fuzzy systems. IEEE Trans Fuzzy Syst 7(6):713–722

Khalil HK (2002) Nonlinear systems. Prentice-Hall, Upper Saddle River

Kim E, Kim D (2001) Stability analysis and synthesis for an affine fuzzy system via LMI and ILMI: discrete case. IEEE Trans Syst Man Cybern 31:132–140

Krstić M, Kanellakopoulos I, Kokotović P (1995) Nonlinear and adaptive control design. Wiley, New York

Lilly JH (2010) Fuzzy control and identification. Wiley, Hoboken

Li N, Li SY (2004) Stability analysis and design of T-S fuzzy control system with simplified linear rule consequent. IEEE Trans Syst Man Cybern 34:788–795

Marino R, Tome P (1995) Nonlinear control design: geometric, adaptive and robust. Prentice-Hall, Englewood Cliffs

Nijmeijer H, Van der Schaft A (1990) Nonlinear dynamical control systems. Spring-Verlag, New York

Passino KM, Yurkovich S (1998) Fuzzy control. Addison Wesley Longman Inc., Reading

Rovatti R (1998) Fuzzy piecewise multilinear and piecewise linear systems as universal approximators in Sobolev norms. IEEE Trans Fuzzy Syst 6:235–249

Tanaka K, Sano M (1994) A robust stabilization problem of fuzzy control systems and its application to backing up control of a truck-trailer. IEEE Trans Fuzzy Syst 2(2):119–134

Tanaka K, Sugeno M (1992) Stability analysis and design of fuzzy control systems. Fuzzy Sets Syst 45:135–156

Tanaka K, Ikeda T, Wang HO (1998) Fuzzy regulators and fuzzy observers: relaxed stability conditions and LMI-based designs. IEEE Trans Fuzzy Syst 6(2):250–265

Wang L-X (1994) Adaptive fuzzy systems and control-design and stability analysis. Prentice-Hall, Englewood Cliffs

Wang L-X (1998) Universal approximation by hierarchical fuzzy systems. Fuzzy Sets Syst 93:223–230

Wang LX, Mendel JM (1992) Fuzzy basis functions, universal approximation, and orthogonal least-square learning. IEEE Trans Neural Netw 3:807–814

Wang HO, Tanaka K, Griffin MF (1996) An approach to fuzzy control of nonlinear systems: stability and design issues. IEEE Trans Fuzzy Syst 4:14–23

Ying H (1994) Sufficient conditions on general fuzzy systems as function approximators. Automatica 30:521–C525

Ying H (1998a) General SISO Takagi–Sugeno fuzzy systems with linear rule consequent are universal approximators. IEEE Trans Fuzzy Syst 28:582–587

Ying H (1998b) Sufficient conditions on uniform approximation of multivariate functions by general Takagi–Sugeno fuzzy systems with linear rule consequent. IEEE Trans Syst Man Cybern 28:515–520

Zeng XJ, Singh MG (1994) Approximation theory of fuzzy systems - SISO case. IEEE Trans Fuzzy Syst 2:162–176

Zeng XJ, Singh MG (1995) Approximation theory of fuzzy systems - MIMO case. IEEE Trans Fuzzy Syst 3(2):219–235

Zhang YJ, Tao G, Chen M (2017) Parametrization and adaptive control of multivariable non-canonical T–S fuzzy systems. IEEE Trans Fuzzy Syst 25(1):156–171

Zhou SS, Feng G, Lam J, Xu SY (2005) Robust H-infinity control for discrete fuzzy systems via basis-dependent Lyapunov functions. Inf Sci 174(3–4):197–217

Chapter 3
Adaptive Control: A Tutorial Introduction

Adaptive control is a well-established and powerful methodology to deal with systems with unknown constant or slowly time-varying parameters. The basic idea of adaptive control is to employ a parameter adaptation scheme to estimate those unknown parameters and replace the unknown parameters in the feedback controller with their estimates. Adaptive control is essentially nonlinear control due to the involvement of dynamic parameter adaptive law, no matter it is applied to linear systems or nonlinear systems.

Traditionally, there are two approaches in adaptive control design. One approach is called *direct adaptive control*, which estimates the controller parameters directly, and the other is called *indirect adaptive control*, which estimates the plant parameters and the controller parameters are calculated from the plant parameter estimates.

When designing the parameter adaptive law, there are also two approaches. One approach is *Lyapunov-based design* and the other is *estimation-based design* (Krstić et al. 1995). The major difference between the two approaches resides in the way to derive the parameter adaptive laws and the corresponding stability and convergence proof. In Lyapunov-based designs, usually a Lyapunov function containing tracking error and parameter estimation error is chosen first, and then the derivative of the Lyapunov function along the tracking error dynamics is derived, based on which the parameter adaptive law is designed such that the derivative of the Lyapunov function is nonpositive. Estimation-based designs treat the parameter estimation as a separate module and guarantee the properties of parameter estimation independent of the controller module. Various parameter identification algorithms such as gradient and least-squares optimization algorithms can be applied in estimation-based designs.

In adaptive control, there is an important design principle called *certainty equivalent principle* (Astrom and Wittenmark 1995). The controller is first designed by assuming all the plant parameters were known. When the plant parameters are

R. Qi et al., *Fuzzy System Identification and Adaptive Control*, Communications and Control Engineering,
https://doi.org/10.1007/978-3-030-19882-4_3

unknown, the control law is computed from the parameter estimates by treating them as if they were the true parameters, which is called a *certainty equivalence controller*. The certainty equivalent controllers have been proven to be satisfactory for adaptive control of linear systems.

This chapter is aimed at providing an introductory overview of adaptive linear and nonlinear control, to help readers understand the basic ideas and design approaches of adaptive control. We use simple single-input single-output (SISO) linear and nonlinear systems as design examples to illustrate the basic ideas of adaptive control. For readers who want to get a deeper understanding of adaptive control, there are many elegant books in this area (Goodwin and Sin 1984; Astrom and Wittenmark 1995; Krstić et al. 1995; Ioannou and Sun 1996; Tao 2003; Landau and Lozano et. al. 2011).

3.1 Adaptive Linear Control

In this section, we introduce the basic ideas of indirect and direct adaptive control, and model reference adaptive control for linear systems. We try to make it easier for beginners to understand how an adaptive control system is designed and what are the key properties it can achieve. We employ a simple SISO linear plant with two unknown parameters as a demonstration example in the following designs.

Consider the following SISO linear plant:

$$\dot{y} = ay + bu, \tag{3.1}$$

where $y \in R$ is the output and $u \in R$ is the input; a and b are two plant parameters. It is assumed that $b \neq 0$, which means the relative degree of (3.1) is one. The control design objective is to make y track a reference signal y_m when a and b are unknown under the following assumption.

Assumption 3.1 The reference signal y_m and its first derivative $\dot{y}_m$ are bounded.

We first present the basic methodology of indirect and direct adaptive control for the plant (3.1).

3.1.1 Indirect Adaptive Control

Let us first consider the nominal case that a and b are known. One simple choice of u is

$$u = \frac{1}{b}[-ay + \dot{y}_m - k(y - y_m)], \tag{3.2}$$

where k is a positive constant, which is a design parameter chosen by the user.

Define the tracking error $e = y - y_m$. Substituting (3.2) into (3.1) yields

$$\dot{e} = -ke, \tag{3.3}$$

which implies that e approaches zero asymptotically (the convergence speed depends on the value of k) and the design objective is achieved.

Now we consider the case that a and b are unknown. In this case, the nominal controller (3.2) cannot be implemented.

Suppose that we can obtain the parameter estimates $(\hat{a},\ \hat{b})$ for $(a,\ b)$, then the controller (3.2) can be implemented by the following indirect adaptive controller:

$$u = \frac{1}{\hat{b}}[-\hat{a}y + \dot{y}_m - k(y - y_m)], \tag{3.4}$$

$$\dot{\hat{a}} = \gamma_1 ey, \tag{3.5}$$

$$\dot{\hat{b}} = \gamma_2 eu, \tag{3.6}$$

where $\gamma_1, \gamma_2 > 0$ are two design parameters. The adaptive control scheme (3.4)–(3.6) is called *indirect* adaptive control because the controller is calculated based on the estimates $(\hat{a},\ \hat{b})$ of the plant parameters $(a,\ b)$.

Applying (3.4) into (3.1), we obtain

$$\dot{e} = -ke - \tilde{a}y - \tilde{b}u, \tag{3.7}$$

where $\tilde{a} = \hat{a} - a$ and $\tilde{b} = \hat{b} - b$.

The stability of the error dynamics (3.7) can be checked by examining the derivative of the Lyapunov function

$$V(e, \tilde{a}, \tilde{b}) = \frac{1}{2}e^2 + \frac{1}{2\gamma_1}\tilde{a}^2 + \frac{1}{2\gamma_2}\tilde{b}^2, \tag{3.8}$$

where $\gamma_1, \gamma_2 > 0$ are two design parameters. The derivative of (3.8) along the error dynamics (3.7) follows

$$\dot{V} = -ke^2 - \tilde{a}ey - \tilde{b}eu + \frac{1}{\gamma_1}\tilde{a}\dot{\tilde{a}} + \frac{1}{\gamma_2}\tilde{b}\dot{\tilde{b}}. \tag{3.9}$$

Since a and b are two constants, we have $\dot{a} = \dot{b} = 0$, and from (3.5)–(3.6), we obtain

$$\dot{\tilde{a}} = \dot{\hat{a}} = \gamma_1 ey,$$
$$\dot{\tilde{b}} = \dot{\hat{b}} = \gamma_2 eu,$$

which are used in (3.9), leading to

$$\dot{V} = -ke^2, \tag{3.10}$$

which indicates $V(e, \tilde{a}.\tilde{b})$ evaluated along (3.7) is a nonincreasing function of time. This proves $e,\ \tilde{a},\ \tilde{b} \in L_\infty$, i.e., bounded for all $t \geq 0$, which also indicates $y,\ \hat{a},\ \hat{b} \in L_\infty$. From (3.4), we have $u \in L_\infty$ when $\hat{b} \neq 0$ (which can be guaranteed by using various parameter projection algorithms). Then, from (3.7), we have $\dot{e} \in L_\infty$. By integrating both sides of (3.10), we have

$$\int_0^\infty e^2 dt = \frac{1}{k}(V(0) - V(\infty)) < \infty, \tag{3.11}$$

which means $e \in L^2$. With $e,\ \dot{e} \in L_\infty$ and $e \in L_2$, it can be concluded from a corollary of Barbălat's lemma that $\lim_{t\to\infty} e(t) = 0$, which means the design objective has been achieved using the indirect adaptive control scheme (3.4)–(3.6).

Barbălat's lemma and its corollary are widely used in proving the signal convergence of an adaptive control system, which are given as follows (Popov 1966).

Lemma 3.1 (Barbălat) *Consider the function $\phi : R_+ \to R$. If ϕ is uniformly continuous and* $\lim_{t\to\infty} \int_0^t \phi(\tau)d\tau$ *exists and is finite, then*

$$\lim_{t\to\infty} \phi(t) = 0. \tag{3.12}$$

Corollary 3.1 *Consider the function $\phi : R_+ \to R$. If $\phi, \dot{\phi} \in L_\infty$, and $\phi \in L_p$ for some $p \in [1, \infty)$, then*

$$\lim_{t\to\infty} \phi(t) = 0. \tag{3.13}$$

Parameter projection. The parameter estimate $\hat{b}$ can be guaranteed bounded away from zero by using various parameter projection algorithms (Ioannou and Sun 1996; Tao 2003). To employ a parameter projection algorithm, some knowledge of the control gain b is needed.

Assumption 3.2 The sign of b is known and $b \in [b_1, b_2]$ for some known constants b_1 and b_2.

Without loss of generality, it is assumed $b > 0$ in the following design. The parameter adaptive law (3.6) for updating $\hat{b}$ is modified as

$$\dot{\hat{b}} = \gamma_2 eu + f(t), \tag{3.14}$$

where $f(t)$ is a projection signal defined as

$$f(t) = \begin{cases} 0 & \text{if } \hat{b} \in (b_1, b_2) \\ & \text{or } \hat{b} = b_1 \text{ and } \gamma_2 eu \geq 0 \\ & \text{or } \hat{b} = b_2 \text{ and } \gamma_2 eu \leq 0, \\ -\gamma_2 eu & \text{otherwise,} \end{cases}$$

which for $\hat{b}(0) \in [b_1, b_2]$ guarantees that $\hat{b}(t) \in [b_1, b_2]$.

Consider the same Lyapunov function (3.8), whose derivative along (3.21), (3.5) and (3.14) is

$$\dot{V} = \begin{cases} -ke^2 & \text{if } \hat{b} \in (b_1, b_2) \\ & \text{or } \hat{b} = b_1 \text{ and } eu \geq 0 \\ & \text{or } \hat{b} = b_2 \text{ and } eu \leq 0, \\ -ke^2 - \tilde{b}eu & \text{otherwise.} \end{cases} \tag{3.15}$$

When $\hat{b} = b_1$ and $eu < 0$, we have $\tilde{b}eu = (\hat{b} - b)eu \geq 0$; when $\hat{b} = b_2$ and $eu > 0$, we have $\tilde{b}eu = (\hat{b} - b)eu \geq 0$ as well. Therefore, from (3.15), we can conclude

$$\dot{V} \leq -ke^2 \leq 0. \tag{3.16}$$

By following the similar analysis for the case without parameter projection, it can be obtained that all the closed-loop signals are bounded and $\lim_{t\to\infty} e(t) = 0$ under the adaptive control law (3.4), (3.5) and (3.14).

3.1.2 Direct Adaptive Control

In Sect. 3.1.1, the parameters of the adaptive controller (3.4) are obtained from the estimates $(\hat{a}, \hat{b})$ of plant parameters (a, b), which is called indirect adaptive control. In this section, we try to estimate the parameters of the adaptive controller directly, which leads to the *direct adaptive control* approach.

First, we parameterize the control law (3.2) as

$$u = -\theta_1 y + \theta_2 \dot{y}_m - k\theta_2 e, \tag{3.17}$$

where $\theta_1 = \frac{a}{b}$ and $\theta_2 = \frac{1}{b}$. With the new parametrization, the plant (3.1) can be written as

$$\dot{y} = \frac{\theta_1}{\theta_2} y + \frac{1}{\theta_2} u. \tag{3.18}$$

When θ_1 and θ_2 are known (from the knowledge of a and b), applying (3.17) into (3.18) yields

$$\dot{e} = -ke, \tag{3.19}$$

which leads to $\lim_{t\to\infty} e(t) = 0$.

When θ_1 and θ_2 are unknown, their estimates $\hat{\theta}_1$ and $\hat{\theta}_2$ are used to build an adaptive controller

$$u = -\hat{\theta}_1 y + \hat{\theta}_2 \dot{y}_m - k\hat{\theta}_2 e. \tag{3.20}$$

To implement (3.20), we need to develop parameter update laws for $\hat{\theta}_1$ and $\hat{\theta}_2$. Generally, there are two approaches to obtain the parameter update laws: one is Lyapunov-based design and the other is estimation-based design. The Lyapunov-based design chooses a Lyapunov function of the tracking error and the parameter estimation error. Then the parameter update laws are designed to make the derivative of the Lyapunov function nonpositive. The estimation-based approach is to design a parameter identifier with guaranteed parameter estimation properties, such as the gradient algorithm and the least-squares algorithms.

Here, we use the Lyapunov-based design to obtain the parameter adaptive laws for $\hat{\theta}_1$ and $\hat{\theta}_2$. First, we need to develop the tracking error dynamics subject to the controller (3.20). Applying (3.20) into (3.18) and with some simple mathematical manipulations, we obtain

$$\dot{e} = -ke - \frac{1}{\theta_2}(\hat{\theta}_1 - \theta_1)y + \frac{1}{\theta_2}(\hat{\theta}_2 - \theta_2)(\dot{y}_m - ke). \tag{3.21}$$

Define $\tilde{\theta}_1 = \hat{\theta}_1 - \theta_1$, $\tilde{\theta}_2 = \hat{\theta}_2 - \theta_2$ and $\omega = \dot{y}_m - ke$, then the error dynamics (3.21) can be written as

$$\dot{e} = -ke - \frac{1}{\theta_2}\tilde{\theta}_1 y + \frac{1}{\theta_2}\tilde{\theta}_2 \omega. \tag{3.22}$$

The error dynamics (3.22) builds the relation between the tracking error and the parameter estimation errors, which is crucial for the Lyapunov-based design.

It can be noticed that the signals $e(t)$, $y(t)$ and $\omega(t)$ are all available at time t. Our parameter adaptive laws will be developed using these signals. Besides, the information on the sign of θ_2, that is, the sign of control gain b, is also needed.

Assumption 3.3 The sign of the control gain b is known.

Assumption 3.3 means the control direction of the linear system (3.1) is known, which can be satisfied in many control applications.

Consider the following Lyapunov function:

$$V(e, \tilde{\theta}_1, \tilde{\theta}_2) = \frac{1}{2}e^2 + \frac{1}{2\gamma_1|\theta_2|}\tilde{\theta}_1^2 + \frac{1}{2\gamma_2|\theta_2|}\tilde{\theta}_2^2, \tag{3.23}$$

where $\gamma_1, \gamma_2 > 0$ are two design parameters.

The derivative of (3.23) along (3.22) is

$$\begin{aligned}\dot{V} &= -ke^2 - \frac{1}{\theta_2}\tilde{\theta}_1 ey + \frac{1}{\theta_2}\tilde{\theta}_2 e\omega + \frac{1}{\gamma_1|\theta_2|}\tilde{\theta}_1\dot{\tilde{\theta}}_1 + \frac{1}{\gamma_2|\theta_2|}\tilde{\theta}_2\dot{\tilde{\theta}}_2 \\ &= -ke^2 + \frac{\tilde{\theta}_1}{|\theta_2|}\left(\frac{1}{\gamma_1}\dot{\tilde{\theta}}_1 - \text{sign}(\theta_2)ey\right) + \frac{\tilde{\theta}_2}{|\theta_2|}\left(\frac{1}{\gamma_2}\dot{\tilde{\theta}}_2 - \text{sign}(\theta_2)e\omega\right) \\ &= -ke^2 + \frac{\tilde{\theta}_1}{|\theta_2|}\left(\frac{1}{\gamma_1}\dot{\hat{\theta}}_1 - \text{sign}(b)ey\right) + \frac{\tilde{\theta}_2}{|\theta_2|}\left(\frac{1}{\gamma_2}\dot{\hat{\theta}}_2 - \text{sign}(b)e\omega\right). \end{aligned} \tag{3.24}$$

If the parameter adaptive laws are designed as

$$\dot{\hat{\theta}}_1 = \gamma_1 \text{sign}(b)ey, \tag{3.25}$$

$$\dot{\hat{\theta}}_2 = \gamma_2 \text{sign}(b)e\omega, \tag{3.26}$$

from (3.24), we have

$$\dot{V} = -ke^2, \tag{3.27}$$

based on which we can conclude e, $\tilde{\theta}_1$ and $\tilde{\theta}_2$ are bounded. By following the similar analysis as we have done in Sect. 3.1.1, it can be obtained that all the closed-loop signals are bounded and $\lim_{t\to\infty} e(t) = 0$.

Since no parameter estimate appears in the denominator of the control law (3.20), no parameter projection is required for the parameter adaptive laws (3.25)–(3.26). It can be observed that the information of sign(b) is required in the parameter adaptive laws (3.25)–(3.26) for the direct adaptive control scheme, while the same information is also required in the modified parameter adaptive law (3.14) for the indirect adaptive control scheme.

3.1.3 Model Reference Adaptive Control

In model reference adaptive control, the expected output is generated by a reference model. For the plant (3.1), suppose the reference model is given by

$$\dot{y}_m = -a_m y_m + b_m r, \tag{3.28}$$

where $y_m \in R$ and $r \in R$ are the output and input of the reference model, respectively; a_m and b_m are the model parameters. To make the reference model stable, a_m satisfies $a_m > 0$. The control objective is to make the plant output y track the reference model output y_m as closely as possible.

First, we consider the nominal case that a and b in (3.1) are known. The nominal controller has the following structure:

$$u = k_1 y + k_2 r, \tag{3.29}$$

where k_1 and k_2 are two parameters to be designed. Applying (3.29) into (3.1) results in the following closed-loop dynamics

$$\dot{y} = (a + bk_1)y + bk_2 r. \tag{3.30}$$

If k_1 and k_2 are chosen to satisfy the following *matching conditions*:

$$a + bk_1 = -a_m, \quad bk_2 = b_m, \tag{3.31}$$

then the closed-loop system becomes

$$\dot{y} = -a_m y + b_m r. \tag{3.32}$$

Define the output tracking error $e = y - y_m$. Subtracting (3.28) from (3.32) yields the following tracking error dynamics:

$$\dot{e} = -a_m e, \tag{3.33}$$

which implies e converges to zero asymptotically.

In this case, the matching conditions (3.31) can be easily satisfied with

$$k_1 = \frac{-a_m - a}{b}, \quad k_2 = \frac{b_m}{b}. \tag{3.34}$$

Thus, the control objective can be achieved perfectly when a and b are known.

Now we consider the case that a and b are unknown. In this case, the controller parameters k_1 and k_2 cannot be calculated from the matching conditions (3.31) and their estimates $\hat{k}_1$ and $\hat{k}_2$ are used to form an adaptive control law

$$u = \hat{k}_1 y + \hat{k}_2 r. \tag{3.35}$$

To develop adaptive laws for $\hat{k}_1$ and $\hat{k}_2$, we need to obtain the tracking error dynamics for the plant (3.1) subject to the adaptive control law (3.35).

Applying (3.35) into (3.1), we have

$$\begin{aligned}\dot{y} &= ay + b\hat{k}_1 y + b\hat{k}_2 r \\ &= (a + bk_1)y + bk_2 r + b\tilde{k}_1 y + b\tilde{k}_2 r,\end{aligned} \tag{3.36}$$

where $\tilde{k}_1 = \hat{k}_1 - k_1$ and $\tilde{k}_2 = \hat{k}_2 - k_2$.

Using the matching conditions (3.31), from (3.36), we have

$$\dot{y} = -a_m y + b_m r + b\tilde{k}_1 y + b\tilde{k}_2 r. \tag{3.37}$$

Subtracting (3.28) from (3.37) yields the following closed-loop error dynamics:

$$\dot{e} = -a_m e + b\tilde{k}_1 y + b\tilde{k}_2 r. \tag{3.38}$$

It can be noticed that the control gain b appears on the right side of (3.38). To develop parameter adaptive laws through Lyapunov-based approach, we need the information of the sign of b, that is, Assumption 3.3.

Consider the following Lyapunov function:

$$V(e, \tilde{k}_1, \tilde{k}_2) = \frac{1}{2}e^2 + \frac{|b|}{2\gamma_1}\tilde{k}_1^2 + \frac{|b|}{2\gamma_2}\tilde{k}_2^2. \tag{3.39}$$

The derivative of (3.39) along (3.38) is

$$\begin{aligned} \dot{V} &= e\dot{e} + \frac{|b|}{\gamma_1}\tilde{k}_1\dot{\tilde{k}}_1 + \frac{|b|}{\gamma_2}\tilde{k}_2\dot{\tilde{k}}_2 \\ &= -a_m e^2 + |b|\tilde{k}_1\left(\text{sign}(b)ey + \frac{1}{\gamma_1}\dot{\tilde{k}}_1\right) + |b|\tilde{k}_2\left(\text{sign}(b)er + \frac{1}{\gamma_2}\dot{\tilde{k}}_2\right). \end{aligned} \tag{3.40}$$

If the parameter adaptive laws are chosen as

$$\begin{aligned} \dot{\hat{k}}_1 &= \dot{\tilde{k}}_1 = -\gamma_1 \text{sign}(b)ey \\ \dot{\hat{k}}_2 &= \dot{\tilde{k}}_2 = -\gamma_2 \text{sign}(b)er, \end{aligned} \tag{3.41}$$

from (3.40), we have

$$\dot{V} = -a_m e^2 \leq 0, \tag{3.42}$$

which means $V(e, \tilde{k}_1, \tilde{k}_2)$ is nonincreasing such that $e,\ \tilde{k}_1, \tilde{k}_2 \in L_\infty$. Hence, $y,\ \hat{k}_1,\ \hat{k}_2 \in L_\infty$ as well. Then, from (3.35), we obtain $u \in L_\infty$. Thus, all the closed-loop signals are bounded.

Furthermore, from (3.38) we have $\dot{e} \in L_\infty$. By integrating both sides of (3.42), we obtain

$$\int_0^\infty e^2 dt = \frac{1}{a_m}(V(0) - V(\infty)) < \infty, \tag{3.43}$$

which means $e \in L_2$. With $e, \dot{e} \in L_\infty$ and $e \in L_2$, it can be concluded based on the Corollary of Barbălat's Lemma that $\lim_{t\to\infty} e(t) = 0$.

3.1.4 Discrete-Time Adaptive Linear Control

Consider the following discrete-time linear plant:

$$y(t+1) = ay(t) + bu(t), \tag{3.44}$$

where $y \in R$ and $u \in R$ are output and input, respectively; a and b ($b \neq 0$) are plant parameters. Our design objective is to make y track a bounded reference signal y_m.

First, we consider how to design a nominal control law for known a and b. Designing the control law as

$$u(t) = \frac{1}{b}[-ay(t) + y_m(t+1)], \tag{3.45}$$

which, when applied to the system (3.44), brings $y(t+1)$ to $y_m(t+1)$ in one step and leads to the closed-loop dynamics

$$y(t) = y_m(t), \forall t > 1. \tag{3.46}$$

Remark 3.1 The nominal control law for known a and b can also be designed as

$$u(t) = \frac{1}{b}[-ay(t) + y_m(t+1) + k(y(t) - y_m(t))], \tag{3.47}$$

where the parameter k is chosen to satisfy $|k| < 1$. Applying (3.47) into (3.44) yields the closed-loop dynamics

$$y(t+1) = y_m(t+1) + k(y(t) - y_m(t)). \tag{3.48}$$

Define the tracking error $e(t) = y(t) - y_m(t)$. From (3.48), we have the tracking error dynamics

$$e(t+1) = ke(t), \tag{3.49}$$

which, due to $|k| < 1$, leads to $\lim_{t\to\infty} e(t) = 0$.

When a and b are unknown, their estimates $\hat{a}(t)$ and $\hat{b}(t)$ at time t are used to construct an adaptive control law

$$u(t) = \frac{1}{\hat{b}(t)}[-\hat{a}(t)y(t) + y_m(t+1)], \tag{3.50}$$

where $\hat{b}$ should be ensured nonzero.

Applying (3.50) into (3.44) and with some simple mathematical manipulations, we obtain the following error dynamics:

$$e(t+1) = -\tilde{a}(t)y(t) - \tilde{b}(t)u(t), \tag{3.51}$$

where $\tilde{a}(t) = \hat{a}(t) - a$ and $\tilde{b}(t) = \hat{b}(t) - b$.

In Sect. 3.1.1, we derive the parameter adaptive laws through a Lyapunov-based design. However, a major difference between discrete-time adaptive control and continuous-time adaptive control is that the Lyapunov-based design is not applicable to the discrete-time case. For discrete-time systems, it is more convenient to derive the parameter adaptive law through an estimation-based approach, that is, employing a parameter identifier to estimate the unknown parameter.

The plant (3.44) can be parameterized as

$$y(t+1) = \theta^T \phi(t), \tag{3.52}$$

where $\theta = [a, \ b]^T$ is a vector of unknown plant parameters and $\phi(t) = [y(t), \ u(t)]^T$ is a regressor vector.

Define the *estimation error* with $\hat{\theta}(t)$ as

$$\varepsilon(t) = y(t+1) - \hat{\theta}^T(t)\phi(t), \tag{3.53}$$

where $\hat{\theta}(t) = [\hat{a}(t), \ \hat{b}(t)]^T$. In view of (3.52), the estimation error (3.53) can be formulated into

$$\varepsilon(t) = -\tilde{\theta}^T(t)\phi(t), \quad \tilde{\theta}(t) = \hat{\theta}(t) - \theta, \tag{3.54}$$

which is a linear parametric error model (linear in the parameter error $\tilde{\theta}$).

Remark 3.2 The linear parametric model (3.54) is a key model in adaptive control and system identification, which builds the relation between the parameter error $\tilde{\theta}$ and estimation error ε. Based on the measurable signals $y(t+1)$ and $\phi(t)$, an adaptive law for updating $\hat{\theta}$ can be derived to ensure $\hat{\theta}$ has some desirable properties, such as gradient and least-squares methods.

Based on (3.53), a *normalized gradient algorithm* can be applied for updating the parameter vector $\hat{\theta}$ to minimize a normalized quadratic cost function

$$J(\hat{\theta}) = \frac{\varepsilon^2(t)}{2m^2(t)}, \tag{3.55}$$

where $m(t)$ is a normalizing signal which does not explicitly depend on $\hat{\theta}(t)$. A preferred choice of $m(t)$ is the one which ensures the boundedness of $\frac{\phi^T(t)\phi(t)}{m^2(t)}$. In this example, we choose

$$m(t) = \sqrt{c + \phi^T(t)\phi(t)}, \tag{3.56}$$

where $c > 0$ is a small design parameter.

$$\hat{\theta}(t+1) = \hat{\theta}(t) + \frac{\gamma(t)\phi(t)\varepsilon(t)}{m^2(t)}, \tag{3.57}$$

where $0 < \gamma_0 < \gamma(t) < 2 - \gamma_0$ is an adaptation gain for some constant $\gamma_0 \in (0, 1)$.

The parameter adaptive law (3.57) has the following properties.

Lemma 3.2 *The parameter adaptive law (3.57) guarantees that*

(i) $\hat{\theta}(t)$ *and* $\frac{\varepsilon(t)}{m(t)}$ *are bounded;*
(ii) $\hat{\theta}(t+1) - \hat{\theta}(t)$ *and* $\frac{\varepsilon(t)}{m(t)}$ *belong to* L_2*; and*
(iii) $\lim_{t\to\infty} \frac{\varepsilon(t)}{m(t)} = 0$.

Proof (i) Consider the following Lyapunov function candidate:

$$V(\tilde{\theta}) = \tilde{\theta}^T\tilde{\theta}. \tag{3.58}$$

The time increment of $V(\tilde{\theta})$ along (3.57) is

$$\begin{aligned} V(\tilde{\theta}(t+1)) - V(\tilde{\theta}(t)) &= \|\hat{\theta}(t+1) - \theta\|^2 - \|\hat{\theta}(t) - \theta\|^2 \\ &= -\gamma(t)\left[2 - \frac{\gamma(t)\phi^T(t)\phi(t)}{c + \phi^T(t)\phi(t)}\right] \cdot \frac{\varepsilon^2(t)}{c + \phi^T(t)\phi(t)}. \end{aligned} \tag{3.59}$$

Since

$$\begin{aligned} -\gamma(t)\left[2 - \frac{\gamma(t)\phi^T(t)\phi(t)}{c + \phi^T(t)\phi(t)}\right] &\le -\gamma(t)\left[2 - \gamma(t)\right] \\ &\le -\gamma_0(2 - \gamma_0), \end{aligned} \tag{3.60}$$

we have the inequality

$$V(\tilde{\theta}(t+1)) - V(\tilde{\theta}(t)) \le -\gamma_0(2 - \gamma_0)\frac{\varepsilon^2(t)}{m^2(t)} \le 0, \tag{3.61}$$

which means $V(\tilde{\theta})$ is a nonincreasing function. Thus, $\tilde{\theta}$ and $\hat{\theta}$ are bounded. From (3.61), we have

$$\gamma_0(2 - \gamma_0)\frac{\varepsilon^2(t)}{m^2(t)} \le V(\tilde{\theta}(t)) - V(\tilde{\theta}(t+1)) < \infty \tag{3.62}$$

which gives $\frac{\varepsilon(t)}{m(t)} \in L_\infty$.

(ii) It follows from (3.61) that

$$\gamma_0(2 - \gamma_0)\sum_{\tau=0}^{t-1} \frac{\varepsilon^2(\tau)}{m^2(\tau)} = V(\tilde{\theta}(0)) - V(\tilde{\theta}(t)) \le V(\tilde{\theta}(0)), \tag{3.63}$$

which means $\frac{\varepsilon(t)}{m(t)} \in L_2$, and from (3.57) that $\hat{\theta}(t+1) - \hat{\theta}(t) \in L_2$.

(iii) With $\frac{\varepsilon(t)}{m(t)} \in L_\infty \bigcap L_2$, it is straightforward to obtain

$$\lim_{t\to\infty} \frac{\varepsilon(t)}{m(t)} = 0,$$

which completes the proof. ∇

To avoid singularity in the adaptive control law (3.50), $\hat{b}$ should be guaranteed nonzero. Similar to the continuous-time case, parameter projection can be applied to make $\hat{b}$ bounded away from zero. Without loss of generality, the following assumption is made on b.

Assumption 3.4 It is assumed that $0 < b_1 \le b \le b_2$ for some known constants b_1 and b_2.

The parameter adaptive law (3.57) is modified as

$$\hat{a}(t+1) = \hat{a}(t) + \frac{\gamma(t)y(t)\varepsilon(t)}{m^2(t)}, \tag{3.64}$$

$$\hat{b}(t+1) = \hat{b}(t) + \frac{\gamma(t)u(t)\varepsilon(t)}{m^2(t)} + f(t), \tag{3.65}$$

where $\hat{b}(0) \in [b_1, b_2], f(t)$ is a projection signal which is defined as

$$f(t) = \begin{cases} 0 & h(t) \in [b_1, b_2] \\ b_2 - h(t) & h(t) > b_2 \\ b_1 - h(t) & h(t) < b_1, \end{cases} \tag{3.66}$$

where $h(t) = \hat{b}(t) + \frac{\gamma(t)u(t)\varepsilon(t)}{m^2(t)}$.

The modified parameter adaptive laws (3.64)–(3.65) guarantee for $\hat{b}(0) \in [b_1, b_2]$ that $\hat{b}(t) \in [b_1, b_2]$ and the desired properties of Lemma 3.2 are still valid, which can be checked by evaluating the increment of the Lyapunov function $V(\tilde{a}, \tilde{b}) = \tilde{a}^2 + \tilde{b}^2$ along (3.64)–(3.65).

Now we consider the property of the tracking error $e(t)$. Based on Lemma 3.2, we have the following desired closed-loop properties.

Theorem 3.1 *The adaptive controller (3.50) with the parameter adaptive laws (3.64)–(3.65), applied to the plant (3.44), can guarantee that all the closed-loop signals are bounded and the tracking error* $\lim_{t\to\infty} e(t+1) = 0$.

Proof Substituting (3.50) into (3.44), we obtain the closed-loop system

$$y(t+1) = \theta^T\phi(t) - \hat{\theta}\phi(t) + y_m(t+1), \tag{3.67}$$

from which we have

$$e(t+1) = -\tilde{\theta}^T(t)\phi(t). \tag{3.68}$$

From (3.54), we have

$$e(t+1) = \varepsilon(t), \tag{3.69}$$

which, divided by $m(t)$ on both sides, gives

$$\frac{e(t+1)}{m(t)} = \frac{\varepsilon(t)}{m(t)}. \tag{3.70}$$

Define $\bar{\varepsilon}(t) = \frac{\varepsilon(t)}{m(t)}$. Then from (3.70), we have

$$e(t+1) = \bar{\varepsilon}(t)m(t). \tag{3.71}$$

From (3.56), we have

$$m(t) \leq \sqrt{c} + |\phi(t)|. \tag{3.72}$$

With $\phi(t) = [y(t), u(t)]^T$, we have

$$\|\phi(t)\| \leq |y(t)| + |u(t)|. \tag{3.73}$$

From (3.50) and Assumption 3.7, we obtain

$$|u(t)| \leq c_1|y(t)| + c_2, \tag{3.74}$$

where c_1 and c_2 are two positive constants.

With (3.72)–(3.74) and $e(t) = y(t) - y_m(t)$, it can be derived that

$$m(t) \leq c_3|e(t)| + c_4, \tag{3.75}$$

where c_3 and c_4 are two positive constants.

With (3.69) and (3.75), it can be derived that

$$|e(t+1)| \leq c_3|\bar{\varepsilon}(t)||e(t)| + c_4|\bar{\varepsilon}(t)|. \tag{3.76}$$

From Lemma 3.2, we have $\lim_{t\to\infty} \bar{\varepsilon}(t) = 0$. Therefore, it can be concluded from (3.76) that $e(t) \in L_\infty$. Hence, $y(t),\ u(t),\ \phi(t) \in L_\infty$ as well.

With $\phi(t) \in L_\infty$ and $\frac{\varepsilon(t)}{m(t)} \in L_2$, it can be concluded from (3.69) that $e(t+1) \in L_2$. With $e(t+1) \in L_\infty \bigcap L_2$, we finally have $\lim_{t\to\infty} e(t+1) = 0$.

3.2 Adaptive Nonlinear Control

In this section, we mainly consider adaptive feedback linearization control for nonlinear systems. We employ a continuous-time example and a discrete-time example to illustrate the use of adaptive control methodology in nonlinear feedback control.

3.2.1 A Continuous-Time Design Example

In this example, we consider the tracking control design problem for the following nth-order continuous-time nonlinear plant:

$$\begin{aligned}\dot{x}_1 &= x_2\\ \dot{x}_2 &= x_3\\ &\vdots\\ \dot{x}_n &= \theta^T f(x) + g(x)u\\ y &= x_1,\end{aligned} \tag{3.77}$$

where $x = [x_1, x_2, \ldots, x_n]^T \in R^n$ is the state vector, $u \in R$ is input, $y \in R$ is output, $f : R^n \to R^l$ and $g : R^n \to R$ are sufficiently smooth in a domain $D \subset R^n$, $\theta \in R^l$ is a constant parameter vector. The design objective is to find an adaptive controller such that the output $y(t)$ tracks a bounded reference signal $y_m(t)$ for unknown parameter vector θ. It is assumed that the reference signal $y_m(t)$ satisfies the following assumption.

Assumption 3.5 $y_m(t),\ \dot{y}_m(t),\ \ldots,\ y_m^{(n)}(t)$ are bounded.

Without loss of generality, we make the following assumption on $g(x)$.

Assumption 3.6 $g(x) \geq g_0 > 0$, where g_0 is a positive constant.

Assumption 3.6 implies the relative degree of (3.77) is n.

We first consider the nominal case that θ is known. Define the tracking error $e(t) = y(t) - y_m(t)$, then the nominal control law can be designed as

$$u = \frac{1}{g(x)}\left(-\theta^T f(x) + y_m^{(n)} - k_{n-1}e^{(n-1)} - \cdots - k_1\dot{e} - k_0 e\right), \tag{3.78}$$

where the parameters $k_1,\ k_2, \ldots, k_{n-1}$ are chosen such that the polynomial

$$s^n + k_{n-1}s^{n-1} + \cdots + k_1 s + k_0 = 0 \tag{3.79}$$

is Hurwitz, i.e., all roots of (3.79) lie in the left-half complex plane.

Applying (3.78) into (3.77) results in the following tracking error dynamics

$$e^{(n)} + k_{n-1}e^{(n-1)} + \cdots + k_1\dot{e} + k_0 e = 0, \tag{3.80}$$

which leads to $\lim_{t\to\infty} e(t) = 0$.

When θ is unknown, its estimate $\hat{\theta}$ is used to implement the following adaptive control law:

$$u = \frac{1}{g(x)}\left(-\hat{\theta}^T f(x) + y_m^{(n)} - k_{n-1}e^{(n-1)} - \cdots - k_1\dot{e} - k_0 e\right). \tag{3.81}$$

Substituting (3.81) into (3.77) yields the following tracking error dynamics:

$$e^{(n)} + k_{n-1}e^{(n-1)} + \cdots + k_1\dot{e} + k_0 e + \tilde{\theta}^T f(x) = 0, \tag{3.82}$$

where $\tilde{\theta} = \hat{\theta} - \theta$.

Define $\underline{e} = [e, \dot{e}, \ldots, e^{(n-1)}]^T$. Then (3.82) can be formulated as

$$\dot{\underline{e}} = A\underline{e} + B\tilde{\theta}^T f(x), \tag{3.83}$$

where

$$A = \begin{bmatrix} 0 & 1 & 0\cdots & 0 \\ 0 & 0 & 1 & 0 \\ \vdots & & \ddots & \vdots \\ 0 & & & 1 \\ -k_0 & -k_1 & \cdots & -k_{n-1} \end{bmatrix}, \quad B = \begin{bmatrix} 0 \\ 0 \\ \vdots \\ 0 \\ 1 \end{bmatrix}. \tag{3.84}$$

Since A is a Hurwitz matrix, for a positive definite matrix Q, there exists a positive matrix P such that

$$A^T P + PA = -Q. \tag{3.85}$$

Consider the following Lyapunov function candidate:

$$V(\underline{e}, \tilde{\theta}) = \frac{1}{2}\underline{e}^T P \underline{e} + \frac{1}{2}\tilde{\theta}^T \Gamma^{-1}\tilde{\theta}, \tag{3.86}$$

where Γ is a diagonal matrix with its diagonal elements being positive constants.

The derivative of (3.86) along (3.82) is given by

$$\begin{aligned} \dot{V} &= -\frac{1}{2}\underline{e}^T Q\underline{e} + \tilde{\theta}^T \underline{e}^T PBf(x) + \tilde{\theta}^T \Gamma^{-1}\dot{\tilde{\theta}} \\ &= -\frac{1}{2}\underline{e}^T Q\underline{e} + \tilde{\theta}^T\left(\underline{e}^T PBf(x) + \Gamma^{-1}\dot{\hat{\theta}}\right). \end{aligned} \tag{3.87}$$

Selecting the parameter adaptive law as

$$\dot{\hat{\theta}} = -\Gamma \underline{e}^T PBf(x), \tag{3.88}$$

it can be obtained from (3.87) that

$$\dot{V} = -\frac{1}{2}\underline{e}^T Q\underline{e} \leq 0, \tag{3.89}$$

which implies $\underline{e}$, $\hat{\theta} \in L_\infty$. Since $e = x_1 - y_m, \dot{e} = x_2 - \dot{y}_m, \ldots e^{(n)} = x_n - y_m^{(n)}$, with Assumption 3.5, we have $x \in L_\infty$. Based on Assumption 3.6 and (3.81), it can be concluded that u are bounded. From (3.82), we obtain $\dot{\underline{e}} \in L_\infty$. Integrating both sides of (3.89), it can be obtained that $\int_0^\infty \underline{e}^T Q\underline{e} < \infty$, that is, $\underline{e} \in L_2$. With Barbălat's lemma, we finally obtain $\lim_{t\to\infty} \underline{e}(t) = 0$.

3.2.2 A Discrete-Time Design Example

In this example, we consider a nonlinear plant described by the following discrete-time model:

$$\begin{aligned} x_1(t+1) &= x_2(t) \\ x_2(t+1) &= x_3(t) \\ &\vdots \\ x_{n-1}(t+1) &= x_n(t) \\ x_n(t+1) &= \theta^T f(x(t)) + g(x(t))u(t) \\ y(t) &= x_n(t), \end{aligned} \tag{3.90}$$

where $x(t) = [x_1(t), x_2(t), \ldots, x_n(t)]^T \in R^n$ is the state vector, $u \in R$ is input, $y \in R$ is output, $f : R^n \to R^l$ and $g : R^n \to R$ are sufficiently smooth in a domain $D \subset R^n$, and $\theta \in R^l$ is a constant parameter vector. The control objective is to design a control law $u(t)$ to make the output $y(t)$ track a reference signal $y_m(t)$ when the parameter vector θ is unknown and under the following assumptions:

Assumption 3.7 $y_m(t)$ is bounded for any $t > 0$.

Assumption 3.8 For $x \in D \subset R^n$, $f(x)$ satisfies $\|f(x)\| \leq L\|x\|$, where L is a positive constant.

Assumption 3.9 $g(x) \geq g_0 > 0$, where g_0 is a positive constant.

Assumption 3.9 implies the relative degree of (3.90) is one. The plant (3.90) can be formulated as

$$y(t+1) = \theta^T f(x(t)) + g(x(t))u(t). \tag{3.91}$$

When θ is known, the nominal control law

$$u(t) = \frac{1}{g(x(t))}\left(-\theta^T f(x(t)) + y_m(t+1)\right), \tag{3.92}$$

applied to (3.91), leads the closed-loop system to

$$y(t) = y_m(t), \quad t > 1. \tag{3.93}$$

When θ is unknown, its estimate $\hat{\theta}$ is used to implement the following adaptive control law:

$$u(t) = \frac{1}{g(x(t))}\left(-\hat{\theta}^T(t) f(x(t)) + y_m(t+1)\right), \tag{3.94}$$

which results in the closed-loop dynamics described by

$$y(t+1) = y_m(t+1) - \tilde{\theta}^T(t) f(x(t)), \tag{3.95}$$

where $\tilde{\theta}(t) = \hat{\theta}(t) - \theta$.

Unlike the continuous-time example, the discrete-time parameter adaptive law cannot be derived from Lyapunov-based design. Various parameter identification algorithms such as gradient algorithms and least-squares algorithms can be applied to update $\hat{\theta}$.

Define the *estimation error* with $\hat{\theta}(t)$ as

$$\varepsilon(t) = y(t+1) - \hat{\theta}^T(t) f(x(t)) - g(x(t))u(t), \tag{3.96}$$

which, in view of (3.91), can be expressed as

$$\varepsilon(t) = -\tilde{\theta}^T(t)\phi(t), \tag{3.97}$$

where the regressor $\phi(t) \triangleq f(x(t))$.

It can be noted that (3.97) has the same form as (3.54). Based on (3.96) and (3.97), the parameter adaptive law (3.57) can be applied to update $\hat{\theta}$ and the desired properties given by Lemma 3.2 also hold.

From (3.95), we have

$$e(t+1) = -\tilde{\theta}^T(t)\phi(t). \tag{3.98}$$

In view of (3.97) and (3.98), we have

$$e(t+1) = \varepsilon(t), \tag{3.99}$$

which, divided by $m(t)$ on both sides, gives

$$\frac{e(t+1)}{m(t)} = \frac{\varepsilon(t)}{m(t)}. \tag{3.100}$$

From (3.56), we have

$$m(t) = \sqrt{c + \phi^T(t)\phi(t)} \leq \sqrt{c} + \|\phi(t)\|, \tag{3.101}$$

where, under Assumption 3.8, $\phi(t)$ satisfies

$$\|\phi(t)\| = \|f(x(t))\| \leq L\|x(t)\|. \tag{3.102}$$

Since $x(t) = [y(t-n+1), \ldots, y(t-1), y(t)]^T$ and $e(t) = y(t) - y_m(t)$, with Assumption 3.7, we have

$$\|x(t)\| \leq \sum_{\tau=0}^{n-1} |e(t-\tau)| + c_1 \leq n \max_{0\leq\tau\leq n-1} |e(t-\tau)| + c_1, \tag{3.103}$$

where c_1 is a positive constant.

With (3.101)–(3.103), we obtain

$$m(t) \leq nL \max 0 \leq \tau \leq n-1 |e(t-\tau)| + c_2, \tag{3.104}$$

where c_2 is a positive constant.

Then from (3.100), we have

$$e(t) \leq |\frac{\varepsilon(t)}{m(t)}| m(t) \leq nL|\frac{\varepsilon(t)}{m(t)}| \max_{0\leq\tau\leq n-1} |e(t-\tau)| + c_2|\frac{\varepsilon(t)}{m(t)}|. \tag{3.105}$$

With Lemma 3.2, we have $\lim_{t\to\infty} \frac{\varepsilon(t)}{m(t)} = 0$. Therefore, the inequality (3.105) implies $e(t) \in L_\infty$. Hence, $y(t)$, $u(t)$, $x(t) \in L_\infty$ as well.

With $x(t) \in L_\infty$ and Assumption 3.8, we have $\phi(t) \in L_\infty$ and $m(t) \in L_\infty$. Furthermore, $\frac{\varepsilon(t)}{m(t)} \in L_2$ by Lemma 3.2. Therefore, it can be concluded from (3.100) that $e(t+1) \in L_2$. With $e(t+1) \in L_\infty \bigcap L_2$, we finally obtain $\lim_{t\to\infty} e(t+1) = 0$.

3.3 Summary

Adaptive control is a sophisticated methodology to deal with parametric uncertainties. Parameterization is the foundation of adaptive control designs, which transforms the uncertain part of a system into a vector of unknown parameters and a regressor with measurable signals. Based on the certainty equivalent principle, the estimates of unknown parameters are used to compute the control signal.

The adaptive law for updating the parameter estimates can be derived either through *Lyapunov-based design* or *estimation-based design*. The Lyapunov-based design is based on a Lyapunov function containing both tracking error and parameter estimate error, and the parameter adaptive law is selected such that the Lyapunov function is nonincreasing. The estimation-based design utilizes various parameter identification algorithms, such as least-squares and gradient algorithms, which can ensure certain desirable properties of parameter adaptation that are independent of control. For discrete-time systems, it is difficult to employ the Lyapunov-based design and the parameter adaptive laws are usually derived through the estimation-based design.

When the adaptive law updates the plant parameter estimates and the controller parameters are calculated from the plant parameter estimates, the corresponding adaptive control scheme is called *indirect adaptive control*. When the adaptive law directly updates the controller parameters, the corresponding adaptive control scheme is called *direct adaptive control*.

The sign of the control gain is usually assumed to be known in adaptive control, which is required when designing the parameter adaptive law. The model reference adaptive control requires certain plant-model matching conditions.

Since there may exist modeling error and disturbances in an adaptive control system, the parameter adaptive law designed for the ideal case with no modeling error and disturbances need to be modified to avoid parameter drift. Interested readers may refer to the books (Ioannou and Sun (1996), Tao (2003)) for more knowledge on robust adaptive control.

References

Astrom K, Wittenmark B (1995) Adaptive control, 2nd edn. Addison-wesley, Reading

Goodwin GC, Sin KS (1984) Adaptive filtering prediction and control. Prentice-Hall, Englewood Cliffs

Ioannou PA, Sun J (1996) Robust adaptive control. Prentice-Hall, Englewood Cliffs

Krstić M, Kanellakopoulos I, Kokotović P (1995) Nonlinear and adaptive control design. Wiley, New York

Landau ID, Lozano R, M'Saad M, Karimi A (2011) Adaptive control algorithms, analysis and applications. Springer, London

Popov VM (1966) Editura Adcademiei Republicü Socialiste România, Bucharest, Hyperstability of Automatic Control Systems (in Romanian)

Tao G (2003) Adaptive control design and analysis. Wiley, New York

Chapter 4
T–S Fuzzy System Identification Using I/O Data

4.1 Introduction

In this chapter, we consider the identification of T–S fuzzy models based on input–output (I/O) data. The identification of T–S fuzzy models includes two major tasks: structure identification and parameter identification (Takagi and Sugeno 1985). Structure identification determines the premise (input) variables, the number of fuzzy rules, and the initial positions of membership functions. Parameter identification determines a feasible set of parameters including antecedent (membership function) parameters and consequent parameters under a given structure.

If an I/O data collection can be obtained in advance, the fuzzy system identification can be carried out using offline identification algorithms. If the data come in an online mode, the fuzzy system can also be identified using online identification algorithms. In this chapter, both offline identification and online identification algorithms will be presented. We intend to give a self-contained and succinct description of the identification algorithms, which is easy to understand and apply for a first-year graduate student. By following the offline fuzzy system identification procedure and online fuzzy system identification procedure in this chapter, one can obtain a T–S fuzzy model totally based on I/O data that is good enough for prediction or control design.

Fuzzy system identification is a research field in progress. There are many research results regarding fuzzy clustering, selection of key premise variables, robust parameter estimation, etc., that are not covered by this chapter. Interested readers may also refer to the books on the topic of fuzzy system identification (Babuska 1998; Nelles 2001; Angelov 2002; Abonyi 2003; Oviedo et al. 2005; Lilly 2010).

R. Qi et al., *Fuzzy System Identification and Adaptive Control*, Communications and Control Engineering,
https://doi.org/10.1007/978-3-030-19882-4_4

4.2 Offline Identification of T–S Fuzzy Systems

Consider a general single-input single-output (SISO) nonlinear plant represented by the following discrete-time nonlinear difference equation:

$$y(t+1) = f(y(t), \ldots, y(t-n+1), u(t), \ldots, u(t-n+1)), \tag{4.1}$$

where $f(\cdot, \ldots, \cdot)$ is an unknown nonlinear function, $y(\cdot)$ is the output, $u(\cdot)$ is the input, $t = 0, 1, 2, \ldots,$ is the discrete time variable, and n is the system order.

Denote $\xi(t) = [\xi_1(t), \xi_2(t), \ldots, \xi_L(t)]^T$, where $\xi_j(t)$, $j = 1, 2, \ldots, L$, are premise variables selected from $\{y(t), y(t-1), \ldots, y(t-n+1), u(t), \ldots, u(t-n+1)\}$ representing the structure information of the nonlinear plant. Then the nonlinear plant (4.1) can be approximated by the following T–S fuzzy model with a group of fuzzy rules and a set of parameters:

$$\begin{aligned} R^i:\ & \text{IF } \xi_1(t) \text{ is } F_1^i \text{ and } \xi_2(t) \text{ is } F_2^i \text{ and } \ldots \text{ and } \xi_L(t) \text{ is } F_L^i \\ & \text{THEN } y(t+1) = \theta_{i,0} + \theta_{i,1}\xi_1(t) + \cdots + \theta_{i,L}\xi_L(t), \end{aligned} \tag{4.2}$$

where R^i denotes the ith fuzzy rule, $\theta_{i,0}, \theta_{i,1}, \ldots, \theta_{i,L}$ are coefficients of the ith subsystem, and F_j^i denotes a fuzzy set associated with which there is a membership function $F_j^i(\xi_j(t))$ to indicate the membership degree of $\xi_j(t)$ in F_j^i, $i = 1, 2, \ldots, N$, $j = 1, \ldots, L$.

Using *singleton fuzzification*, *product inference*, and *weighted average*, the fuzzy model (4.2) can be transformed into the following global model:

$$y(t+1) = \sum_{i=1}^{N} \mu_i(\xi)\bar{\xi}^T(t)\theta_i, \tag{4.3}$$

where $\bar{\xi}(t) = [1, \xi(t)^T]^T$, $\theta_i = [\theta_{i0}, \theta_{i1}, \ldots, \theta_{iL}]^T$, $\mu_i(\xi)$ is the normalized firing strength of the ith rule, satisfying

$$\begin{aligned} \mu_i(\xi) &= \frac{\lambda_i(\xi)}{\sum_{i=1}^{N}\lambda_i(\xi)}, \quad \lambda_i(\xi) = \prod_{j=1}^{L} F_j^i(\xi_j) \\ \mu_i(\xi) &\geq 0, \quad \sum_{i=1}^{N}\mu_i(\xi) = 1. \end{aligned} \tag{4.4}$$

The membership functions F_j^i can be chosen as typical fuzzy membership functions such as triangular, trapezoidal, Gaussian functions, etc.

The objective of offline identification is to identify a T–S fuzzy model described by (4.2) from a collection of input/output (I/O) data.

Identification of a T–S fuzzy model requires both structure identification and parameter identification, which includes the following four tasks:

1. identification of premise variables;
2. identification of number of rules;
3. estimation of consequent parameters; and
4. adjustment of membership parameters.

The first two tasks belong to *structure identification* while the last two belong to *parameter identification*. In principle, structure identification and parameter identification are not totally independent since to identify the model structure, we need an initial group of parameters for model evaluation. On the other hand, the number of parameters to be identified relies on the model structure.

4.2.1 Identification of Premise Variables

In most identification methods, the input and output variables of the plant are assumed known (Ljung 1987). However, if no *a prior* knowledge can be obtained about the structure of the plant, including its order, the premise (input) variables can be selected from a group of candidates. The complexity of the problem is exponential, that is, if there are p possible candidates, 2^p possibilities need to be evaluated. In reality, the exhaustive search of 2^p possibilities is not practical. Some shortcuts to guide the search have been proposed such as the search tree method (Sugeno and Yasukawa 1993) and the genetic selection method (Espinosa et al. 2005). Here, the search tree (ST) method is employed to identify the premise variables. The ST method is a heuristic search method which is easy to understand and can greatly reduce the search complexity. Figure 4.1 shows an ST example, where there are three nodes in Level 1 which represent three candidate premise variables x_1, x_2, x_3. Each branch starting

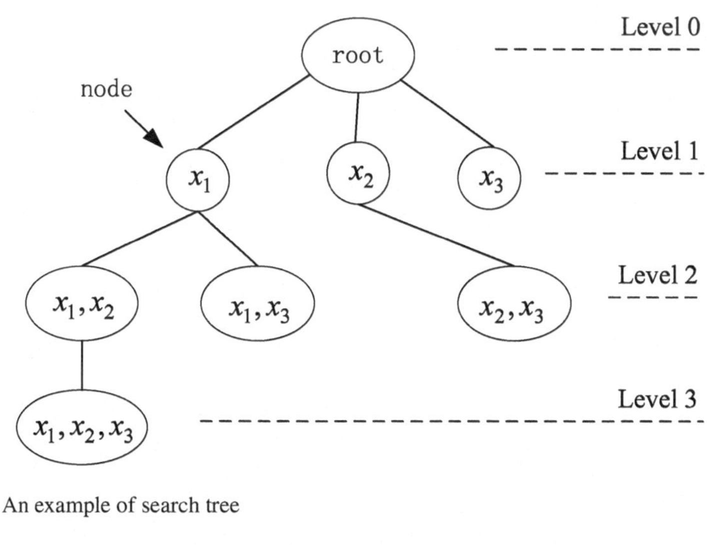

Fig. 4.1 An example of search tree

from each node in Level 1 forms a new node in Level 2 by combining two candidate premise variables. Following the similar way, Level 3 is formed by combining three candidate premise variables. If there are p nodes in Level 1, the ST would have a total of $2^p - 1$ nodes. In the ST algorithm, all the nodes in Level 1 are first evaluated and an optimal node is selected. Then, the search would go along the branches starting from the optimal node in Level 1 and the branches from other nodes would be neglected. Through this way, the search complexity is greatly reduced and at most $p(p+1)/2$ nodes are evaluated.

In the following study, we will use a simple example to show how to apply the ST method to select premise variables.

Example 4.1 Consider a motor-driven single robotic link described by the following equation (Lilly 2010):

$$\ddot{\psi} = -64 \sin \psi - 5\dot{\psi} + 400u, \tag{4.5}$$

where $\psi \in [-\pi/2, \pi/2]$ is the angle of link from the vertical-down position and u is the current input to the motor. The output is the measured link angle $y = \psi$. The problem is to identify a T–S fuzzy model (4.2) using the I/O data from (4.5).

Collecting I/O data. Using MATLAB/Simulink, the differential equation (4.5) is solved with a fourth-order Runge–Kutta (RK4) integration algorithm by a fixed step size of 0.001s. The input signal u is chosen as a uniform random sequence within $[-0.5, 0.5]$. The input and output signals are shown in Fig. 4.2. The I/O data pairs $\{u(kT), y(kT)\}$ are collected with a sampling time $T = 0.01$s for $t = 1, 2, \ldots, 1000$, that is, the input and output signals are saved every 10 steps through the RK4 integration routine. The input and output data are shown in Fig. 4.2.

Constructing a search tree. In this example, the candidate premise variables are selected as $u(t), y(t), y(t-1)$, which are the nodes in Level 1 of the search tree (ST), as shown in Fig. 4.3.

To select an optimal node from Level 1, we need to evaluate the approximation capability of the fuzzy models formed from each node. For example, with $y(t)$ being the premise variable, the fuzzy rule has the following form:

$$R^i : \text{IF } y(t) \text{ is } F_1^i, \quad \text{THEN } \hat{y}(t+1) = \theta_{i,0} + \theta_{i,1} y(t), \tag{4.6}$$

where $i = 1, 2, \ldots, N$, $\hat{y}(t+1)$ is the model output.

Building a simplest fuzzy model. To evaluate its approximation capability, we also need to determine the rule number N and the membership functions F_1^i for the fuzzy sets F_1^i. Since there is only one premise variable $y(t)$, the number of membership functions will determine the number of rules. From the I/O data, we know the value of y is within $[-1.57, 1.57]$. To form a simplest fuzzy model, we need at least two membership functions F_1^1 and F_1^2 on $[-1.57, 1.57]$, as shown in Fig. 4.4, where the membership function type is chosen as triangular. Then, the simplest fuzzy model has the following two rules:

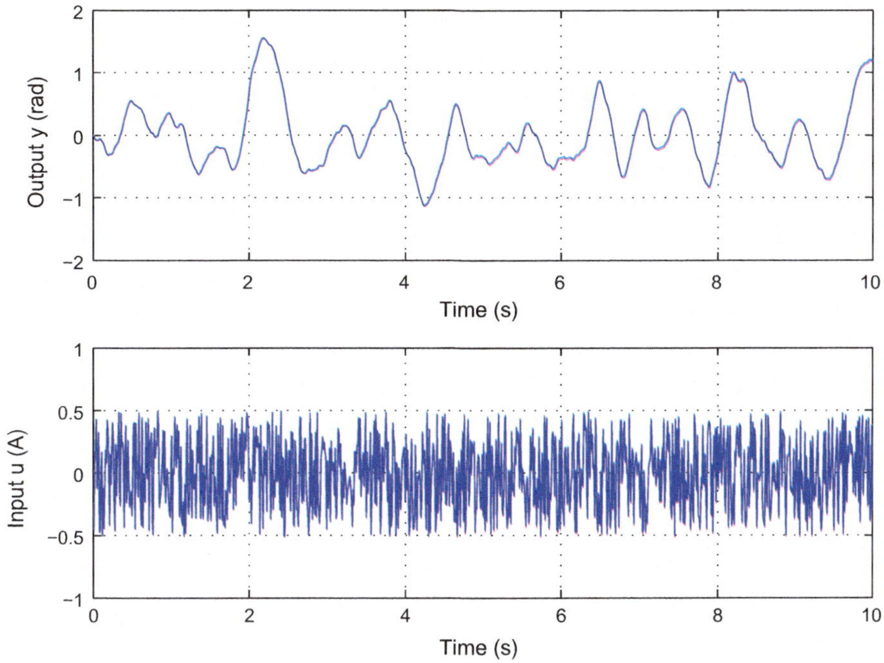

Fig. 4.2 Output and input data

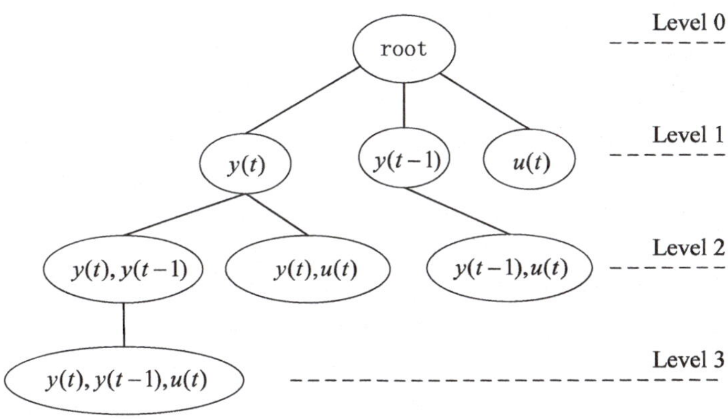

Fig. 4.3 Search tree of Example 4.1

$$
\begin{aligned}
R^1 &: \text{ IF } y(t) \text{ is } F_1^1, \quad \text{THEN } \hat{y}(t+1) = \theta_{1,0} + \theta_{1,1} y(t), \\
R^2 &: \text{ IF } y(t) \text{ is } F_1^2, \quad \text{THEN } \hat{y}(t+1) = \theta_{2,0} + \theta_{2,1} y(t).
\end{aligned} \tag{4.7}
$$

The overall output of the fuzzy model is

$$\hat{y}(t+1) = \mu_1(y(t))(\theta_{1,0} + \theta_{1,1}y(t)) + \mu_2(y(t))(\theta_{2,0} + \theta_{2,1}y(t), \tag{4.8}$$

where $\mu_1(y(t)) = F_1^1(y(t))$ and $\mu_2(y(t)) = F_1^2(y(t))$.

Estimating consequent parameters. Equation (4.8) can be formulated into the following parameterized model:

$$\hat{y}(t+1) = \phi^T(t)\theta, \tag{4.9}$$

where

$$\phi(t) = \begin{bmatrix} \mu_1(y(t)) \\ \mu_1(y(t))y(t) \\ \mu_2(y(t)) \\ \mu_2(y(t))y(t) \end{bmatrix}, \quad \theta = \begin{bmatrix} \theta_{1,0} \\ \theta_{1,1} \\ \theta_{2,0} \\ \theta_{2,1} \end{bmatrix}. \tag{4.10}$$

The parameter vector θ can be identified using either the batch least squares (BLS) algorithm (4.26) or recursive least squares (RLS) algorithm (4.27).

Now, we have finished the identification task with $y(t)$ being the premise variable. Following the similar procedure, we can identify the T–S fuzzy model with the other two nodes, $y(t-1)$ and $u(t)$ in Level 1. The membership functions of $y(t-1)$ are chosen the same as those of $y(t)$, as shown in Fig. 4.4. The membership functions of $u(t)$ are given in Fig. 4.5. Each model has two rules, with $y(t-1)$ and $u(t)$ being the premise variables, respectively.

Calculating RC value. To compare the model quality with different premise variables, we split the I/O data into two groups A and B, which are used for model identification and crossover checking. The identification and crossover checking procedures are shown in Fig. 4.6. Models A and B are identified by data from groups A and B, respectively, and then their approximation capability is evaluated by feeding them with input data from the other group, which is called *crossover checking*.

In crossover checking, the following regularity criterion (RC) is introduced (Barada and Singh 1998):

$$\mathrm{RC} = \left[\sum_{j=1}^{N_A} (Y_j^A - \hat{Y}_j^{AB})^2 / N_A + \sum_{j=1}^{N_B} (Y_j^B - \hat{Y}_j^{BA})^2 / N_B \right], \tag{4.11}$$

where

N_A	number of data in group A
N_B	number of data in group B
Y_j^A	the jth output data in group A
Y_j^B	the jth output data in group B
$\hat{Y}_j^{AB}$	the jth output of model B subject to the jth input in group A
$\hat{Y}_j^{BA}$	the jth output of model A subject to the jth input in group B.

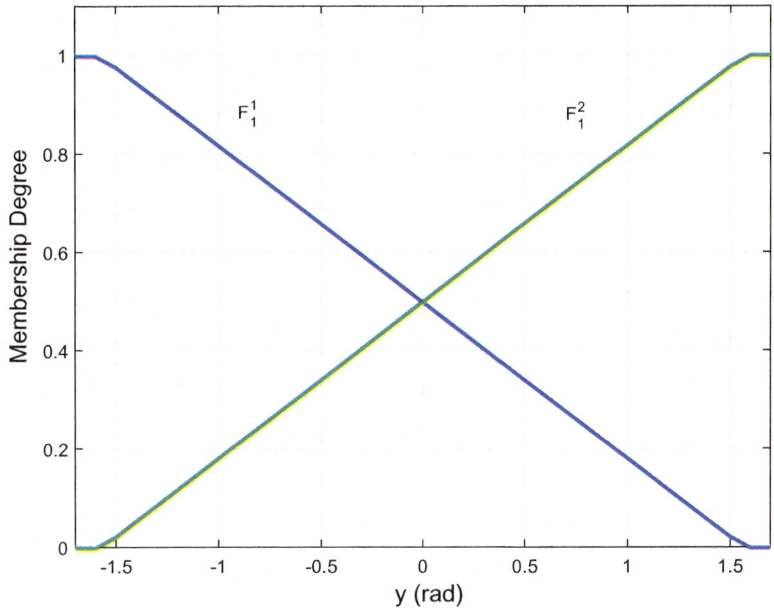

Fig. 4.4 Membership functions of $y(t)$

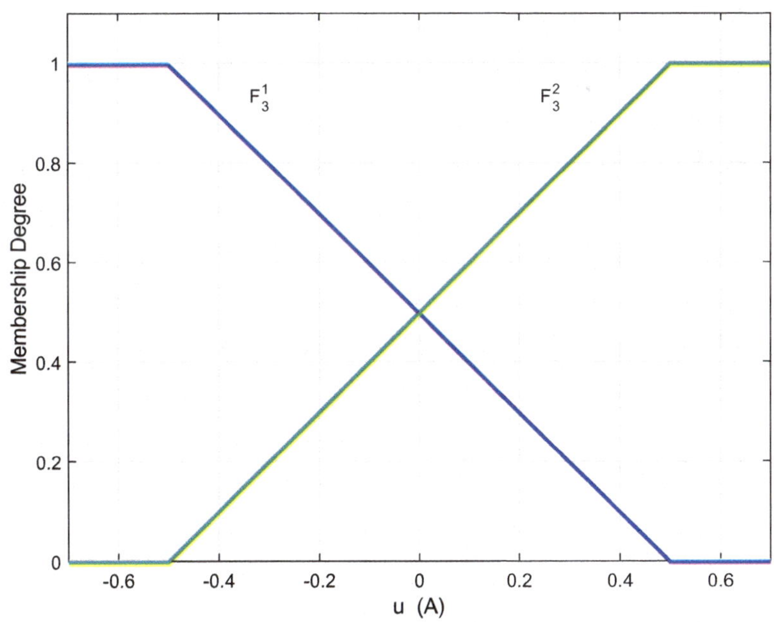

Fig. 4.5 Membership functions of $u(t)$

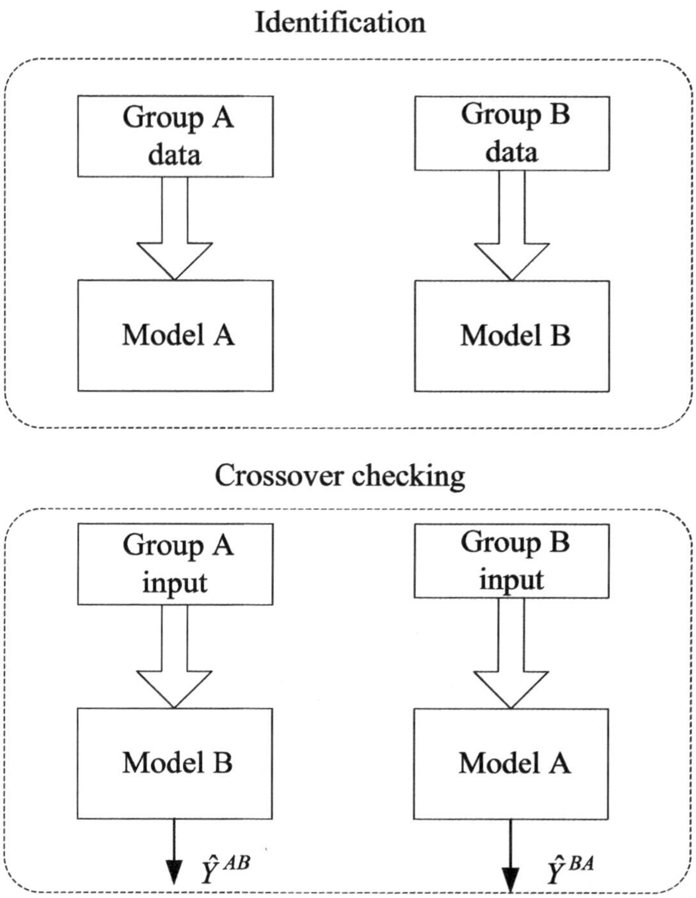

Fig. 4.6 Identification and crossover checking using two data groups A and B

Determining the optimal node. By calculating the RC values of all the nodes of Level 1 in Fig. 4.3, we find the local optimal node $y(t)$ with the minimum RC value, as shown in Table 4.1. Then we continue to calculate the RC values of the nodes starting from $y(t)$ on Level 2. For each node in Level 2, there are two candidate variables. For example, the first node in Level 2 is $\{y(t), y(t-1)\}$. The membership functions of $y(t)$ and $y(t-1)$ are chosen the same as those for the nodes in Level 1. Since each premise variable has two membership functions, the rule number is $2^2 = 4$:

$$\begin{aligned} R^i :\ & \text{IF } y(t) \text{ is } F_1^i \text{ AND } y(t-1) \text{ is } F_2^i, \\ & \text{THEN } \hat{y}(t+1) = \theta_{i,0} + \theta_{i,1} y(t) + \theta_{i,2} y(t-1), \quad i = 1, 2, 3, 4. \end{aligned} \tag{4.12}$$

Table 4.1 RC calculation in Level 1 of Example 4.1

Tree level	Input variables	No. of rules	RC
Level 1	$y(t)$	2	2.80×10^{-3}*
	$y(t-1)$	2	0.011
	$u(t)$	2	0.523

Table 4.2 RC calculation in Levels 2 and 3 of Example 4.1

Tree level	Input variables	No. of rules	RC
Level 2	$y(t), y(t-1)$	4	1.30×10^{-4}**
	$y(t), u(t)$	4	2.70×10^{-3}
Level 3	$y(t), y(t-1), u(t)$	8	6.55×10^{-5}***

The overall T–S fuzzy model can also be formulated into a parametrized model in the form similar to (4.9) and the consequent parameter vector θ can be identified by the BLS algorithm (4.26) or RLS algorithm (4.27).

The two T–S fuzzy models constructed from the nodes in Level 2 are compared through crossover checking and it turns out that the node $\{y(t), y(t-1)\}$ has the minimum RC value. Then, we move to Level 3 from the node $\{y(t), y(t-1)\}$.

Finally, we find the node with the minimum RC value is $\{y(t), y(t-1), u(t)\}$. The RC calculation results of Levels 2 and 3 are shown in Table 4.2. Therefore, the premise variables are selected as $\{y(t), y(t-1), u(t)\}$, and at the same time we have obtained an initial fuzzy model with the premise variables $\{y(t), y(t-1), u(t)\}$.

In Example 4.1, the number and parameters of membership functions for each premise variable are chosen manually, which also determine the number of fuzzy rules. Actually, the number and parameters of membership functions can also be identified from I/O data through fuzzy clustering methods. In the following section, we will introduce how to apply offline fuzzy clustering methods to identify the number of fuzzy rules and the parameters of membership functions.

4.2.2 *Identification of Number of Rules*

Fuzzy clustering is a way to partition the data space into a number of overlapped subspaces based on similarities between data. Fuzzy clustering algorithms can be employed to obtain the number of rules and the parameters of membership functions simultaneously. Various offline fuzzy clustering algorithms have been proposed depending on the assumed model structure of the identification. Popular fuzzy clustering algorithms include fuzzy c-means (FCM) algorithm (Bezdek 1976, 1981), Gustafson–Kessel (GK) algorithm (Gustafson and Kessel 1979), and Gath–Geva (GG) algorithm (Gath and Geva 1989), subtractive clustering algorithm (Chiu 1994).

The identification of the number of rules and membership functions from I/O data corresponds to the partition of I/O data into a set of subspaces. The number of subspaces determines the number of rules. The centers of the subspaces and the overlap among subspaces determine the parameters of the membership functions.

In the following study, as a representative, Gaussian function is employed as membership function, which has the following form:

$$F_j^i(\xi_j) = \exp\left\{-\frac{(\xi_j - c_j^i)^2}{2\sigma_j^{i2}}\right\}, \tag{4.13}$$

where c_j^i and σ_j^i, $j = 1, 2, \ldots, L$, $i = 1, 2, \ldots, N$, are the centers and radii of Gaussian functions, respectively. c_j^i denotes the center for the Gaussian function of the jth premise variable in the subspace defined by the ith rule, and σ_j^i denotes the radius of Gaussian function of the jth premise variable in the subspace defined by the ith rule.

Recall the general T–S fuzzy model (4.2) with the following rules:

$$\begin{aligned} R^i :\ & \text{IF } \xi_1(t) \text{ is } F_1^i \text{ and } \xi_2(t) \text{ is } F_2^i \text{ and } \ldots \text{ and } \xi_L(t) \text{ is } F_L^i \\ & \text{THEN } y(t+1) = \theta_{i,0} + \theta_{i,1}\xi_1(t) + \cdots + \theta_{i,L}\xi_L(t). \end{aligned} \tag{4.14}$$

Define $\xi(t) = [\xi_1(t), \xi_2(t), \ldots, \xi_L(t)]^T \in R^L$ as the vector of the premise variables. In Example 4.1, the premise variables are identified as $u(t), y(t), y(t-1)$, and then we have $\xi(t) = [u(t), y(t), y(t-1)]^T$.

Suppose the I/O data are given by

$$X = \begin{bmatrix} \xi(1)^T \\ \xi(2)^T \\ \vdots \\ \xi(N_D)^T \end{bmatrix}, \quad Y = \begin{bmatrix} y(2) \\ y(3) \\ \vdots \\ y(N_D+1) \end{bmatrix}, \tag{4.15}$$

where $X \in R^{N_D \times L}$ and $Y \in R^{N_D}$ represent the input and output data, respectively, and N_D is the number of data.

Subtractive clustering. Define $z(t) = [\xi^T(t), y(t+1)]^T \in R^{L+1}$ as the tth I/O data point. The subtractive clustering approach considers each data point $z(t)$ as a candidate cluster center. The capability of a data point $z(t)$ to be a cluster center is evaluated through its potential. The subtractive clustering algorithm includes the following steps (Chiu 1994):

(1) The potential of the data point $z(t)$ is computed as

$$P(z(t)) = \sum_{l=1, l \neq t}^{N_D} \exp\{-\alpha \|z(t) - z(l)\|^2\}, \quad (4.16)$$

where $\alpha = 4/r_a^2$, $r_a = r \left\| \bar{v} - \underline{v} \right\|$, $\bar{v}$ and $\underline{v}$ are vectors consisting of the upper and lower bounds of each component in $z(t)$, $t \in [1, N_D]$, r effectively defines the radius of a neighborhood and determines the number of the cluster centers. r needs to be carefully chosen to avoid averaging (r is too large) or over-fitting (r is too small). A recommended choice is $r \in [0.3, 0.5]$.

(2) After computing the potentials of all the data points, the data point with the highest potential is selected as the first cluster center v_1^* and

$$P(v_1^*) = \max_{t=1}^{N_D} P(z(t)), \quad (4.17)$$

where $P(v_1^*)$ denotes the potential of the first cluster center.

(3) The potential of the data point $z(t)$ is updated by subtracting an amount proportional to the potential of the chosen center v_i^* and inversely proportional to the distance to this center:

$$P(z(t)) := P(z(t)) - P(v_i^*) \exp\{-\beta \left\| z(t) - v_i^* \right\|^2\}, \quad (4.18)$$

where $\beta = 4/r_b^2$, r_b is a positive constant and a recommended choice is $r_b = 1.5r_a$.

(4) Define $P_{\max}^*$ as the highest potential of all existing cluster centers. With the updated potential (4.18), the following criteria are used to determine whether a data point should be accepted as a new cluster center:

1. *If* $P(z(t)) > \varepsilon_1 P_{\max}^*$, $z(t)$ *is accepted as a new cluster center.*
2. *If* $P(z(t)) < \varepsilon_2 P_{\max}^*$, $z(t)$ *is rejected as a new cluster center.*
3. *If* $\varepsilon_2 P_{\max}^* \leq P(z(t)) \leq \varepsilon_1 P_{\max}^*$, *let* $d_{\min} = \min_{i=1}^{c} \|z(t) - v_i^*\|$ *be the shortest distance between* $z(t)$ *to all the previously found* c *cluster centers and if*

$$\frac{P(z(t))}{P_{\max}^*} + \frac{d_{\min}}{r_a} \geq 1 \quad (4.19)$$

then $z(t)$ *is accepted as a new cluster center.*

The two positive constants ε_1 and ε_2 define the boundaries for potential evaluation. The recommended values for them are $\varepsilon_1 = 0.5$ and $\varepsilon_2 = 0.15$.

With the above procedure, a set of c cluster centers $\{v_1^*, v_2^*, \ldots, v_c^*\}$ can be found in the $(L+1)$-dimensional space. The number of the obtained cluster centers determines the number of fuzzy rules ($N = c$). The obtained cluster centers can be used to form the membership functions. For example, the first L elements in $v_i^* = [v_{i,1}^*, v_{i,2}^*, \ldots, v_{i,L+1}^*]^T$ can be directly taken as the estimates of the centers $\hat{c}^i = [\hat{c}_1^i, \hat{c}_2^i, \ldots, \hat{c}_L^i]^T$ in the Gaussian membership functions (4.28), that is

$$\hat{c}_j^i = v_{i,j}^*, \tag{4.20}$$

and the estimates of the radii $\hat{\sigma}_j^i$ can be calculated from

$$\hat{\sigma}_j^i = 1/\sqrt{2\alpha}, \tag{4.21}$$

where $j = 1, 2, \ldots, L; i = 1, 2, \ldots, N$.

The subtractive clustering algorithm can be used as an independent clustering approach or used to provide the estimation of the initial values of cluster centers in iterative optimization-based clustering algorithms such as fuzzy C-means. It relies on the idea that each cluster center is representative of a characteristic behavior of the plant so that the resulting cluster centers can be used as parameters of the membership functions defining the focal points of the rules of the T–S model.

4.2.3 Estimation of Consequent Parameters

Once the membership functions and number of rules are fixed, the model (4.8) can be formulated into the following linearly parameterized model:

$$y(t+1) = \phi^T(t)\theta, \tag{4.22}$$

where $\bar{\xi}(t) = [1, \xi(t)^T]^T \in R^{L+1}, \theta_i = [\theta_{i,0}, \theta_{i,1}, \ldots, \theta i, L]^T \in R^{L+1}$

$$\phi(t) = \begin{bmatrix} \mu_1(\xi(t))\bar{\xi}(t) \\ \mu_2(\xi(t))\bar{\xi}(t) \\ \vdots \\ \mu_N(\xi(t))\bar{\xi}(t) \end{bmatrix} \in R^{N(L+1)}, \quad \theta = \begin{bmatrix} \theta_1 \\ \theta_2 \\ \vdots \\ \theta_N \end{bmatrix} \in R^{N(L+1)}. \tag{4.23}$$

The parameter vector θ can be identified from the I/O data. Define

$$\Phi = \begin{bmatrix} \phi(1)^T \\ \phi(2)^T \\ \vdots \\ \phi(N_D)^T \end{bmatrix} \in R^{N_D \times N(L+1)}, \quad Y = \begin{bmatrix} y(2) \\ y(3) \\ \vdots \\ y(N_D+1) \end{bmatrix} \in R^{N_D}, \tag{4.24}$$

where N_D is the number of training data. Then, the parameter estimation problem can be formulated into solving the following least squares problem:

$$\min_{\theta} \|Y - \Phi\theta\|^2. \tag{4.25}$$

Batch least squares (BLS). If the matrix $\Phi^T\Phi$ is invertible, the solution can be derived by the following *batch least squares (BLS)* algorithm:

$$\hat{\theta} = (\Phi^T\Phi)^{-1}\Phi^T Y, \tag{4.26}$$

which shows that we can directly calculate the least square estimate $\hat{\theta}$ from the batch of N_D data given by Φ and Y. If the inputs to system can make the system "sufficiently excited" (Ljung 1987), then it can be guaranteed that $\Phi^T\Phi$ is invertible.

The BLS algorithm can successfully provide the parameter estimates as long as there are enough I/O data. However, it needs all the data are collected before the algorithm can be applied. In addition, if the number of data N_D is very large, the computation of the inverse of $\Phi^T\Phi$ may become prohibitive. To deal with those problems, a recursive version of BLS is derived which allows the parameter estimate $\hat{\theta}$ to be updated recursively when a new data point comes, and does not need to compute the inverse of $\Phi^T\Phi$.

Recursive least squares (RLS). The consequent parameters can also be estimated by the following *recursive least squares (RLS)* algorithm (Takagi and Sugeno 1985):

$$\begin{aligned} \hat{\theta}(t) &= \hat{\theta}(t-1) + C_t\phi(t-1)(y(t) - \phi(t-1)^T\hat{\theta}(t-1)) \\ C_t &= C_{t-1} - \frac{C_{t-1}\phi(t-1)\phi(t-1)^T C_{t-1}}{1 + \phi(t-1)^T C_{t-1}\phi(t-1)}, \end{aligned} \tag{4.27}$$

where $\hat{\theta}(t) \in R^{L+1}$ is the parameter estimate based on the data at time t, C_t is a $N(L+1) \times N(L+1)$-dimensional covariance matrix. The initial conditions are set as $\hat{\theta}(0) = 0$ and $C_0 = \Omega I$, where Ω is a large positive number, I is an identity matrix.

For linearly parametrized consequent parameters, it is possible to use BLS or RLS to estimate them in a batch or recursive mode. However, for those parameters in membership functions, which cannot be formulated into a linearly parametrized form, cannot be tuned by BLS or RLS. In the next section, we will introduce how to tune the membership parameters based on gradient method.

4.2.4 Adjustment of Membership Parameters

In Sect. 4.2.2, we determine the centers and radii of Gaussian functions using the information generated from the clustering procedure, which does not guarantee those parameters have achieved their best approximation. The estimates of those parameters can be further tuned. In this section, we will derive update laws for membership parameters using gradient method.

The membership functions used in this section are still chosen as Gaussian functions, as a representative. Recall that

$$F_j^i(\xi_j) = \exp\left\{-\frac{(\xi_j - c_j^i)^2}{2\sigma_j^{i2}}\right\}, \tag{4.28}$$

where c_j^i and $\sigma_j^i, j = 1, 2, \ldots, L, i = 1, 2, \ldots, N$, are the centers and radii of Gaussian functions, respectively. Recall that c_j^i denotes the center for the Gaussian function of the jth premise variable in the subspace defined by the ith rule, and σ_j^i denotes the radius of Gaussian function of the jth premise variable in the subspace defined by the ith rule.

Consider the training data given by (4.15) and suppose you are given the tth training data pair $(\xi(t), y(t+1))$. Let $\hat{y}(t+1)$ be the predicted output with the input $\xi(t)$ based on the T–S fuzzy model (4.3) and using the parameter estimates $\hat{c}_j^i$, $\hat{\sigma}_j^i$ and $\hat{\theta}_i$, that is

$$\hat{y}(t+1) = \sum_{i=1}^{N} \mu_i(\xi(t)|\hat{c}_j^i, \hat{\sigma}_j^i)\bar{\xi}^T(t)\hat{\theta}_i. \tag{4.29}$$

Recall that $\bar{\xi} = [1, \xi(t)^T]^T$ and

$$\mu_i(\xi) = \frac{\lambda_i(\xi)}{\sum_{i=1}^{N} \lambda_i(\xi)}, \quad \lambda_i(\xi) = \prod_{j=1}^{L} F_j^i(\xi_j) \tag{4.30}$$

satisfying $\mu_i(\xi) \geq 0$ and $\sum_{i=1}^{N} \mu_i(\xi) = 1$.

Let $\varepsilon(t+1)$ be the error between the model output $\hat{y}(t+1)$ and the plant output $y(t+1)$ and define the following *objective function*:

$$J(t+1) = \frac{1}{2}\varepsilon(t+1)^2 = \frac{1}{2}[\hat{y}(t+1) - y(t+1)]^2. \tag{4.31}$$

Our goal is to minimize (4.31) by tuning the parameters $\hat{c}_j^i$ and $\hat{\sigma}_j^i$ in (4.29).

First, let we consider how to tune the parameter $\hat{c}_j^i$ to minimize J. One effective way is to use the following update formula:

$$\hat{c}_j^i(t+1) = \hat{c}_j^i(t) - \gamma_c \frac{\partial J(t+1)}{\partial \hat{c}_j^i(t)}, \tag{4.32}$$

which is a gradient descent approach. Gradient descent is a first-order iterative optimization algorithm for finding a minimum of a function by taking steps proportional to the negative of the gradient. The parameter $\gamma_c > 0$ characterizes the step size, which determines how big a step to take down the J surface. A small γ_c indicates a slow tuning speed of $\hat{c}_j^i$. A big γ_c will lead to fast convergence. However, if γ_c is chosen too big, the minimum value of J may be skipped. In practice, γ_c should be carefully selected to give moderate convergence speed yet avoid missing a minimum.

With (4.29)–(4.31), the gradient term $\frac{\partial J(t+1)}{\partial \hat{c}_j^i(t)}$ in (4.32) can be derived using the following chain rule:

$$\frac{\partial J(t+1)}{\partial \hat{c}_j^i(t)} = \frac{\varepsilon(t+1)\left(\bar{\xi}^T(t)\hat{\theta}_i(t) - y(t+1)\right)}{\sum_{i=1}^N \lambda_i(\xi(t))} \cdot \frac{\partial \lambda_i(\xi(t))}{\partial \hat{c}_j^i(t)}, \tag{4.33}$$

$$\frac{\partial \lambda_i(\xi(t))}{\partial \hat{c}_j^i(t)} = \frac{(\xi_j(t) - \hat{c}_j^i(t))}{(\hat{\sigma}_j^i(t))^2} \cdot \lambda_i(\xi(t)). \tag{4.34}$$

Similarly, the update law for tuning $\hat{\sigma}_j^i$ can be derived as follows:

$$\hat{\sigma}_j^i(t+1) = \hat{\sigma}_j^i(t) - \gamma_\sigma \frac{\partial J(t+1)}{\partial \hat{\sigma}_j^i(t)}, \tag{4.35}$$

$$\frac{\partial J(t+1)}{\partial \hat{\sigma}_j^i(t)} = \frac{\varepsilon(t+1)\left(\bar{\xi}^T(t)\hat{\theta}_i(t) - y(t+1)\right)}{\sum_{i=1}^N \lambda_i(\xi(t))} \cdot \frac{\partial \lambda_i(\xi(t))}{\partial \hat{\sigma}_j^i(t)}, \tag{4.36}$$

$$\frac{\partial \lambda_i(\xi(t))}{\partial \hat{\sigma}_j^i(t)} = \frac{(\xi_j(t) - \hat{c}_j^i(t))^2}{(\hat{\sigma}_j^i(t))^3} \lambda_i(\xi(t)), \tag{4.37}$$

where $\gamma_\sigma > 0$ determines the step size for adjusting $\hat{\sigma}_j^i$.

The initial values $\hat{c}_j^i(0)$ and $\hat{\sigma}_j^i(0)$ in the parameter update laws (4.32)–(4.37) can be set as those values obtained from the initially manually designed membership functions or offline clustering.

The parameter update laws (4.32)–(4.37) are derived for Gaussian membership functions. If other types of membership functions are employed, the parameter update laws can also be derived by minimizing the objective function (4.31) using the gradient descent approach.

The gradient method can be used to tune the consequent parameters as well. However, compared with RLS, the gradient method has a slower convergence speed. Therefore, RLS is usually applied to estimate the consequent parameters while the gradient method is employed to estimate the membership parameters in the identification of T–S fuzzy models.

4.2.5 Procedure for Offline Identification

In the previous sections, we have presented how to identify a T–S fuzzy model by identifying the premise variables, the number of rules, the consequent parameters, and the membership parameters based on offline I/O data. In this section, the procedure for offline identification of T–S fuzzy model is summarized as follows.

Step 1: Select a set of candidate premise variables and build up an ST.

Step 2: Choose the type of membership function and determine the range of each candidate premise variable.

Step 3: Start from the first node in the ST, construct a T–S fuzzy model by manually designing the membership functions and determining the number of rules, or using subtractive clustering to obtain the number of rules and membership parameters simultaneously.

Step 4: Split the data into two groups: training data and checking data; estimate the consequent parameters using BLS (4.26) or RLS(4.27); calculate the RC value of each node in Level 1 according to (4.11) to find the node with the minimum RC value.

Step 5: Start from the node with the minimum RC value and continue to calculate the RC values of the nodes in the next level. Repeat this step until finding the node with the minimum RC value in the last level.

Step 6: Compare the minimum RC values of different levels, find the node with the minimum RC value of the whole ST, which determines the premise variables.

Step 7: With the identified premise variables, vary the radius r in subtractive clustering to find a suitable number of rules.

Step 8: With fixed premise variables and number of rules, further adjust the consequent parameters using RLS and the membership parameters using the gradient method.

Till now, we have finished the task of offline identification of T–S fuzzy models based on I/O data collection. The offline identification method assumes that all the I/O data are available before the identification. If the offline data are rich enough to cover all the important dynamic characteristics of the plant to be identified, we can obtain a good model through offline identification. In reality, however, the data collection is usually limited, which can only partially reflect the dynamic behaviors of the plant to be identified. When the plant works online, new data are produced

Table 4.3 MSE and no. of rules in Simulation 1

r	No. of Rules	MSE
0.5	1	1.2119e-05
0.4	1	1.2119e-05
0.3	2	1.0672e-05
0.2	2	1.0817e-05
0.15	3	1.1412e-05
0.14	4	1.0296e-05*
0.13	6	1.1469e-05
0.1	8	1.1188e-05

which may contain some new dynamic characteristic of the plant. In Sect. 4.3, we will study how to identify T–S fuzzy models when the data are collected continuously online.

4.2.6 Simulation Study

In this section, we use two simulation examples to demonstrate the performance of the offline identification algorithm. The first example is Example 4.1. In Example 4.1, we manually set the membership function parameters and the number of rules. In this section, the number of rules and the membership function parameters will be obtained from offline subtractive clustering. The second example is a discrete-time second-order nonlinear plant (Narendra and Parthasarathy 1990).

Simulation 1. The offline I/O data have been collected from Example 4.1. With the identified premise variables $u(t), y(t), y(t-1)$, the number of the cluster centers will be affected by the parameter r which defines the radius of a neighborhood.

With a specific r, after identifying the whole T–S fuzzy model based on the offline I/O data, select another input $u_1(t) = 0.02\sin 0.1\pi t + 0.15\sin(\pi t) + 0.2\sin(10\pi t) + 0.2\sin(100\pi t)$, which consists of sinusoidal signals of four frequencies. Then we can obtain the output from the differential equation (4.5) and the output from the T–S fuzzy model within 20 s separately. To choose a suitable r, the errors between y and $\hat{y}$ under different situations need to be calculated and compared. The following mean squared error (MSE) is introduced:

$$MSE = \frac{\sum_{k=1}^{N_D} (y(t) - \hat{y}(t))^2}{N_D}. \tag{4.38}$$

As can be seen in Table 4.3, we find that when r is set as 0.14, the MSE index is smallest and the number of rules is suitable. With this value of r, the input and output signals are shown in Fig. 4.7.

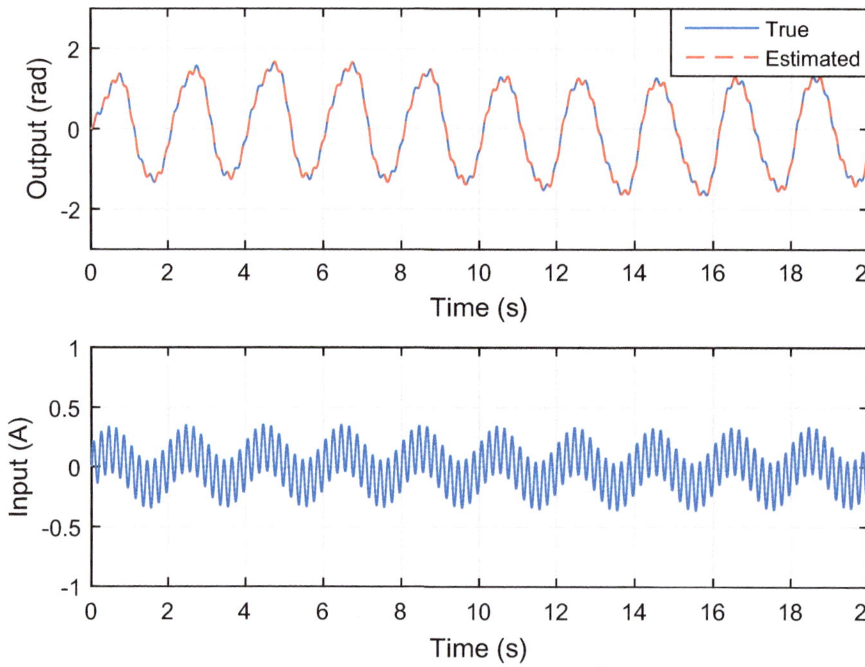

Fig. 4.7 Estimated and true outputs in Simulation 1

Simulation 2. In this simulation example, a discrete-time second-order nonlinear plant is identified based on offline collected I/O data.

Consider a second-order nonlinear plant described by the following difference equation (Narendra and Parthasarathy 1990):

$$y(t+1) = g[y(t), y(t-1)] + u(t), \tag{4.39}$$

where the nonlinear function

$$g[y(t), y(t-1)] = \frac{y(t)y(t-1)[y(t)+2.5]}{1+y(t)^2+y(t-1)^2} \tag{4.40}$$

is assumed to be unknown. A T–S fuzzy model is employed to approximate $g[y(t),$ $y(t-1)]$ and our task to identify this T–S fuzzy model based on the I/O data collected from the plant. 1000 data points are simulated from the discrete nonlinear function, using a compound sinusoidal input signal $u(t) = 0.3\sin(2\pi t/25) + 0.3\sin(2\pi t/250)$. The parameter r is selected as $r = 0.3$. By following the offline procedure summarized in Sect. 4.2.5, we obtain a T–S fuzzy system with three rules. The comparison between the estimated output and the true output is shown in Fig. 4.8.

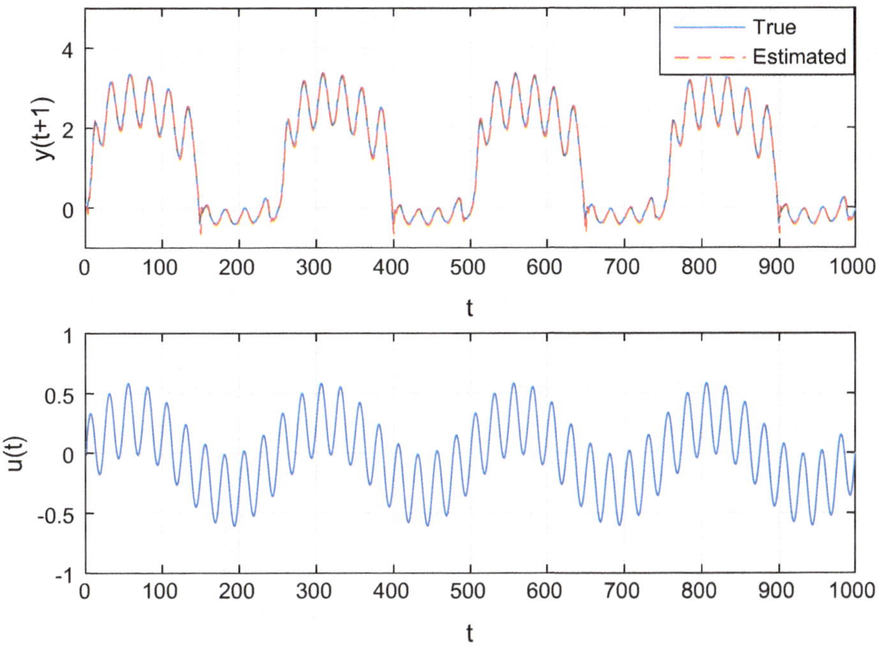

Fig. 4.8 Estimated and true outputs in Simulation 2

4.3 Online Identification of T–S Fuzzy Systems

In online identification, the data are collected continuously, rather than being a fixed set. Some of the new data reinforce and confirm the information contained in the old data. Other data, however, bring new information, which could indicate a change in operating conditions, development of a fault or simply a more significant change in the dynamics of the process. These data may possess enough new information to add a new rule or to modify an existing rule.

In this section, we assume the premise variables and the type of membership functions are fixed and try to identify other parts (number of rules, consequent parameters and membership parameters) in an online mode. The subtractive clustering algorithm (Chiu 1994) used in Sect. 4.2.2 will be developed into an online version, which can add a new rule, replace an existing rule, or delete a redundant rule based on the online data, leading to a self-evolving T–S fuzzy model. The parameter learning algorithms are also modified so that they can adjust parameters online. This online clustering algorithm uses the concept of evolving rule-based (eR) modeling proposed in Angelov (2002). This evolution mechanism ensures greater generality of structural changes and inheritance of the structure information.

4.3.1 Online Fuzzy Clustering Algorithm

In the following study, as a representative, Gaussian functions are employed as membership functions, which are given in (4.28).

Define $\xi = [\xi_1, \xi_2, \ldots, \xi_L]^T$ as the vector of the premise variables and define $z(t) = [z_1(t), z_2(t), \ldots, z_{L+1}(t)]^T = [\xi^T(t-1), y(t)]^T$ as a data point at time t. Each data point can be considered as a candidate prototype cluster center. The online clustering algorithm can start with an existing group of cluster centers or start with the first data point.

If the online clustering algorithm starts with the first data point $z(1)$, $z(1)$ is treated as the first clustering center v_1^* and its potential is set as 1:

$$v_1^* = z(1), \quad P_1(v_1^*) = 1. \tag{4.41}$$

If the online clustering algorithm starting with a group of c cluster centers, $\{v_1^*, v_2^*, \ldots, v_c^*\}$, which can be obtained through a priori knowledge or from offline fuzzy clustering algorithms such as subtractive clustering introduced in Sect. 4.2.2, their initial potentials are set as

$$P_1(v_i^*) = 1, \quad i = 1, 2, \ldots, c. \tag{4.42}$$

At time t $(t \geq 2)$, when the new data $z(t)$ comes, the potential of $z(t)$ is calculated through a potential function (Angelov and Filev 2004):

$$P(z(t)) = \left(1 + \frac{1}{t-1}\sum_{m=1}^{t-1}\sum_{j=1}^{L+1}(d_{mt}^j)^2\right)^{-1}, \tag{4.43}$$

where $d_{mt}^j = |z_j(m) - z_j(t)|$ denotes the projected distance between the data $z(m)$ and $z(t)$ on the z_j axis.

The reason to choose the potential function (4.43) is that it can be formulated into the following recursive form:

$$P(z(t) = \frac{t-1}{(t-1)(\vartheta_t + 1) + \sigma_t - 2\upsilon_t}, \tag{4.44}$$

where $\vartheta_t = \sum_{j=1}^{L+1}(z_j(t))^2$, $\sigma_t = \sum_{m=1}^{t-1}\sum_{j=1}^{L+1}(z_j(m))^2$, $\upsilon_t = \sum_{j=1}^{L+1} z_j(t)\beta_t^j$ and $\beta_t^j = \sum_{m=1}^{t-1} z_j(m)$. The two terms ϑ_t and υ_t can be calculated from the new data $z(t)$ at time t while the other two terms σ_t and β_t^j can be recursively updated by

$$\begin{aligned} \sigma_t &= \sigma_{t-1} + \sum_{j=1}^{L+1}(z_j(t-1))^2, \\ \beta_t^j &= \beta_{t-1}^j + z_j(t-1). \end{aligned} \tag{4.45}$$

The new data $z(t)$ also changes the potentials of all the existing cluster centers v_i^*, which are updated as

$$P_t(v_i^*) = \frac{(t-1)P_{t-1}(v_i^*)}{k-2+P_{t-1}(v_i^*)+(t-1)P_{t-1}(v_i^*)\sum_{j=1}^{L+1}(d_{t(t-1)}^j)^2}, \quad (4.46)$$

where $P_t(v_i^*)$ denotes the updated potential of v_i^* at time t.

The potential of the new data point $P(z(t))$ is compared with the updated potentials of the existing cluster centers $P_t(v_i^*)$ to determine whether to accept this new data point as a new cluster center according to the following criterion.

If the potential of the new data point is higher than the potential of all existing cluster centers and the new data point is not very close to any existing cluster center:

$$P(z(t)) > \max_{i=1}^{c}\{P_t(v_i^*)\} \quad (4.47)$$

$$\frac{P(z(t))}{\max_{i=1}^{c} P_t(v_i^*)} - \frac{\min_{i=1}^{c}\left\|z(t)-v_i^*\right\|}{r_a} < 1 \quad (4.48)$$

then the data point $z(t)$ is accepted as a new cluster center and ***added*** *to the set of cluster centers.*

If only the condition (4.47) *is satisfied while the condition* (4.48) *is not satisfied, the data point $z(t)$* ***replaces*** *the cluster center that is closest to it.*

In (4.48), the parameter $r_a = r_1 \left\|\bar{v} - \underline{v}\right\|$, $\bar{v} = [\bar{v}_1, \bar{v}_2, \ldots, \bar{v}_{L+1}]^T$, $\underline{v} = [\underline{v}_1, \underline{v}_2, \ldots, \underline{v}_{L+1}]^T$, $\bar{v}_j$ and $\underline{v}_j$ represent the upper bound and lower bound of $z_j(t)$, respectively, $j = 1, 2, \ldots, L+1$, $r_1 > 0$ effectively defines the radius of a neighborhood and determines the number of the cluster centers. r_1 needs to be carefully chosen to avoid averaging (r_1 is too large) or over-fitting (r_1 is too small). If any element of $z(t)$, $z_j(t)$, exceeds the range of $[\underline{v}_j, \bar{v}_j]$, $\underline{v}_j$ or $\bar{v}_j$ is reset as $z_j(t)$ and $r_a = r_1 \|\bar{v} - \underline{v}\|$ is recalculated using the new boundaries.

To make the rule base compact and to improve the computation efficiency, redundant rules need to be deleted. In clustering, it means deleting redundant cluster centers.

Let $d_{\min}^*$ be the shortest distance between two existing cluster centers and v_j^* be the center with a smaller potential. If the following condition is satisfied:

$$\frac{d_{\min}^*}{r_b} + \frac{P_t(v_j^*)}{\max_{i=1}^{c} P_t(v_i^*)} < 1, \quad (4.49)$$

then v_j^* is ***deleted*** *from the set of cluster centers.*

In (4.49), $r_b = r_2 \left\|\bar{v} - \underline{v}\right\|$ with $r_2 \in (0, 1)$.

The above online clustering approach provides a dynamic and evolving rule base by upgrading it when incoming new data bring new information. The number of the cluster centers determines the number of the fuzzy rules, and the values of them are used to construct the membership functions of the input variables.

Determining membership parameters. Based on the obtained cluster centers, the parameters of Gaussian membership functions are set as follows:

$$c_j^i = v_i^*, \quad \delta_j^i = \frac{r_3}{c}(\bar{v}_j - \underline{v}_j), \tag{4.50}$$

where $i = 1, 2, \ldots, c, j = 1, 2, \ldots, L$, where $r_3 \in (0, 1)$ is a design parameter.

4.3.2 Estimation of Consequent Parameters

For a fixed number of rules, $N = c$, and membership parameters, (c_j^i, δ_j^i), the estimation of consequent parameters can be transformed into a least squares problem by transforming the overall T–S model (4.3) into a linearly parametrized form (4.22):

$$y(t+1) = \theta^T \phi(t). \tag{4.51}$$

Recall that

$$\begin{aligned}
\theta &= [\theta_1^T, \theta_2^T, \ldots, \theta_N^T]^T \\
\phi(t) &= [\phi_1^T(t), \phi_2^T(t), \ldots, \phi_N^T(t)]^T \\
\phi_i(t) &= [\mu_i, \mu_i\xi_1(t), \mu_i\xi_2(t), \ldots, \mu_i\xi_L(t)]^T \\
\theta_i &= [\theta_{i0}, \theta_{i1}, \ldots, \theta_{iL}]^T.
\end{aligned} \tag{4.52}$$

The Eq. (4.51) with θ unknown and $\phi(t)$ known is a regression form with a linear parametrization for which many parameter estimation algorithms can be adopted, such as the RLS algorithm (4.27) in Sect. 4.2.3. The initial estimation $\hat{\theta}(0)$ takes the value obtained from offline identification.

Evolution of rule base. At time t, when a new cluster center is generated, a new rule is **added** to the rule base as well. The estimated parameter vector at time $t-1$ is expanded by

$$\hat{\theta}(t-1) := [\hat{\theta}_1^T(t-1), \cdots, \hat{\theta}_N^T(t-1), \hat{\theta}_{N+1}^T(t-1)]^T \tag{4.53}$$

with $\hat{\theta}_{N+1}^T(t-1) = 0$ and $\hat{\theta}_1^T(t-1), \cdots, \hat{\theta}_N^T(t-1)$ inherited from the previous time step. The covariance matrix for RLSE are reset as

$$C_{t-1} := \begin{bmatrix} \eta C_{t-1} & 0 \\ 0 & \Omega I \end{bmatrix}, \tag{4.54}$$

where $\eta > 1$ is a positive coefficient and a recommended choice of η is $\eta = (N^2 + 1)/N^2$. Then, with the expanded $\hat{\theta}(t-1)$ and C_{t-1}, the parameter estimate at time t is calculated by (4.27).

When an existing cluster center is **replaced** by a new cluster center, the membership function parameters of the corresponding rule are updated using the updated cluster center based on (4.50) while its consequent parameters are inherited from the replaced rule.

When an existing cluster center is **deleted**, the corresponding rule is deleted as well. The parameter estimate $\hat{\theta}(t-1)$ and the covariance matrix C_{t-1} are modified by deleting the corresponding elements in them and then updated by (4.27).

4.3.3 Procedure for Online Identification

The procedure of online identification includes the following steps:

Step 1: Initialize the rule-base structure, including premise variables, the membership function type, and the number of rules. The premise variables are obtained from *a prior* knowledge or identified by the offline identification approach. The membership function type is set as Gaussian function. The initial number of rules (cluster centers) can be determined from offline identification. The initial potentials of existing cluster centers are set by (4.42). If the online clustering algorithm starts with the first data point, it is treated as the first clustering center and its potential is set according to (4.41).

Step 2: At time t, when the tth new data point comes, its potential is calculated by (4.43) and the potentials of existing cluster centers are updated by (4.46). The potential of the new data point is compared with the potentials of old cluster centers and the set of clusters centers is updated based on the criteria (4.47)–(4.49).

Step 3: Update the rule-base structure based on the updated set of cluster centers (if a new rule is added, its membership parameters are determined by (4.50)), and modify the consequent parameter estimate vector if needed (if a new rule is added, the consequent parameter estimate vector is expanded according to (4.53)).

Step 4: Update the consequent parameters according to (4.27).

Step 5: Repeat *Step* 2–4 for $t := t + 1$.

4.3.4 Simulation Study

In this section, the online identification algorithms are applied to two simulation examples, which have been used in offline identification. The first example is a motor-driven single robot link introduced in Example 4.1, and the second example is a discrete-time second-order nonlinear plant (Narendra and Parthasarathy 1990).

Simulation 1. Different from the offline identification, the online identification starts from the first data point, and the I/O data are collected with a sampling time $T = 0.01$s. RLS is employed to update parameter estimates recursively and the parameter Ω is set as $\Omega = 1000$. The parameter r_3 for calculating the membership function

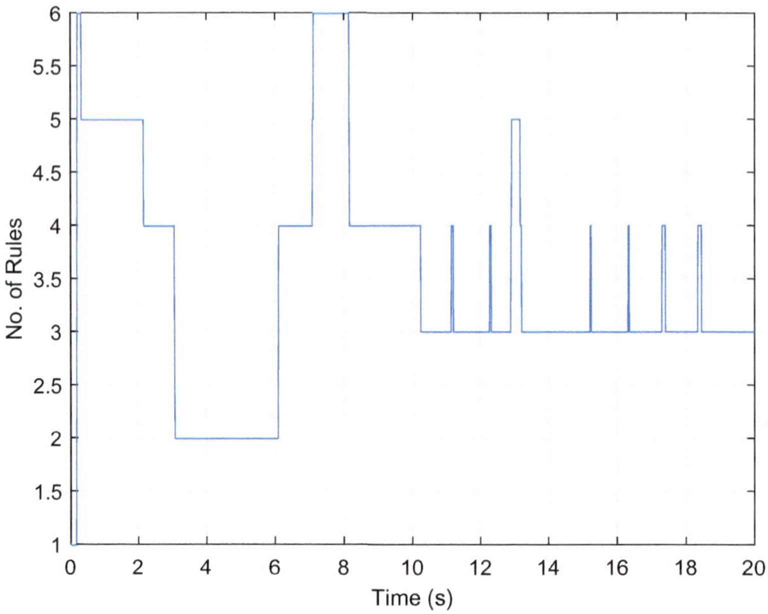

Fig. 4.9 Evolution of the number of rules in Simulation 1

parameters is set as $r_3 = 0.9$. The parameters r_1 and r_2 affect the criteria for adding or deleting rules, which play critical roles in rule-base evolution. By trying various values of r_1 and r_2, finally, we select $r_1 = 0.95$ and $r_2 = 0.95$ which produce good approximation performance and a compact rule base. The evolution of the number of rules is shown in Fig. 4.9. The comparison between the true and estimated outputs is shown in Fig. 4.10. The estimated parameters are shown in Figs. 4.11 and 4.12.

Simulation 2. In this example, we identify the discrete-time second-order nonlinear plant (4.39) in an online mode. The input signal is selected as $u(t) = 0.3\sin(2\pi t/50) + 0.3\sin(2\pi t/250)$. The online clustering algorithm starts with the first data point with $\Omega = 1000$ and $r_3 = 0.9$. For each value of r_1 from 0.9 to 0.1, we vary the value of r_2 from 0.9 to 0.1 and observe the corresponding MSE. For each value of r_1, the value of r_2 with the minimum MSE is shown in Table 4.4. It can be observed that when r_1 is set as 0.8 and r_2 as 0.2, the MSE index is the smallest. However, the maximal number of rules is 20. When r_1 is set as 0.4 and r_2 as 0.9, the MSE index is small and the number of rules is suitable.

With $r_1 = 0.4$ and $r_2 = 0.9$, the online evolution of the rule number is shown in Fig. 4.13. The comparison between the estimated output from the T–S fuzzy system and the true output from the plant is shown in Fig. 4.14. The estimated parameters are shown in Fig. 4.15.

In online identification, the parameters r_1 and r_2 greatly affect the procedure of rule-base evolution. Currently, we use *trial-and-error* to select the values of r_1 and

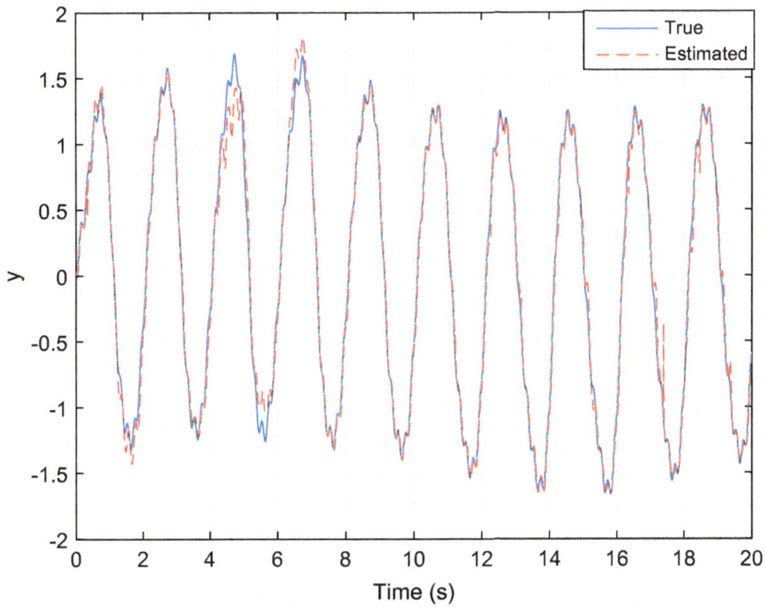

Fig. 4.10 True and estimated outputs in Simulation 1

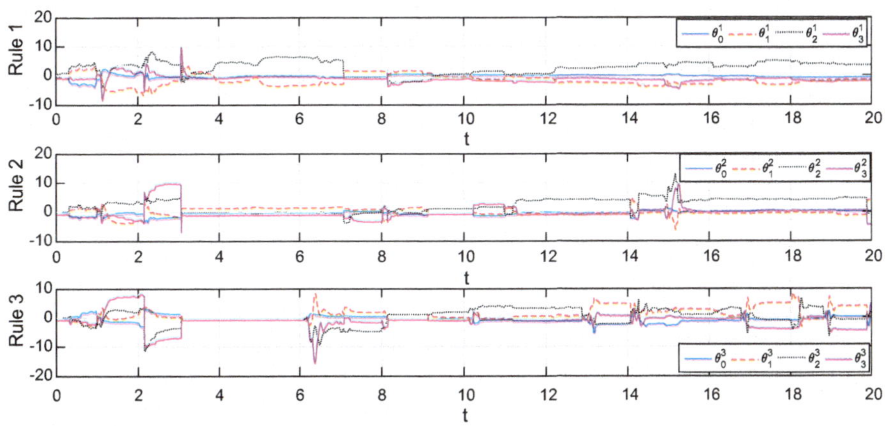

Fig. 4.11 Estimated parameters of rules 1–3 in Simulation 1

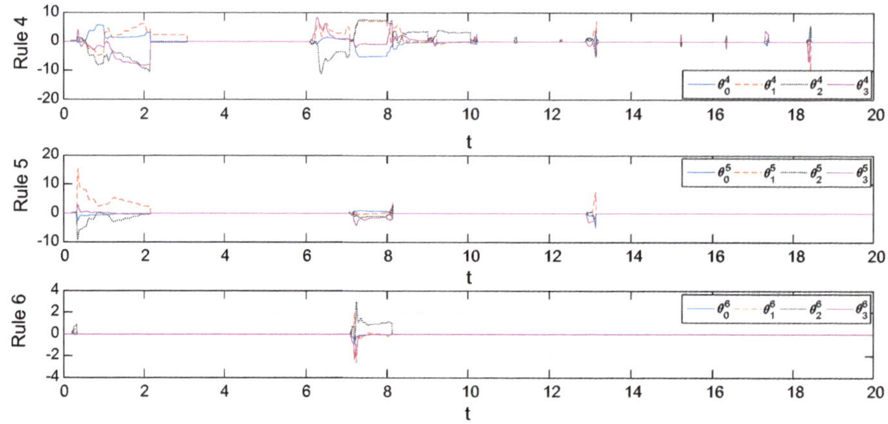

Fig. 4.12 Estimated parameters of rules 4–6 in Simulation 1

Table 4.4 MSE in Simulation 2

r_1	r_2	Maximal no. of rules	MSE
0.9	0.1	23	0.0146
0.8	0.2	20	0.0124
0.7	0.2	17	0.0368
0.6	0.7	7	0.0925
0.5	0.1	30	0.0126
0.4	0.9	4	0.0141
0.3	0.9	9	0.0158
0.2	0.8	9	0.0170
0.1	0.8	9	0.0170

r_2 which are not very efficient. Further study should be done on how to optimize the values of r_1 and r_2 through certain well-designed mechanism.

4.4 Summary

Identification of T–S fuzzy systems consists of four tasks: (1) identification of premise variables; (2) identification of the number of rules; (3) estimation of consequent parameters; and (4) adjustment of membership parameters. The first two tasks belong to structure identification and the last two belong to parameter identification.

In practice, the structure identification and the parameter identification cannot be completely separated. To identify premise variables and the number of rules, the evaluation of the identification performance relies on a group of consequent

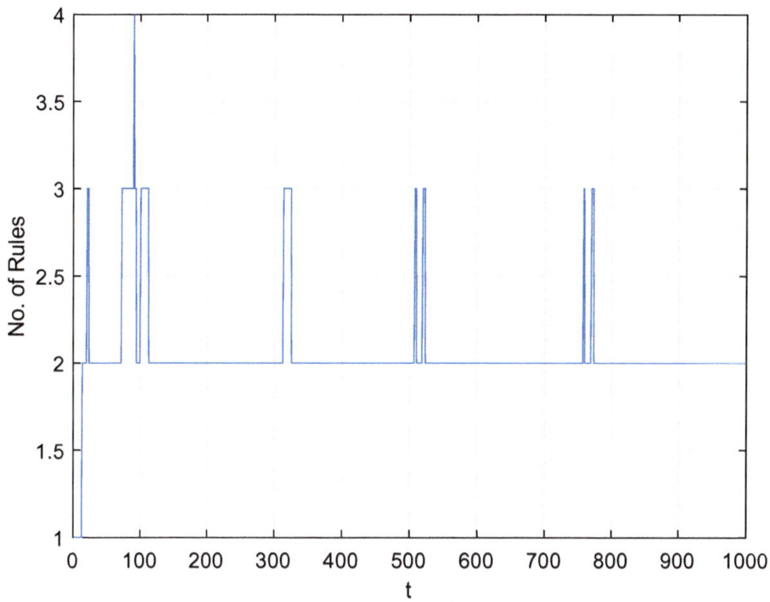

Fig. 4.13 Evolution of the number of rules in Simulation 2

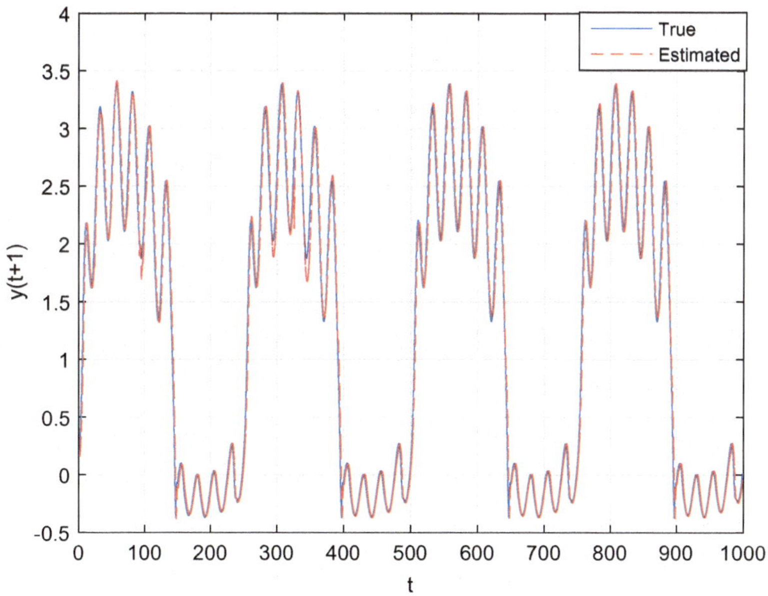

Fig. 4.14 True and estimated outputs in Simulation 2

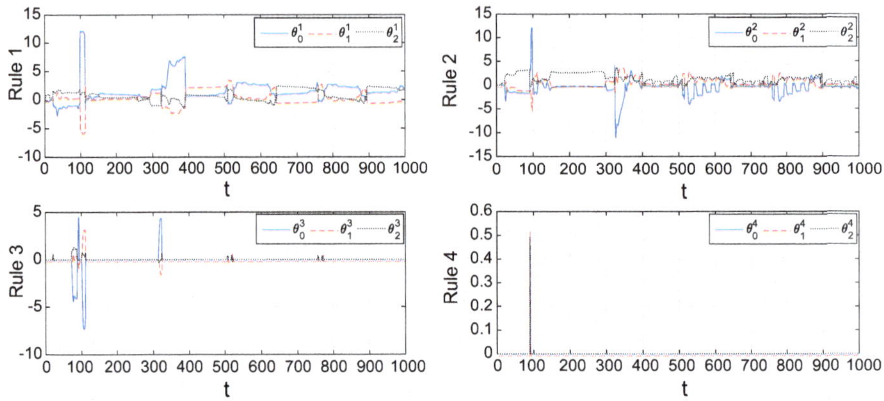

Fig. 4.15 Estimated parameters in Simulation 2

parameters and membership parameters. To optimize parameters, we need to fix the premise variables and the number of rules.

Identification of T–S fuzzy systems can be carried out in either an offline mode or an online mode. In the offline mode, based on the data collection, the premise variables are identified using a search tree algorithm and the model quality subject to different premise variables is compared based on the RC value. The number of rules can be manually selected or determined through offline fuzzy clustering. The consequent parameters can be estimated using either the batch least squares estimation algorithm or the recursive least squares estimation algorithm. The membership parameters can be adjusted based on gradient methods.

In the online mode, the data come recursively. A potential function is defined to evaluate the potential of a new data point as a cluster center, which can be calculated in a recursive way. The potentials of all existing cluster centers are also updated when a new data point comes. Then the potential of the new data point is compared with the updated potentials of existing cluster centers to determine whether the new data point can be accepted as a new cluster center or replace an existing cluster center, and whether any existing cluster center should be deleted from the set of cluster centers. The number of rules is determined by the number of cluster centers, and the parameters of membership functions are determined from the coordinates of cluster centers. The consequent parameters are estimated using the recursive least squares estimation algorithm.

By following the offline and online identification procedures introduced in this chapter, it is easy for a graduate student to obtain a T–S fuzzy model using I/O data. However, more advanced topics in system identification such as how to handle the noise in the I/O data, how to conduct closed-loop identification, etc., are not addressed in this chapter. The obtained fuzzy model can serve as an initial model for control design and the model parameters can be further adjusted by parameter adaptive laws developed from adaptive control schemes which will be presented in the following chapters.

References

Abonyi J (2003) Fuzzy model identification for control. Birkhäuser, Boston
Angelov PP (2002) Evolving rule-based models: a tool for design of flexible adaptive systems. Springer, Heidelberg
Angelov PP, Filev DP (2004) An approach to online identification of Takagi-Sugeno fuzzy models. IEEE Trans Syst Man Cybern - Part B: Cybern 34(1):484–498
Babuska B (1998) Fuzzy modeling for control. Kluwer Academic Publishers, Boston
Barada S, Singh H (1998) Generating optimal adaptive fuzzy-neural models of dynamical systems with applications to control. IEEE Trans Syst Man Cybern - Part C: Appl Revi 28(3):297–313
Bezdek J (1976) A physical interpretation of fuzzy isodata. IEEE Trans Syst Man and Cybern 387–389
Bezdek J (1981) Pattern recognition with fuzzy objective function. Plenum Press, New York
Chiu SL (1994) Fuzzy model identification based on cluster estimation. J Intell Fuzzy Syst 2:267–278
Espinosa J, Vandewalle J, Wertz V (2005) Fuzzy logic, identification and predictive control. Springer, London
Gath I, Geva AB (1989) Unsupervised optimal fuzzy clustering. IEEE Trans. Pattern Mach Intell 7:773–781
Gustafson DE, Kessel WC (1979) Fuzzy clustering with a fuzzy covariance matrix. In: Proceedings of IEEE CDC, San Diego, CA, pp 761–766
Lilly JH (2010) Fuzzy control and identification. Wiley, Hoboken
Ljung L (1987) System identification: theory for the user. Prentice-Hall, Englewood Cliffs
Narendra KS, Parthasarathy K (1990) Identification and control of dynamical systems using neural networks. IEEE TransNeural Netw 1:4–27
Nelles O (2001) Nonlinear system identification: from classical approaches to neural networks and fuzzy models. Springer, Berlin
Oviedo JJE, Vandewalle JPL, Wertz V (2005) Fuzzy logic, identification and predictive control. Springer, London
Sugeno M, Yasukawa T (1993) A Fuzzy-logic-based approach to qualitative modeling. IEEE Trans Fuzzy Syst 1:7–31
Takagi T, Sugeno M (1985) Fuzzy identification of systems and its applications to modeling and control. IEEE Trans Syst Man Cybern 15(1):116–132

Chapter 5
Adaptive T–S Fuzzy State Tracking Control Using State Feedback

In this chapter, we will focus on adaptive state tracking control designs for T–S fuzzy systems. As is well-known, for model reference adaptive control of linear systems, there are some matching conditions between the plant and the reference model which should be satisfied such that state tracking can be achieved. For T–S fuzzy systems, there also exist such plant-model matching conditions which will be developed in this chapter. The reference model for a linear system is usually a stable linear system. For a T–S fuzzy system, the reference model can be chosen as either a stable linear system or a stable T–S fuzzy system, which leads to different matching conditions. On the other hand, for T–S fuzzy systems with different system structures (the forms of A_i and B_i), different controller structures can be proposed which also lead to different plant-model matching conditions. We will carry out detailed studies on these issues.

When the parameters of T–S fuzzy systems are unknown, the controller parameters cannot be calculated directly from the matching conditions. A parameter estimation scheme is essential for developing an adaptive controller for a T–S fuzzy system. We will study how to parametrize the T–S fuzzy system and the state tracking controller and how to derive a key estimation model that can be used for developing parameter adaptive laws.

We will also discuss some other important issues in adaptive state tracking control of T–S fuzzy systems, i.e., how to guarantee the input matrix with estimated parameters nonsingular in the adaptive controller.

5.1 Problem Statement

In this section, we describe the state tracking control problem for T–S fuzzy systems in a canonical form and in a general form.

Consider the following nonlinear dynamic system:

$$\dot{x}(t) = f(x(t), u(t)), \tag{5.1}$$

R. Qi et al., *Fuzzy System Identification and Adaptive Control*, Communications and Control Engineering,
https://doi.org/10.1007/978-3-030-19882-4_5

where $x = [x_1, x_2, \ldots, x_n]^T \in R^n$ is the state vector, $u \in R^m$ is the input, and $f(\cdot, \cdot)$ is an n-dimensional smooth nonlinear function. As described in Sect. 2.2.1, a T–S fuzzy system is capable of representing a nonlinear system by fuzzily blending a group of linearized models and enables more convenient analysis and controller design by employing sophisticated linear control techniques.

Consider a T–S fuzzy system representation for the nonlinear system (5.1) with the following rules:

$$\begin{aligned}&\text{IF} \quad \xi_1(t) \ \text{is} \ F_1^i \ \text{and} \ \ldots \ \text{and} \ \xi_L(t) \ \text{is} \ F_L^i,\\ &\text{THEN} \quad \dot{x}(t) = A_i x(t) + B_i u(t),\end{aligned} \tag{5.2}$$

where $i = 1, 2, \ldots, N$, F_j^i is a fuzzy set associated with which there is a membership function $F_j^i(\xi_j(t))$ to indicate the degree of membership of $\xi_j(t)$ in F_j^i.

Using the standard technique of *singleton fuzzification*, *product inference* and *weighted average*, we obtain the following global T–S fuzzy model:

$$\dot{x}(t) = \sum_{i=1}^{N} \mu_i (A_i x(t) + B_i u(t)), \tag{5.3}$$

where μ_i is the normalized firing strength:

$$\lambda_i = \prod_{j=1}^{L} F_j^i(\xi_j(t)), \ \mu_i = \frac{\lambda_i}{\sum_{i=1}^{N} \lambda_i}, \ \mu_i \geq 0, \ \sum_{i=1}^{N} \mu_i = 1. \tag{5.4}$$

Control objective. The control objective is to design a state feedback controller $u(t)$ for the global version of the fuzzy system (5.3) with unknown parameters (A_i, B_i) to ensure closed-loop signal boundedness and asymptotic tracking of a reference state signal $x_m(t)$ provided by a reference model by the system state $x(t)$.

It is well-known that some matching conditions between the plant model and the reference model are needed so that state tracking can be achieved for linear systems. For T–S fuzzy systems, such matching conditions are also required. We will show in the following sections what the matching conditions are and how the conditions can be relaxed by choosing different reference models and different controller structures.

We will start with the adaptive state tracking design for T–S fuzzy systems in a canonical form, and then develop our adaptive state tracking control schemes for T–S fuzzy systems in a general form.

5.2 Design for T–S Fuzzy Systems in Canonical Form

Many practical systems have certain special structures and their dynamic equations have certain canonical forms. When T–S fuzzy systems are used to approximate such nonlinear systems, they are also in certain canonical forms. This section considers state tracking control design for T–S fuzzy systems in a canonical form.

5.2.1 Plant Model and Reference System

If the nonlinear dynamic system (5.1) can be formulated into the following canonical form:

$$\begin{cases} \dot{x}_i(t) = x_{i+1}(t), \quad i = 1, 2, \ldots, n-1 \\ \dot{x}_n(t) = f_n(x(t), u(t)), \end{cases} \tag{5.5}$$

where $x = [x_1, x_2, \ldots, x_n]^T \in R^n$ is the state vector, $u \in R$ is the input, and $f_n(x, u)$ is a smooth function, the T–S fuzzy system to approximate the nonlinear system (5.5) also has the canonical form.

T–S fuzzy system in canonical form. In this case, the system matrices A_i and B_i in (5.2) are in the following canonical form:

$$A_i = \begin{bmatrix} 0 & 1 & \cdots & 0 & 0 \\ \vdots & \vdots & & \vdots & \vdots \\ 0 & 0 & \cdots & 0 & 1 \\ a_1^i & a_2^i & \cdots & a_{n-1}^i & a_n^i \end{bmatrix}, \quad B_i = \begin{bmatrix} 0 \\ \vdots \\ 0 \\ b^i \end{bmatrix}. \tag{5.6}$$

Linear reference model. Consider the following linear reference model:

$$\dot{x}_m(t) = A_m x_m(t) + B_m r(t), \tag{5.7}$$

where $x_m = [x_{m1}, x_{m2}, \ldots, x_{mn}]^T \in R^n$, $r \in R$, $A_m \in R^{n\times n}$ is a Hurwitz matrix and $B_m \in R^n$ with the following forms

$$A_m = \begin{bmatrix} 0 & 1 & \cdots & 0 & 0 \\ \vdots & \vdots & & \vdots & \vdots \\ 0 & 0 & \cdots & 0 & 1 \\ a_{m1} & a_{m2} & \cdots & a_{m,n-1} & a_{mn} \end{bmatrix}, \quad B_m = \begin{bmatrix} 0 \\ \vdots \\ 0 \\ b_m \end{bmatrix}. \tag{5.8}$$

The control objective is to make the T–S fuzzy system states track the states of the reference model (5.7).

5.2.2 Nominal Controller and Matching Conditions

Before developing an adaptive control scheme for (5.3) with unknown parameters, it is important to make sure there exists a solution to meet the desired control objective when all the parameters are known. For each local linear model, the local controller is designed as

$$\begin{aligned} &\text{IF } \xi_1(t) \text{ is } F_1^i \text{ and } \ldots \text{ and } \xi_L(t) \text{ is } F_L^i, \\ &\text{THEN } u(t) = k_1^{iT} x(t) + k_2^i r(t), \end{aligned} \tag{5.9}$$

where $i = 1, 2, \ldots, N, k_1^i = [k_{11}^i, k_{12}^i, \ldots, k_{1n}^i]^T \in R^n$ and $k_2^i \in R$ are design parameters.

It can be noted that we use the similar state feedback controller structure for linear systems to design each local controller. This is a natural choice since each local system in a T–S fuzzy system is a linear system. However, how to form the overall controller from the local controllers is an interesting problem. Actually we have many different choices which may lead to different design conditions. For example, a usual choice is to build the overall controller by combining all the local controllers in the same way to build the overall T–S fuzzy system, as given in (5.3). That is,

$$u(t) = \sum_{i=1}^{N} \mu_i (k_1^{iT} x(t) + k_2^i r(t)). \tag{5.10}$$

Substituting (5.10) into (5.3) yields

$$\dot{x}(t) = \sum_{i=1}^{N} \sum_{j=1}^{N} \mu_i \mu_j \left[(A_i + B_i k_1^{jT}) x(t) + B_i k_2^j r(t) \right]. \tag{5.11}$$

Comparing (5.11) with (5.7), it can be obtained that if k_1^j and k_2^j satisfy the following matching conditions:

$$A_i + B_i k_1^{jT} = A_m, \quad B_i k_2^j = B_m, \quad i, j = 1, 2, \ldots, N, \tag{5.12}$$

then the closed-loop fuzzy system (5.11) becomes

$$\dot{x}(t) = A_m x(t) + B_m r(t), \tag{5.13}$$

which leads to $\lim_{t \to \infty} (x(t) - x_m(t)) = 0$. However, it can be observed that the matching conditions (5.12) are rather difficult to satisfy since they require that for the ith subsystem parameters (A_i, B_i), not only ith controller parameters k_1^i and k_2^i but also all the other controller parameters k_1^j and k_2^j, $j \neq i$, $j \in [1, N]$ need to satisfy the matching conditions (5.12).

To remove the coupling between the system parameters and controller parameters for different local models, we propose a different method to build the overall controller from local controllers based on the following assumptions:

Assumption 5.1 The signs of b^i in B_i are known and the same for all $i = 1, 2, \ldots, N$, and $|b^i| \geq b_{i0}^*$, where $b_{i0}^* > 0$.

Remark 5.1 Assumption 5.1 requires the control direction of each local system is known and the same, which is reasonable for many practical systems. With the properties of μ_i in (5.4), Assumption 5.1 indicates $\sum_{i=1}^{N} \mu_i b^i \neq 0$.

If the signs of b^i, $i = 1, 2, \ldots, N$ are not the same for each local system, Assumption 5.1 can be replaced by the following one.

Assumption 5.2 The signs of b^i in B_i are known and there exist known lower and upper bounds $\underline{b}^i$ and $\bar{b}^i$ such that for $b^i \in [\underline{b}^i, \bar{b}^i]$, $\sum_i^N \mu_i b^i \neq 0$.

With Assumption 5.1 or 5.2, we propose the following nominal controller.

Nominal controller. The nominal controller can be designed as

$$u(t) = \frac{\sum_{i=1}^{N} \mu_i b^i (k_1^{iT} x(t) + k_2^i r(t))}{\sum_{i=1}^{N} \mu_i b^i}, \tag{5.14}$$

where the overall controller is a weighted-average of all the local controllers using $\mu_i b_i$ rather than μ_i. The benefit to use the controller structure as (5.14) will be clearly shown by the following much more relaxed matching conditions.

Matching conditions. To achieve perfect state tracking, the parameters in (5.14) should satisfy the matching conditions given in the following proposition.

Proposition 5.1 *For the fuzzy system (5.3) subject to the controller (5.14), if there exist $k_1^j \in R^n$, $k_2^j \in R$ satisfying the following matching conditions:*

$$b^i k_1^{iT} + a^{iT} = a_m^T, \quad b^i k_2^i = b_m, \tag{5.15}$$

where $i = 1, 2, \ldots, N$, $a^i = [a_1^i, a_2^i, \ldots, a_n^i]^T$ and $a_m = [a_{m1}, a_{m2}, \ldots, a_{mn}]^T$, then the closed-loop fuzzy system becomes

$$\dot{x}(t) = A_m x(t) + B_m r(t). \tag{5.16}$$

Proof Substituting (5.14) into (5.3) yields

$$\dot{x}_n(t) = \sum_{i=1}^{N} \mu_i (a^{iT} + b^i k_1^{iT}) x(t) + \mu_i b^i k_2^i r(t). \tag{5.17}$$

With (5.15) and (5.4), the Eq. (5.17) becomes

$$\dot{x}_n(t) = a_m^T x(t) + b_m r(t), \tag{5.18}$$

which can be equivalently written as

$$\dot{x}(t) = A_m x(t) + B_m r(t). \tag{5.19}$$

∇

Remark 5.2 Comparing (5.15) with (5.12), there is no coupling between (A_i, B_i) and (k_1^j, k_2^j) for $j \neq i$ in (5.15). Therefore, the matching conditions (5.15) can be satisfied for achieving state tracking in reality, which means the controller (5.14) is more practical than the controller (5.10) which requires very strong matching conditions for state tracking.

With A_i and B_i known, the nominal controller parameters k_1^i and k_2^i in (5.14) can be solved from (5.15). If A_i and B_i are unknown, an indirect adaptive scheme will be introduced to estimate the controller parameters.

5.2.3 Adaptive Control Scheme

In this section, we design an adaptive parameter estimation algorithm to estimate the unknown controller parameters, develop an indirect adaptive control law and analyze the closed-loop system performance.

First, we use the matching conditions (5.15) to transform the nominal control law (5.14) into the following form:

$$u(t) = \frac{\sum_{i=1}^{N} \mu_i (a_m^T x(t) - a^{iT} x(t) + b_m r(t))}{\sum_{i=1}^{N} \mu_i b^i}, \tag{5.20}$$

which only contains the plant parameters a^i and b^i.

Adaptive control law. Since a^i and b^i are unknown, the controller (5.20) cannot be implemented. Using parameter estimation, an indirect adaptive control law as the adaptive version of (5.20) can be implemented as

$$u(t) = \frac{\sum_{i=1}^{N} \mu_i (a_m^T x(t) - \hat{a}^{iT} x(t) + b_m r(t))}{\sum_{i=1}^{N} \mu_i \hat{b}^i}, \tag{5.21}$$

where $\hat{a}^i$ and $\hat{b}^i$ are the estimates of a^i and b^i, respectively.

Define the *state tracking error*

$$e(t) = x(t) - x_m(t). \tag{5.22}$$

Then the reference model (5.7) can be written as

$$\dot{x}_m(t) = -A_m e(t) + A_m x(t) + B_m r(t), \tag{5.23}$$

which gives

$$\dot{x}_{mn}(t) = -a_m^T e(t) + a_m^T x(t) + b_m r(t). \tag{5.24}$$

It can be obtained from (5.3) and (5.6) that

$$\dot{x}_n(t) = a^{iT} x(t) + b^i u(t), \tag{5.25}$$

which can be formulated as

$$\dot{x}_n(t) = \sum_{i=1}^{N} \mu_i a^{iT} x(t) - \sum_{i=1}^{N} \mu_i \tilde{b}^i u(t) + \sum_{i=1}^{N} \mu_i \hat{b}^i u(t), \tag{5.26}$$

where $\tilde{b}^i = \hat{b}^i - b^i$.

Applying the adaptive controller (5.21) into (5.26), we have

$$\dot{x}_n(t) = -\sum_{i=1}^{N} \mu_i \tilde{a}^{iT} x(t) - \sum_{i=1}^{N} \mu_i \tilde{b}^i u(t) + a_m^T x(t) + b_m r(t), \tag{5.27}$$

where $\tilde{a}^i = \hat{a}^i - a^i$.

Subtracting (5.24) from (5.27) produces

$$\dot{e}_n(t) = a_m^T e(t) - \sum_{i=1}^{N} \mu_i \tilde{a}^i x(t) - \sum_{i=1}^{N} \mu_i \tilde{b}^i u(t). \tag{5.28}$$

Tracking error dynamics. From (5.28), we have the tracking error dynamics

$$\dot{e}(t) = A_m e(t) - \sum_{i=1}^{N} \mu_i (\tilde{A}^i x(t) + \tilde{B}^i u(t)), \tag{5.29}$$

where $\tilde{A}_i = \hat{A}_i - A_i$, $\tilde{B}_i = \hat{B}_i - B_i$, with $\hat{A}_i$ and $\hat{B}_i$ being the estimates of A_i and B_i, respectively.

Parameter adaptive laws. To obtain the parameter estimates $\hat{a}^i$ and $\hat{b}^i$ in (5.21), we design the following parameter adaptive laws:

$$\dot{\hat{a}}^i = \gamma_{1i} \mu_i x P_n e, \tag{5.30}$$

$$\dot{\hat{b}}^i = \begin{cases} \gamma_{2i} \mu_i \hat{u} P_n e & \text{if } \left|\hat{b}^i\right| > b_{i0}^* \\ & \text{or } \left|\hat{b}^i\right| = b_{i0}^* \text{ and } \operatorname{sign}(b^i)\hat{u} P_n e \geq 0 \\ 0 & \text{otherwise,} \end{cases} \tag{5.31}$$

where γ_{1i}, $\gamma_{2i} > 0$ are design parameters, b_{i0}^* is the lower bound of $|b^i|$, which is assumed to be known, and $\hat{b}^i(0)$ satisfies $\hat{b}^i(0)\operatorname{sign}(b^i) \geq 0$, P_n is the last row of a positive definite symmetrical matrix $P \in R^{n \times n}$, which is the solution of the following equation:

$$A_m^T P + P A_m = -Q, \tag{5.32}$$

where $Q \in R^{n\times n}$ is a positive definite symmetrical matrix.

Based on the adaptive control law (5.21) and the parameter adaptation laws (5.30)–(5.31), the closed-loop stability and asymptotic tracking results are proved and summarized in the following theorem.

Theorem 5.1 *For the T–S fuzzy system (5.3), under Assumptions 5.1–5.2, the adaptive controller (5.21) with the parameter adaptive laws (5.30)–(5.31) guarantees:*

(i) all closed-loop system signals are bounded; and
(ii) the state tracking error $e(t) \to 0$ when $t \to \infty$.

Proof Consider the following Lyapunov function candidate

$$V(e, \tilde{a}, \tilde{b}) = e^T P e + \sum_{i=1}^{N} \gamma_{1i}^{-1} \tilde{a}^{iT} \tilde{a}^i + \sum_{i=1}^{N} \gamma_{2i}^{-1} (\tilde{b}^i)^2, \tag{5.33}$$

where $\tilde{a}^i = \hat{a}^i - a^i$ and $\tilde{b}^i = \hat{b}^i - b^i$.

The differentiation of V along the error dynamics (5.29) is

$$\begin{aligned}\dot{V} &= \dot{e}^T P e + e^T P \dot{e} + 2\sum_{i=1}^{N} \gamma_{1i}^{-1} \dot{\tilde{a}}^{iT} \tilde{a}^i + 2\sum_{i=1}^{N} \gamma_{2i}^{-1} \dot{\tilde{b}}^i \tilde{b}^i \\ &= e^T A_m^T P e + e^T P A_m e - 2\sum_{i=1}^{N} \mu_i x^T \tilde{A}^{iT} P e - 2\sum_{i=1}^{N} \mu_i \hat{u}^T \tilde{B}^i P e \\ &\quad + 2\sum_{i=1}^{N} \gamma_{1i}^{-1} \tilde{a}^{iT} \dot{\tilde{a}}^i + 2\sum_{i=1}^{N} \gamma_{2i}^{-1} \tilde{b}^i \dot{\tilde{b}}^i. \end{aligned} \tag{5.34}$$

Applying (5.32) into (5.34) and with simple manipulations, we have

$$\begin{aligned}\dot{V} &= -e^T Q e - 2\sum_{i=1}^{N} \mu_i \tilde{a}^{iT} x P_n e - 2\sum_{i=1}^{N} \mu_i \tilde{b}^i \hat{u} P_n e \\ &\quad + 2\sum_{i=1}^{N} \gamma_{1i}^{-1} \tilde{a}^{iT} \dot{\tilde{a}}^i + 2\sum_{i=1}^{N} \gamma_{2i}^{-1} \tilde{b}^i \dot{\tilde{b}}^i. \end{aligned} \tag{5.35}$$

Using the parameter adaptive laws (5.30)–(5.31), we can obtain from (5.35) that

$$\dot{V} = -e^T Q e \le 0, \tag{5.36}$$

which means e, $\tilde{a}_i$, $\tilde{b}_i \in L_\infty$. From (5.22) and (5.21), we can have x, $u \in L_\infty$ as well.

From (5.29), we have $\dot{e}(t) \in L_\infty$. It follows from (5.36) and the boundedness of V that

$$\int_0^\infty e^T(\tau)Qe(\tau)d\tau = -\int_0^\infty \dot{V}d\tau = V(0) - V(\infty) < \infty, \quad \forall t \geq 0, \quad (5.37)$$

that is, $e(t) \in L_2$. With Barbălat's lemma, we have the desired convergence property: $\lim_{t\to\infty} e(t) = 0$. ∇

Thus far, we have considered the adaptive state tracking control for a T–S fuzzy system in a canonical form. We have designed a nominal controller structure (5.14) which leads to matching conditions (5.15) that can be easily satisfied in practice. Next, we will consider the adaptive state tracking control for T–S fuzzy systems in a general form.

5.3 Design for T–S Fuzzy Systems in General Form ($m \leq n$)

In this section, we consider the case when A_i and B_i are general matrices and the number of inputs m is less than or equal to the number of states n.

5.3.1 *Plant Model and Reference System*

Consider the T–S fuzzy system (5.2) with the following rules:

$$\begin{aligned} &\text{IF } \xi_1(t) \text{ is } F_1^i \text{ and } \ldots \text{ and } \xi_L(t) \text{ is } F_L^i, \\ &\text{THEN } \dot{x}(t) = A_i x(t) + B_i u(t), \end{aligned} \quad (5.38)$$

where $x \in R^n$, $u \in R^m$, $i = 1, 2, \ldots, N$.

T–S fuzzy system in general form. The system matrices A_i and B_i in (5.38) are in the following general form:

$$A_i = \begin{bmatrix} a_{11}^i & a_{12}^i & \cdots & a_{1,n-1}^i & a_{1,n}^i \\ a_{21}^i & a_{22}^i & \cdots & a_{2,n-1}^i & a_{2,n}^i \\ \vdots & \vdots & & \vdots & \vdots \\ a_{n,1}^i & a_{n,2}^i & \cdots & a_{n,n-1}^i & a_{n,n}^i \end{bmatrix}, \quad B_i = \begin{bmatrix} b_{11}^i & b_{12}^i & \cdots & b_{1,m}^i \\ b_{21}^i & b_{22}^i & \cdots & b_{2,m}^i \\ \vdots & \vdots & \cdots & \vdots \\ b_{n,1}^i & b_{n,2}^i & \cdots & b_{n,m}^i \end{bmatrix}. \quad (5.39)$$

The global T–S fuzzy model is the same as (5.3):

$$\dot{x}(t) = \sum_{i=1}^{N} \mu_i (A_i x(t) + B_i u(t)), \quad (5.40)$$

where μ_i has the same meaning as in (5.4).

In this section, we consider two kinds of reference models: linear reference model and T–S fuzzy reference model.

Linear reference model. The linear reference model is given by

$$\dot{x}_m(t) = A_m x(t) + B_m r(t), \tag{5.41}$$

where $x_m \in R^n$, $r \in R^m$, $A_m \in R^{n\times n}$ is a Hurwitz matrix and $B_m \in R^{n\times m}$.

T–S fuzzy reference model. The T–S fuzzy reference model is defined by the following rule:

$$\begin{aligned} &\text{IF } \ \xi_1(t) \ \text{ is } \ F_1^i \ \text{ and } \ \ldots \ \text{ and } \ \xi_L(t) \ \text{ is } \ F_L^i, \\ &\text{THEN } \ \dot{x}_m(t) = A_{mi} x_m(t) + B_{mi} r(t), \ \ i = 1, 2, \ldots, N, \end{aligned} \tag{5.42}$$

where $x_m \in R^n$, $r \in R^m$, $A_{mi} \in R^{n\times n}$ and $B_{mi} \in R^{n\times m}$.

The overall T–S fuzzy reference model is inferred as

$$\dot{x}_m(t) = \sum_{i=1}^{N} \mu_i (A_{mi} x_m(t) + B_{mi} r(t)), \tag{5.43}$$

where μ_i has the same meaning as in (5.4).

In (5.42), the matrices A_{mi} are chosen to satisfy the following condition:

(C.1). A common positive definite symmetrical matrix P exists for some chosen $Q = Q^T > 0$, which is the solution of the following linear matrix inequalities:

$$A_{mi}^T P + P A_{mi} \le -Q, \quad i = 1, 2, \ldots, N. \tag{5.44}$$

The condition (C.1) is a sufficient condition that ensures that the T–S fuzzy reference system (5.43) is stable (Tanaka and Sugeno 1992).

5.3.2 *Nominal Controller and Matching Conditions*

Based on the idea of parallel distributed control (PDC), we can use linear control techniques to design a local linear state feedback controller for each rule:

$$\begin{aligned} &\text{IF } \ \xi_1(t) \ \text{ is } \ F_1^i \ \text{ and } \ \ldots \ \text{ and } \ \xi_L(t) \ \text{ is } \ F_L^i, \\ &\text{THEN } \ u(t) = K_{1i} x(t) + K_{2i} r(t), \end{aligned} \tag{5.45}$$

where $K_{1i} \in R^{m\times n}$ and $K_{2i} \in R^{m\times m}$.

The PDC controller shares the same fuzzy sets with the fuzzy model (5.38) to construct its premise parts. The resulting overall controller, which is nonlinear in general, is a fuzzy blending of each individual linear controller. The overall output of the fuzzy controller is inferred by integrating the local controllers (5.45) using the weighted-average method:

$$u(t) = \sum_{i=1}^{N} \mu_i (K_{1i}x(t) + K_{2i}r(t)), \tag{5.46}$$

where μ_i is the normalized firing strength defined in (5.4).

Applying (5.46) to the fuzzy system (5.40) and with some manipulations, we can have the closed-loop fuzzy control system:

$$\dot{x}(t) = \sum_{i=1}^{N} \sum_{l=1}^{N} \mu_i \mu_j [(A_i + B_i K_{1j})x(t) + B_i K_{2j} r(t)]. \tag{5.47}$$

Matching conditions for linear reference model. To meet the state $x(t)$ of the system (5.47) tracking the state $x_m(t)$ of the reference model (5.41), the following matching conditions need to be satisfied:

$$A_i + B_i K_{1j} = A_m, \quad B_i K_{2j} = B_m, \tag{5.48}$$

for $\forall i, j = 1, 2, \ldots, N$.

Applying (5.48) into (5.47) results in the closed-loop fuzzy system:

$$\dot{x} = A_m x(t) + B_m r(t), \tag{5.49}$$

which has the same dynamics as the reference model (5.41).

Remark 5.3 The matching conditions (5.48) for the T–S fuzzy system (5.40) to achieve state tracking control under the controller (5.46), however, are difficult to satisfy even when all the system parameters are known. There are two ways to relax the matching conditions: (i) change the reference model; (ii) change the controller structure. We will try to choose a different reference model to relax the matching conditions first, and then try to choose a different controller structure.

Matching conditions for T–S fuzzy reference model. With the T–S fuzzy reference model (5.43), the matching conditions for state tracking control become

$$A_i + B_i K_{1j} = A_{mi}, \quad B_i K_{2j} = B_{mi} \tag{5.50}$$

for $\forall i, j = 1, 2, \ldots, N$.

With (5.40), we have

$$\dot{x}(t) = \sum_{i=1}^{N} \mu_i A_i x(t) + \sum_{i=1}^{N} \mu_i B_i u(t). \tag{5.51}$$

With (5.46) and (5.50), and after some simple mathematical manipulations, we obtain

$$\begin{aligned}\sum_{i=1}^{N}\mu_i B_i u(t) &= \sum_{i=1}^{N}\mu_i B_i \sum_{j=1}^{N}\mu_i (K_{1j}x(t) + K_{2j}r(t)) \\ &= \sum_{i=1}^{N}\mu_i [(A_{mi} - A_i)x(t) + B_{mi}r(t)]. \end{aligned} \tag{5.52}$$

Substituting (5.52) into (5.51) yields

$$\dot{x}(t) = \sum_{i=1}^{N}\mu_i (A_{mi}x(t) + B_{mi}r(t)) \tag{5.53}$$

which has the same dynamics as the fuzzy reference model (5.43).

Remark 5.4 Comparing the matching conditions (5.50) with (5.48), we can see that in (5.48) for $\forall i, j = 1, 2, \ldots, N$, A_i, B_i, K_{1j} and K_{2j} need to satisfy the quite conservative relationship with respect to a common A_m and a common B_m, while in (5.50), the relationship is less conservative since one can choose different A_{mi} and B_{mi} now. Therefore, using a T–S fuzzy reference model, the matching conditions can be relaxed. However, it should be noted that the relaxed matching conditions (5.50) still requires that the T–S fuzzy system and the T–S fuzzy reference model have some special structures and only when they have such structures, state tracking by the controller (5.46) can be achieved.

Till now, we have developed the nominal controller for state tracking control of T–S fuzzy systems when all the parameters are known. In the next section, we will solve state tracking problem for unknown system parameters.

5.3.3 Adaptive Control Scheme

In this section, we shall design an adaptive parameter estimation algorithm to estimate the unknown controller parameters, develop an adaptive control law, and analyze the closed-loop system stability and tracking performance.

For an adaptive control design for the fuzzy systems, it is crucial to parametrize the controlled plant in terms of its input signal $u(t)$ and state vector $x(t)$.

Plant parametrization. The T–S fuzzy system (5.40) can be formulated into the following form:

$$\dot{x}(t) = \sum_{i=1}^{N}\mu_i A_{mi}x(t) + \sum_{i=1}^{N}\mu_i [(A_i - A_{mi})x(t) + B_i u(t)]. \tag{5.54}$$

With the matching conditions (5.50), we have $B_i = B_{mi}K_{2i}^{-1}$, $A_{mi} - A_i = B_i K_{1i} = B_{mi}K_{2i}^{-1}K_{1i}$, with which we can obtain from (5.54) that

$$\begin{aligned}\dot{x}(t) &= \sum_{i=1}^{N} \mu_i A_{mi}x(t) + \sum_{i=1}^{N} \mu_i (B_{mi}K_{2i}^{-1}K_{1i}x(t) + B_{mi}K_{2i}^{-1}u(t)) \\ &= \sum_{i=1}^{N} \mu_i [A_{mi}x(t) + B_{mi}(\Theta_{1i}x(t) + \Theta_{2i}u(t))],\end{aligned} \tag{5.55}$$

where $\Theta_{1i} = K_{2i}^{-1}K_{1i} \in R^{m\times n}$ and $\Theta_{2i} = K_{2i}^{-1} \in R^{m\times m}$.

The parametrized model (5.55) is indeed in terms of the input $u(t)$ and the state vector $x(t)$, crucial for constructing an estimator with $u(t)$ and $x(t)$, used for designing parameter adaptive laws.

Estimation model. Let $\hat{\Theta}_{1i}$ and $\hat{\Theta}_{2i}$ be the estimates of Θ_{1i} and Θ_{2i} in (5.55). Then we can construct the following *estimation model*:

$$\dot{\hat{x}}(t) = \sum_{i=1}^{N} \mu_i [A_{mi}\hat{x}(t) + B_{mi}(\hat{\Theta}_{1i}x(t) + \hat{\Theta}_{2i}u(t))]. \tag{5.56}$$

Define the *estimation model state error*

$$e_x(t) = \hat{x}(t) - x(t). \tag{5.57}$$

With (5.56) and (5.55), we can derive the estimation model state error dynamics:

$$\begin{aligned}\dot{e}_x(t) &= \sum_{i=1}^{N} \mu_i A_{mi}e_x(t) + \sum_{i=1}^{N} \mu_i B_{mi}[(\hat{\Theta}_{1i} - \Theta_{1i})x(t) + (\hat{\Theta}_{2i} - \Theta_{2i})u(t)] \\ &= \sum_{i=1}^{N} \mu_i A_{mi}e_x(t) + \sum_{i=1}^{N} \mu_i B_{mi}[\tilde{\Theta}_{1i}x(t) + \tilde{\Theta}_{2i}u(t)],\end{aligned} \tag{5.58}$$

where $\tilde{\Theta}_{1i} = \hat{\Theta}_{1i} - \Theta_{1i}$ and $\tilde{\Theta}_{2i} = \hat{\Theta}_{2i} - \Theta_{2i}$.

Parameter adaptive laws. Based on the estimation model state error dynamics (5.58), we design the parameter adaptive laws for estimating $\hat{\Theta}_{1i}$ and $\hat{\Theta}_{2i}$ as

$$\dot{\hat{\Theta}}_{1i}(t) = -\mu_i \Gamma_{1i} B_{mi}^T P e_x(t) x(t)^T \tag{5.59}$$

$$\dot{\hat{\Theta}}_{2i}(t) = -\mu_i \Gamma_{2i} B_{mi}^T P e_x(t) u(t)^T, \quad i = 1, 2, \ldots, N \tag{5.60}$$

where $\Gamma_{1i} \in R^{m\times m}$ and $\Gamma_{2i} \in R^{m\times m}$ are two diagonal matrices with their diagonal elements being positive constants, $P = P^T > 0$ is chosen as the solution of linear matrix inequalities (5.44).

Adaptive controller. With the parameter estimates $\hat{\Theta}_{1i}$ and $\hat{\Theta}_{2i}$, we propose the following adaptive controller

$$u(t) = (\hat{\Omega}^T \hat{\Omega})^{-1} \left(-\hat{\Omega}^T \sum_{i=1}^{N} \mu_i B_{mi} \hat{\Theta}_{1i} x(t) + \hat{\Omega}^T \sum_{i=1}^{N} \mu_i B_{mi} r(t) \right), \tag{5.61}$$

where $\hat{\Omega} = \sum_{i=1}^{N} \mu_i B_{mi} \hat{\Theta}_{2i}$, which is is an $n \times m$ matrix.

Multiplying $\hat{\Omega}^T$ on both sides of (5.56) and applying (5.61) to it yield

$$\hat{\Omega}^T \dot{\hat{x}}(t) = \hat{\Omega}^T \sum_{i=1}^{N} \mu_i (A_{mi} \hat{x}(t) + B_{mi} r(t)). \tag{5.62}$$

Introducing $\hat{e}(t) = \hat{x}(t) - x_m(t)$, with (5.43) and (5.62), we have $\Omega^T \dot{\hat{e}}(t) = \Omega^T \sum_{i=1}^{N} \mu_i A_{mi} \hat{e}(t)$. Multiplying $\hat{\Omega}$ on both sides of (5.63), we have $\hat{\Omega}\hat{\Omega}^T \dot{\hat{e}}(t) = \hat{\Omega}\hat{\Omega}^T \sum_{i=1}^{N} \mu_i A_{mi} \hat{e}(t)$. If $\hat{\Omega}\hat{\Omega}^T$ is nonsingular which can be guaranteed by the parameter projection algorithm (see (5.64)–(5.65)), it can be guaranteed that $\hat{\Omega}\hat{\Omega}^T$ is nonsingular. Therefore, it can be obtained that

$$\dot{\hat{e}}(t) = \sum_{i=1}^{N} \mu_i A_{mi} \hat{e}(t). \tag{5.63}$$

With the condition (C.1), we obtain $\hat{e}(t) \in L_\infty$ and $\lim_{t\to\infty} \hat{e}(t) = 0$.

To implement the adaptive control law (5.61), it should be guaranteed that the $\hat{\Omega}\hat{\Omega}^T$ and $\hat{\Omega}^T\hat{\Omega}$ are nonsingular.

Define $\Omega = \sum_{i=1}^{N} \mu_i B_{mi} \Theta_{2i}$, which is an $n \times m$ matrix. Let $\Theta_{2i} = (\theta_{j,k}^{2i})_{m\times m}$. We have the following assumption on Θ_{2i}.

Assumption 5.3 $\Omega\Omega^T$ and $\Omega^T\Omega$ are nonsingular for $\theta_{j,k}^{2i} \in [\underline{\theta}_{j,k}^{2i}, \overline{\theta}_{j,k}^{2i}]$, where $\underline{\theta}_{j,k}^{2i}$ and $\overline{\theta}_{j,k}^{2i}$ are the known lower and upper bounds of $\theta_{j,k}^{2i}$.

Parameter projection. If the parameter estimate $\hat{\theta}_{j,k}^{2i}$ could be restricted within the interval $[\underline{\theta}_{j,k}^{2i}, \overline{\theta}_{j,k}^{2i}]$, it can be guaranteed that $\hat{\Omega}\hat{\Omega}^T$ and $\hat{\Omega}^T\hat{\Omega}$ are nonsingular. We apply parameter projection method to make $\hat{\theta}_{j,k}^{2i} \in [\underline{\theta}_{j,k}^{2i}, \overline{\theta}_{j,k}^{2i}]$. The projection algorithm for $\hat{\Theta}_{2i}$ is designed as

$$\dot{\hat{\Theta}}_{2i}(t) = -\mu_i \Gamma_{2i} B_{mi}^T P e_x(t) u(t)^T + F_{2i}(t), \tag{5.64}$$

where $F_{2i}(t) = (f_{j,k}^{2i}(t))_{m\times m}$, $j, k = 1, 2, \ldots, m$ is the projection operator with $f_{j,k}^{2i}$ defined as

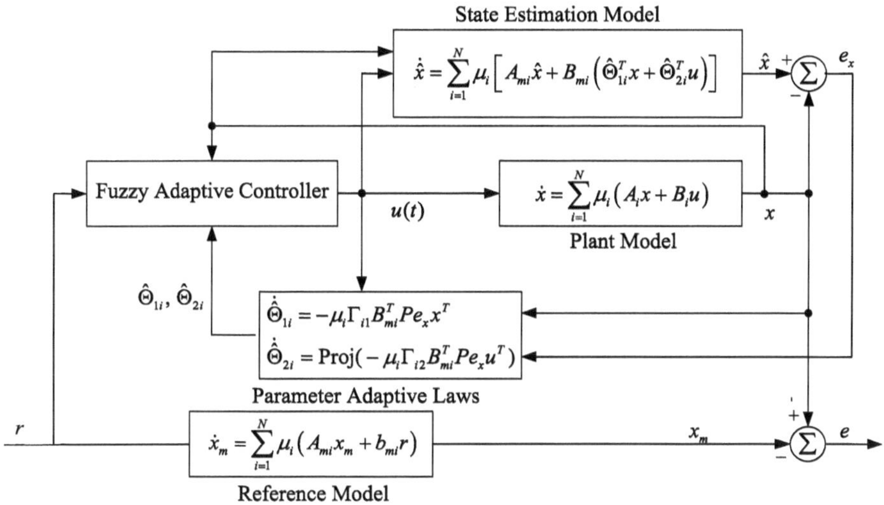

Fig. 5.1 Block diagram of overall control scheme

$$f_{j,k}^{2i}(t) = \begin{cases} 0 & \text{if } \hat{\theta}_{j,k}^{2i} \in [\underline{\theta}_{j,k}^{2i}, \overline{\theta}_{j,k}^{2i}] \\ & \text{or } \hat{\theta}_{j,k}^{2i} = \underline{\theta}_{j,k}^{2i} \text{ and } g_{j,k}^{2i}(t) \geq 0 \\ & \text{or } \hat{\theta}_{j,k}^{2i} = \overline{\theta}_{j,k}^{2i} \text{ and } g_{j,k}^{2i}(t) \leq 0 \\ -g_{j,k}^{2i}(t) & \text{otherwise} \end{cases} \tag{5.65}$$

with $g_{j,k}^{2i}(t)$ being an element in the following matrix:

$$G_{2i}(t) = -\mu_i \Gamma_{2i} B_{mi} P e_x(t) u(t)^T = (g_{j,k}^{2i}(t))_{m\times m}. \tag{5.66}$$

The paramter projection algorithm (5.64)–(5.65) ensures that for any $t \geq 0$, the parameter estimates $\hat{\theta}_{j,k}^{2i} \in [\underline{\theta}_{j,k}^{2i}, \overline{\theta}_{j,k}^{2i}]$ and with Assumption 5.3, $\hat{\Omega}\hat{\Omega}^T$ and $\hat{\Omega}^T\hat{\Omega}$ are nonsingular.

Now, we have finished the design of our fuzzy adaptive control scheme. The overall control scheme is shown in Fig. 5.1.

Define the *state tracking error*

$$e(t) = x(t) - x_m(t). \tag{5.67}$$

The properties of the proposed control scheme are given in the following theorem:

Theorem 5.2 *For the T–S fuzzy system (5.40), under Assumption 5.3, the adaptive controller (5.61) with the adaptive laws (5.59) and (5.64)–(5.65) guarantees:*

(i) all closed-loop system signals are bounded;

(ii) the state tracking error $e(t) \to 0$ when $t \to \infty$.

Proof Consider the following Lyapunov function candidate:

$$V(e_x, \tilde{\Theta}_{1i}, \tilde{\Theta}_{2i}^T) = e_x^T P e_x + \sum_{i=1}^{N} \mathrm{tr}(\tilde{\Theta}_{1i}^T \Gamma_{1i}^{-1} \tilde{\Theta}_{1i}) + \mathrm{tr}\left(\sum_{i=1}^{N} \tilde{\Theta}_{2i}^T \Gamma_{2i}^{-1} \tilde{\Theta}_{2i}\right), \quad (5.68)$$

where $\mathrm{tr}(A)$ denotes the trace of a matrix A. We obtain the time derivative of V along the error dynamics (5.58) under the condition (C.1): $A_{mi}^T P + P A_{mi} \leq -Q$ as follows:

$$\begin{aligned}
\dot{V} &= \dot{e}_x^T P e_x + e_x^T P \dot{e}_x + 2\sum_{i=1}^{N} \mathrm{tr}(\tilde{\Theta}_{1i}^T \Gamma_{1i}^{-1} \dot{\tilde{\Theta}}_{1i}) + 2\sum_{i=1}^{N} \mathrm{tr}(\tilde{\Theta}_{2i}^T \Gamma_{2i}^{-1} \dot{\tilde{\Theta}}_{2i}) \\
&= \sum_{i=1}^{N} \mu_i [e_x^T (A_{mi}^T P + P A_{mi}) e_x + 2 e_x^T P B_{mi} \tilde{\Theta}_{1i} x + 2 e_x^T P B_{mi} \tilde{\Theta}_{2i} u] \\
&\quad + 2\sum_{i=1}^{N} \mathrm{tr}(\tilde{\Theta}_{1i}^T \Gamma_{1i}^{-1} \dot{\tilde{\Theta}}_{1i}) + 2\sum_{i=1}^{N} \mathrm{tr}(\dot{\tilde{\Theta}}_{2i}^T \Gamma_{2i}^{-1} \tilde{\Theta}_{2i}) \\
&\leq -e_x^T Q e_x + 2\sum_{i=1}^{N} \mu_i e_x^T P B_{mi} \tilde{\Theta}_{1i} x + 2\sum_{i=1}^{N} \mu_i e_x^T P B_{mi} \tilde{\Theta}_{2i} u \\
&\quad + 2\sum_{i=1}^{N} \mathrm{tr}(\dot{\hat{\Theta}}_{1i}^T \Gamma_{1i}^{-1} \tilde{\Theta}_{1i}) + 2\sum_{i=1}^{N} \mathrm{tr}(\dot{\hat{\Theta}}_{2i}^T \Gamma_{2i}^{-1} \tilde{\Theta}_{2i}).
\end{aligned} \quad (5.69)$$

Using (5.59) and (5.64) with $f_{j,k}^{2i} = 0$ in (5.65), (5.69) becomes

$$\begin{aligned}
\dot{V} &\leq -e_x^T Q e_x + 2\sum_{i=1}^{N} \mu_i e_x^T P B_{mi} \tilde{\Theta}_{1i}^T x + 2\sum_{i=1}^{N} \mu_i e_x^T P B_{mi} \tilde{\Theta}_{2i} u \\
&\quad - 2\sum_{i=1}^{N} \mathrm{tr}(\mu_i x e_x^T P B_{mi} \tilde{\Theta}_{1i}) + 2\sum_{i=1}^{N} \mathrm{tr}(\mu_i u e_x^T P B_{mi} \tilde{\Theta}_{2i}) \\
&= -e_x^T Q e_x + 2\sum_{i=1}^{N} \mu_i e_x^T P B_{mi} \tilde{\Theta}_{1i}^T x + 2\sum_{i=1}^{N} \mu_i e_x^T P B_{mi} \tilde{\Theta}_{2i} u \\
&\quad - 2\sum_{i=1}^{N} \mathrm{tr}(\mu_i e_x^T P B_{mi} \tilde{\Theta}_{1i} x) + 2\sum_{i=1}^{N} \mathrm{tr}(\mu_i e_x^T P B_{mi} \tilde{\Theta}_{2i} u) \\
&= -e_x^T Q e_x.
\end{aligned} \quad (5.70)$$

Using (5.59) and (5.64) with $f_{j,k}^{2} = -g_{j,k}^{2i}$ in (5.65), (5.69) becomes

$$
\begin{aligned}
\dot{V} &\leq -e_x^T Q e_x + 2\sum_{i=1}^{N} \mu_i e_x^T P B_{mi} \tilde{\Theta}_{2i} u \\
&= -e_x^T Q e_x - 2\sum_{i=1}^{N}\sum_{j=1}^{m}\sum_{k=1}^{m} \Gamma_{2i}^{-1} \tilde{\theta}_{j,k}^{2i} g_{j,k}^{2i}.
\end{aligned} \tag{5.71}
$$

Since $\theta_{j,k}^{2i} \in [\underline{\theta}_{j,k}^{2i}, \overline{\theta}_{j,k}^{2i}]$, when $\hat{\theta}_{j,k}^{2i} = \underline{\theta}_{j,k}^{2i}$, $\tilde{\theta}_{j,k}^{2i} = \hat{\theta}_{j,k}^{2i} - \theta_{j,k}^{2i} \leq 0$, with $g_{j,k}^{2i} < 0$, we have $\dot{V} \leq -e_x^T Q e_x$ from (5.71); when $\hat{\theta}_{j,k}^{2i} = \overline{\theta}_{j,k}^{2i}$, $\tilde{\theta}_{j,k}^{2i} = \hat{\theta}_{j,k}^{2i} - \theta_{j,k}^{2i} \geq 0$, with $g_{j,k}^{2i} > 0$, we have $\dot{V} \leq -e_x^T Q e_x$ from (5.71) as well.

Hence, with (5.70) and (5.71), we have

$$
\dot{V} \leq -e_x^T Q e_x \leq 0. \tag{5.72}
$$

With (5.68) and (5.71), we can obtain that $e_x(t)$, $\tilde{\Theta}_{1i}$ and $\tilde{\Theta}_{2i}$ are all bounded, which indicates $\hat{\Theta}_{1i}$ and $\hat{\Theta}_{2i}$ are bounded as well. Since the state tracking error

$$
e(t) = x(t) - x_m(t) = x(t) - \hat{x}(t) + \hat{x}(t) - x_m(t) = -e_x(t) + \hat{e}(t), \tag{5.73}
$$

with $e_x(t) \in L_\infty$ and $\hat{e}(t) \in L_\infty$, we can have $e(t) \in L_\infty$. With the boundedness of $e(t)$ and $x_m(t)$, we can have $x(t) \in L_\infty$. From (5.61), we can obtain $u(t) \in L_\infty$. Hence, from (5.59) and (5.64), we obtain $\dot{\hat{\Theta}}_{1i}$, $\dot{\hat{\Theta}}_{1i} \in L_\infty$, and from (5.58), we obtain $\dot{e}_x(t) \in L_\infty$. It follows from (5.35) and the boundedness of V that

$$
\int_0^\infty e_x^T(\tau) Q e_x(\tau) d\tau \leq -\int_0^\infty \dot{V} d\tau = V(0) - V(\infty) < \infty, \tag{5.74}
$$

which means $e_x(t) \in L_2$. Then with Barbălat's lemma, we have the desired convergence property: $\lim_{t\to\infty} e_x(t) = 0$. Since it has been proven that $\lim_{t\to\infty} \hat{e}(t) = 0$ based on (5.63), from (5.73), we obtain $e(t) \to 0$ as $t \to \infty$. ∇

Remark 5.5 The closed-loop stability and asymptotic state tracking properties given in Theorem 5.2 are the basic adaptive system properties only for the case when there are only parameterized uncertainties. Those properties are important since the most important feature (advantage) of adaptive control is its ability to deal with parameterized uncertainties. In the presence of unparameterized errors (modeling errors), the adaptive laws (5.59) and (5.64) need to be modified with robustifying signals (Ioannou and Sun 1996; Tao 2003).

5.4 Design for T–S Fuzzy Systems in General Form ($m = n$)

In Sect. 5.3, we have designed an adaptive state tracking controller for T–S fuzzy systems in a general form. The controller is designed based on the PDC idea and a T–S fuzzy reference model is introduced to help relax the needed plant-model matching conditions for achieving state tracking. However, the matching conditions (5.50) are still quite strong, which requires the T–S fuzzy system and the fuzzy reference model have certain special structures and parameters.

In this section, we consider the case when the system has the same number of inputs as states, i.e., $m = n$ and present a controller structure which can further relax the matching conditions for achieving state tracking.

5.4.1 Nominal Controller and Matching Conditions

For the T–S fuzzy system (5.40) with $m = n$, the following assumption is made:

Assumption 5.4 The input gain matrix $\sum_{i=1}^{N} \mu_i(\xi) B_i$ is nonsingular for $\xi \in \Omega_\xi$, where Ω_ξ is a compact set representing the operating region of the fuzzy system (5.40).

Based on Assumption 5.4, we propose the following nominal controller structure for state tracking.

Nominal controller. The nominal controller is constructed by integrating all the local controllers (5.9) in the following way:

$$u(t) = \left(\sum_{i=1}^{N} \mu_i B_i \right)^{-1} \left(\sum_{i=1}^{N} \mu_i B_i K_{1i} x(t) + \sum_{i=1}^{N} \mu_i B_i K_{2i} r(t) \right), \tag{5.75}$$

where $K_{1i} \in R^{n \times n}$ and $K_{2i} \in R^{n \times n}$.

Remark 5.6 Comparing (5.75) with (5.46), we can notice that (5.75) uses $\mu_i B_i$ in the weighted-average integration while (5.46) uses μ_i. Such a modification will lead to much more relaxed matching conditions for model reference state tracking.

Matching conditions for linear reference model. For the linear reference model (5.41), the needed matching conditions for state tracking are

$$A_i + B_i K_{1i} = A_m, \quad B_i K_{2i} = B_m, \tag{5.76}$$

for $\forall i = 1, 2, \ldots, N$.

Matching conditions for T–S fuzzy reference model. For the T–S fuzzy reference model (5.43), the needed matching conditions for state tracking are

$$A_i + B_i K_{1i} = A_{mi}, \quad B_i K_{2i} = B_{mi}, \tag{5.77}$$

for $\forall i = 1, 2, \ldots, N$

Remark 5.7 Comparing (5.76) and (5.77) with (5.48) and (5.50), it can be observed that the matching conditions (5.76) and (5.77) are much easier to satisfy since there are no couplings between local models (represented by (A_i, B_i)) and local controllers (represented by (K_{1j}, K_{2j}), $j \neq i$), and one can compute K_{1i} and K_{2i} independently for each local model represented by A_i and B_i from (5.76) or (5.77).

One can easily verify that with the matching conditions (5.76) or (5.77), the T–S fuzzy system (5.40) under the controller (5.75) has the same dynamics as the linear reference model (5.41) or the T–S fuzzy reference model (5.43).

5.4.2 Adaptive Control Scheme

Now we develop an adaptive control scheme for the system (5.40) with unknown parameters A_i and B_i. Here we only consider adaptive state tracking using a T–S fuzzy reference model. The adaptive state tracking control scheme using a linear reference model can be similarly derived by following the same design idea.

Using the matching conditions (5.77), the nominal controller can be formulated as

$$u(t) = \left(\sum_{i=1}^{N} \mu_i B_i\right)^{-1} \left(-\sum_{i=1}^{N} \mu_i A_i x(t) + \sum_{i=1}^{N} \mu_i A_{mi} x(t) + \sum_{i=1}^{N} \mu_i B_{mi} r(t)\right). \tag{5.78}$$

Adaptive controller. When the system parameters A_i and B_i are uncertain, we cannot implement the nominal controller (5.78) directly. By replacing A_i and B_i by their estimates $\hat{A}_i$ and $\hat{B}_i$, we design the adaptive controller as

$$u(t) = \left(\sum_{i=1}^{N} \mu_i \hat{B}_i\right)^{-1} \left(-\sum_{i=1}^{N} \mu_i \hat{A}_i x(t) + \sum_{i=1}^{N} \mu_i A_{mi} x(t) + \sum_{i=1}^{N} \mu_i B_{mi} r(t)\right). \tag{5.79}$$

It can be seen from (5.79) that to make the adaptive controller nonsingular, the parameter adaptation process for $\hat{B}_i$ must ensure that $\sum_{i=1}^{N} \mu_i \hat{B}_i$ is nonsingular.

Applying the adaptive controller (5.79) to (5.40), we obtain the closed-loop system as

$$\dot{x}(t) = \sum_{i=1}^{N} \mu_i A_{mi} x(t) + \sum_{i=1}^{N} \mu_i B_{mi} r(t) - \sum_{i=1}^{N} \mu_i \tilde{A}_i x(t) - \sum_{i=1}^{N} \mu_i \tilde{B}_i u(t), \tag{5.80}$$

where $\tilde{A}_i = \hat{A}_i - A_i$ and $\tilde{B}_i = \hat{B}_i - B_i$.

Tracking error dynamics. Subtracting (5.43) from (5.80) yields the state tracking error dynamics:

$$\dot{e}(t) = \sum_{i=1}^{N} \mu_i A_{mi} e(t) - \sum_{i=1}^{N} \mu_i \tilde{A}_i x(t) - \sum_{i=1}^{N} \mu_i \tilde{B}_i u(t), \tag{5.81}$$

where $e(t) = x(t) - x_m(t)$.

Before we develop the parameter adaptive laws, the following assumption on $B_i = (b^i_{jk})_{m\times m}$, $i \in [1, N]$, $j, k \in [1, m]$, is made:

Assumption 5.5 There exist known upper and lower bounds for each element of B_i such that for $\underline{b}^i_{jk} \leq b^i_{jk} \leq \bar{b}^i_{jk}$, $\sum_{i=1}^{N} \mu_i(\xi) B_i$ is nonsingular for $\xi \in \Omega_\xi$, where Ω_ξ is a compact set representing the operating region of the fuzzy system (5.40).

Parameter adaptive laws. Based on the tracking error dynamics (5.81), we design the parameter adaptive laws:

$$\dot{\hat{A}}_i(t) = \gamma_{1i} \mu_i e(t) x(t)^T \tag{5.82}$$

$$\dot{\hat{B}}_i(t) = P_{\hat{B}_i}\left[\gamma_{2i} \mu_i e(t) u(t)^T\right], \quad i = 1, 2, \ldots, N \tag{5.83}$$

where γ_{1i}, $\gamma_{2i} > 0$, $P_{\hat{B}_i}$ is a projection operator to ensure that each element of $\hat{B}_i = (\hat{b}^i_{jk})_{m\times m}$ remains within $[\underline{b}^i_{jk}, \bar{b}^i_{jk}]$. The projection operator $P_{\hat{B}_i}(\cdot)$ can be designed by following the similar idea used in (5.64).

With the adaptive controller (5.79) and the parameter adaptive laws (5.82)–(5.83), we establish the following theorem, which shows the properties of the adaptive control scheme.

Theorem 5.3 *For the T–S fuzzy system (5.40) and the T-S fuzzy reference model (5.43), the adaptive controller (5.79) with the parameter adaptive laws (5.82)–(5.83) guarantees that all the closed-loop signals bounded and* $\lim_{t\to\infty} e(t) = 0$.

Proof Consider the following Lyapunov function candidate:

$$V = e^T P e + \sum_{i=1}^{N} \operatorname{tr}\left(\frac{\tilde{A}_i^T P \tilde{A}_i}{\gamma_{1i}}\right) + \sum_{i=1}^{N} \operatorname{tr}\left(\frac{\tilde{B}_i^T P \tilde{B}_i}{\gamma_{2i}}\right), \tag{5.84}$$

where $P = P^T > 0$ is the solution of (5.44)

We obtain the derivative of V along the trajectory of the tracking error dynamics (5.81) as

$$\begin{aligned}
\dot{V} &= \dot{e}^T P e + e^T P \dot{e} + \sum_{i=1}^{N} \operatorname{tr}\left(\frac{\dot{\tilde{A}}_i^T P \tilde{A}_i}{\gamma_{1i}} + \frac{\tilde{A}_i^T P \dot{\tilde{A}}_i}{\gamma_{1i}}\right) + \sum_{i=1}^{N} \operatorname{tr}\left(\frac{\dot{\tilde{B}}_i^T P \tilde{B}_i}{\gamma_{2i}} + \frac{\tilde{B}_i^T P \dot{\tilde{B}}_i}{\gamma_{2i}}\right) \\
&= \sum_{i=1}^{N} \mu_i e^T (A_{mi}^T P + P A_{mi}) e - 2\sum_{i=1}^{N} \mu_i e^T P \tilde{A}_i x - 2\sum_{i=1}^{N} \mu_i e^T P \tilde{B}_i u
\end{aligned}$$

$$
\begin{aligned}
&+2\sum_{i=1}^{N}\mathrm{tr}\left(\frac{\tilde{A}_i^T P \dot{\hat{A}}_i}{\gamma_{1i}}\right)+2\sum_{i=1}^{N}\mathrm{tr}\left(\frac{\tilde{B}_i^T P \dot{\hat{B}}_i}{\gamma_{2i}}\right)\\
&=\sum_{i=1}^{N}\mu_i e^T(A_{mi}^T P + P A_{mi})e + 2\sum_{i=1}^{N}\mathrm{tr}\left(\frac{\tilde{A}_i^T P \dot{\hat{A}}_i}{\gamma_{1i}} - \mu_i \tilde{A}_i^T P e x^T\right)\\
&+2\sum_{i=1}^{N}\mathrm{tr}\left(\frac{\tilde{B}_i^T P \dot{\hat{B}}_i}{\gamma_{2i}} - \mu_i \tilde{B}_i^T P e u^T\right).
\end{aligned}
\tag{5.85}
$$

Using (5.44) and (5.83), we obtain from (5.85) that

$$
\dot{V} \le -e^T Q e \le 0,
\tag{5.86}
$$

which means the tracking error $e(t)$ and the parameter estimates $\hat{A}_i$ and $\hat{B}_i$ are all bounded. Since $e(t) = x(t) - x_m(t)$ and $x_m(t) \in L_\infty$, we have $x(t) \in L_\infty$. Then with (5.79), we have $u \in L_\infty$. From (5.81), we have $\dot{e} \in L_\infty$. Hence, all the closed-loop signals are bounded. From (5.86), we also have

$$
\int_0^\infty e^T(\tau) Q e(\tau) d\tau \le -\int_0^\infty \dot{V} d\tau = V(0) - V(\infty) < \infty,
\tag{5.87}
$$

which means $e(t) \in L_2$. Then with Barbălat's lemma, we have the desired convergence property: $\lim_{t\to\infty} e(t) = 0$. ∇

Remark 5.8 Due to the existence of $\mu_i(\xi)$ in $\sum_{i=1}^{N}\mu_i(\xi)B_i$, it may be difficult to specify the upper and lower bounds $\underline{b}_{jk}^i$ and $\bar{b}_{jk}^i$ to satisfy Assumption 5.5 in practice. For B_i in a general form, how to guarantee the nonsingularity of $\sum_{i=1}^{N}\mu_i\hat{B}_i$ during the parameter adaptation process with more relaxed conditions is still an open problem. For B_i with some special forms, this problem can be handled effectively without requiring strong conditions.

5.4.3 Adaptive Control Design: Special Cases

In this subsection, we consider the adaptive control design for two special cases. Each of them can represent a class of nonlinear systems and special design procedures can be developed to guarantee the control gain matrix nonsingular during parameter adaptation.

Special Case 1: $B_i = B,\ i = 1, 2, \ldots, N$. In this case, all the local models have different system matrices A_i but the common input matrix B. The overall T–S fuzzy system (5.40) becomes

$$
\dot{x}(t) = \sum_{i=1}^{N}\mu_i A_i x(t) + Bu(t).,
\tag{5.88}
$$

where $B \in R^{m\times m}$ is nonsingular. This kind of T–S fuzzy system (5.88) can be used to approximate a class of nonlinear systems described by $\dot{x}(t) = f(x(t)) + Bu(t)$ where the input matrix is a constant matrix.

In the following design, it is assumed that B satisfies the following assumption:

Assumption 5.6 All the leading principle minors Δ_i, $i = 1, 2, \ldots, m$, of B are nonzero and their signs are known.

With Assumption 5.6, the matrix B has the following SDU decomposition.

SDU decomposition of B. The matrix $B \in R^{m\times m}$ with all its principal minors $\Delta_i \neq 0, i = 1, 2, \ldots, m$, has a nonunique decomposition

$$B = SDU, \tag{5.89}$$

where $S \in R^{m\times m}$ is a symmetric and positive definite matrix, $U \in R^{m\times m}$ is an upper triangular matrix, and

$$\begin{aligned} D &= \text{diag}\,\{d_1, d_2, \ldots, d_m\} \\ &= \text{diag}\left\{\text{sign}[\Delta_1]d_1^*, \text{sign}\left[\frac{\Delta_2}{\Delta_1}\right]d_2^*, \ldots, \text{sign}\left[\frac{\Delta_m}{\Delta_{m-1}}\right]d_m^*\right\} \end{aligned} \tag{5.90}$$

with $d_i^* > 0, i = 1, 2, \ldots, m$.

For D, we have the following assumption:

Assumption 5.7 Some lower bounds $\underline{d}_i > 0$ of d_i^*, that is, $d_i^* \geq \underline{d}_i, i = 1, 2, \ldots, m$, are known.

Remark 5.9 If the matrix B satisfies Assumption 5.6, B has a nonunique SDC decomposition. Based on the SDU decomposition of B, the design conditions for adaptive control can be relaxed. By following certain design procedure to be presented, the nonsingularity of the adaptive controller can be guaranteed in practice.

Design Based on SDU Decomposition of B. Using the SDU decomposition (5.90), the model (5.88) can be formulated as

$$\dot{x}(t) = \sum_{i=1}^{N} \mu_i A_i x(t) + SDUu(t). \tag{5.91}$$

Multiplying S^{-1} on both sides of (5.91) yields

$$\begin{aligned} S^{-1}\dot{x}(t) &= S^{-1}\sum_{i=1}^{N} \mu_i A_i x(t) + DUu(t) \\ &= S^{-1}\sum_{i=1}^{N} \mu_i A_i x(t) + Du - D(I - U)u. \end{aligned} \tag{5.92}$$

Multiplying S^{-1} on both sides of (5.43) yields

$$S^{-1}\dot{x}_m(t) = S^{-1}\sum_{i=1}^{N}\mu_i A_{mi}x_m(t) + S^{-1}\sum_{i=1}^{N}\mu_i B_{mi}r(t). \tag{5.93}$$

Based on (5.92) and (5.93), the *nominal controller* structure is designed as

$$u(t) = D^{-1}\left(-S^{-1}\sum_{i=1}^{N}\mu_i A_i x(t) + S^{-1}\sum_{i=1}^{N}\mu_i A_{mi}x(t) + S^{-1}\sum_{i=1}^{N}\mu_i B_{mi}r(t)\right) \\ +(I-U)u(t), \tag{5.94}$$

which can be parameterized as

$$u(t) = D^{-1}\left(-\sum_{i=1}^{N}\mu_i \Phi_{1i}x(t) + \sum_{i=1}^{N}\mu_i \Phi_{2i}r(t) + \Phi_0 u(t)\right), \tag{5.95}$$

where

$$\Phi_{1i} = S^{-1}(A_i - A_{mi}),\ \ \Phi_{2i} = S^{-1}B_{mi},\ i = 1, 2, \ldots, N \tag{5.96}$$

and Φ_0 is in a special form

$$\Phi_0 = D(I-U) = \begin{bmatrix} 0 & \phi^0_{12} & \phi^0_{13} & \cdots & \phi^0_{1m} \\ 0 & 0 & \phi^0_{23} & \cdots & \phi^0_{2m} \\ \vdots & \vdots & \vdots & \vdots & \vdots \\ 0 & 0 & \cdots & 0 & \phi^0_{m-1,m} \\ 0 & \cdots & \cdots & 0 & 0 \end{bmatrix}. \tag{5.97}$$

Let $\hat{D}$, $\hat{\Phi}_0$, $\hat{\Phi}_{1i}$ and $\hat{\Phi}_{2i}$ be the estimates of D, Φ_0, Φ_{1i} and Φ_{2i}, respectively. From the parameterized nominal controller (5.95), we propose the adaptive controller

$$u(t) = \hat{D}^{-1}\left(-\sum_{i=1}^{N}\mu_i \hat{\Phi}_{1i}x(t) + \sum_{i=1}^{N}\mu_i \hat{\Phi}_{2i}r(t) + \hat{\Phi}_0 u(t)\right), \tag{5.98}$$

where $\hat{D}$ has the same structure as D, that is, $\hat{D}$ is also a diagonal matrix; $\hat{\Phi}_0$ has the same structure as Φ_0, that is, $\hat{\Phi}_0$ is an upper triangular matrix with all the diagonal elements being 0.

With (5.96), the system model (5.92) and the reference model (5.93) can be formulated as

$$S^{-1}\dot{x}(t) = \sum_{i=1}^{N} \mu_i \Phi_{1i} x(t) + \sum_{i=1}^{N} \mu_i A_{mi} x(t) + Du(t) - \Phi_0 u(t) \tag{5.99}$$

and

$$S^{-1}\dot{x}_m(t) = S^{-1} \sum_{i=1}^{N} \mu_i A_{mi} x_m(t) + \sum_{i=1}^{N} \mu_i \Phi_{2i} r(t), \tag{5.100}$$

respectively.

Subtracting (5.100) from (5.99) results in

$$\begin{aligned} S^{-1}\dot{e}(t) = & \sum_{i=1}^{N} \mu_i \Phi_{1i} x(t) + S^{-1} \sum_{i=1}^{N} \mu_i A_{mi} e(t) - \sum_{i=1}^{N} \mu_i \Phi_{2i} r(t) \\ & + Du(t) - \Phi_0 u(t). \end{aligned} \tag{5.101}$$

Applying (5.98) into (5.101) yields

$$\begin{aligned} S^{-1}\dot{e}(t) = & S^{-1} \sum_{i=1}^{N} \mu_i A_{mi} e(t) - \sum_{i=1}^{N} \mu_i \tilde{\Phi}_{1i} x(t) + \sum_{i=1}^{N} \mu_i \tilde{\Phi}_{2i} r(t) \\ & - \tilde{D} u(t) + \tilde{\Phi}_0 u(t), \end{aligned} \tag{5.102}$$

which can be equivalently written into the following tracking error dynamics

$$\begin{aligned} \dot{e}(t) = & \sum_{i=1}^{N} \mu_i A_{mi} e(t) - \sum_{i=1}^{N} \mu_i S \tilde{\Phi}_{1i} x(t) + \sum_{i=1}^{N} \mu_i S \tilde{\Phi}_{2i} r(t) \\ & - S\tilde{D} u(t) + S\tilde{\Phi}_0 u(t), \end{aligned} \tag{5.103}$$

where $\tilde{\Phi}_0 = \hat{\Phi}_0 - \Phi_0$, $\tilde{\Phi}_{1i} = \hat{\Phi}_{1i} - \Phi_{1i}$, $\tilde{\Phi}_{2i} = \hat{\Phi}_{2i} - \Phi_{2i}$, $\tilde{D} = \hat{D} - D$.

With the error model (5.103), the parameter adaptive laws are designed as

$$\dot{\hat{\Phi}}_0(t) = P_{\hat{\Phi}_0}[-\gamma_0 P e(t) u(t)^T] \tag{5.104}$$

$$\dot{\hat{\Phi}}_{1i}(t) = \gamma_{3i} \mu_i P e(t) x(t)^T \tag{5.105}$$

$$\dot{\hat{\Phi}}_{2i}(t) = -\gamma_{4i} \mu_i P e(t) r(t)^T \tag{5.106}$$

$$\dot{\hat{D}}(t) = P_{\hat{D}}[\gamma_D P e(t) u(t)^T], \tag{5.107}$$

where $\gamma_0, \gamma_{3i}, \gamma_{4i}, \gamma_D > 0, i = 1, 2, \ldots, N$, $P_{\hat{\Phi}_0}[\cdot]$ and $P_{\hat{D}}[\cdot]$ are parameter projection operators. Since $\hat{\Phi}_0 = (\hat{\phi}^0_{jk})_{m\times m}$ and $\hat{D} = (\hat{d}_{jk})_{m\times m}$ should keep the same structures as Φ_0 and D, the following parameter projection operations for $P_{\hat{\Phi}_0}[\cdot]$ and $P_{\hat{D}}[\cdot]$ are applied:

$$\dot{\hat{\phi}}^0_{jk}(t) = \begin{cases} \varphi^0_{jk}(t) & \text{if } \ j < k \\ 0 & \text{otherwise} \end{cases} \tag{5.108}$$

$$\dot{\hat{d}}_{jj}(t) = \begin{cases} \upsilon_{jj}(t) & \text{if } \ |\hat{d}_{jj}(t)| > \underline{d}_j \\ & \text{or } \ |\hat{d}_{jj}(t)| = \underline{d}_j \ \text{ and } \ \text{sign}[d_j(t)]\upsilon_{jj}(t) \geq 0 \\ 0 & \text{otherwise} \end{cases} \tag{5.109}$$

$$\dot{\hat{d}}_{jk}(t) = 0 \quad \text{for} \quad j \neq k, \tag{5.110}$$

where $\varphi^0_{jk}(t)$ is the jkth element in the matrix $-\gamma_0 Pe(t)u(t)^T$, $\upsilon_{jj}(t)$ is the jth diagonal element in the matrix $\gamma_D Pe(t)u(t)^T$, $j, k = 1, 2, \ldots, m$. The initial parameters of $\hat{\Phi}_0$ and $\hat{D}$ are set as

$$\begin{aligned} \hat{\phi}^0_{jk}(0) &= 0 \quad \text{for} \quad j \geq k \\ \hat{d}_{jk}(0) &= 0 \quad \text{for} \quad j \neq k. \end{aligned} \tag{5.111}$$

With the parameter projection scheme (5.108)–(5.110), we can ensure $\hat{\Phi}_0$ and $\hat{D}$ keep the same structures as Φ_0 and D, and $\hat{D}$ is nonsingular.

Let

$$u(t) = \begin{bmatrix} u_1(t) \\ u_2(t) \\ \vdots \\ u_m(t) \end{bmatrix}, \quad \hat{\Phi}_0 = \begin{bmatrix} 0 & \hat{\phi}^0_{12} & \hat{\phi}^0_{13} & \cdots & \hat{\phi}^0_{1m} \\ 0 & 0 & \hat{\phi}^0_{23} & \cdots & \hat{\phi}^0_{2m} \\ \vdots & \vdots & \vdots & \vdots & \vdots \\ 0 & 0 & \cdots & 0 & \hat{\phi}^0_{m-1,m} \\ 0 & \cdots & \cdots & 0 & 0 \end{bmatrix}, \quad \hat{\Phi}_{1i} = \begin{bmatrix} \hat{\phi}^{1i}_1 \\ \hat{\phi}^{1i}_2 \\ \vdots \\ \hat{\phi}^{1i}_m \end{bmatrix} \quad \hat{\Phi}_{2i} = \begin{bmatrix} \hat{\phi}^{2i}_1 \\ \hat{\phi}^{2i}_2 \\ \vdots \\ \hat{\phi}^{2i}_m \end{bmatrix},$$

$$\hat{D} = \text{diag}\{\hat{d}_1, \hat{d}_2, \ldots, \hat{d}_m\}, \tag{5.112}$$

where $\hat{\phi}^{1i}_j, \hat{\phi}^{2i}_j \in R^{1\times m}, i = 1, 2, \ldots, N, j = 1, 2, \ldots, m$. Then the controller (5.98) can be implemented as

$$\begin{aligned} u_m(t) &= \hat{d}_m^{-1}\left(-\sum_{i=1}^{N}\mu_i\hat{\phi}^{1i}_m x(t) + \sum_{i=1}^{N}\mu_i\hat{\phi}^{2i}_m r(t)\right) \\ u_{m-1}(t) &= \hat{d}_{m-1}^{-1}\left(-\sum_{i=1}^{N}\mu_i\hat{\phi}^{1i}_{m-1}x(t) + \sum_{i=1}^{N}\mu_i\hat{\phi}^{2i}_{m-1}r(t) + \hat{\phi}^0_{m-1,m}u_m(t)\right) \\ &\vdots \\ u_j(t) &= \hat{d}_j^{-1}\left(-\sum_{i=1}^{N}\mu_i\hat{\phi}^{1i}_j x(t) + \sum_{i=1}^{N}\mu_i\hat{\phi}^{2i}_j r(t) + \sum_{k=j+1}^{m}\hat{\phi}^0_{jk}u_k(t)\right) \\ &\vdots \\ u_1(t) &= \hat{d}_1^{-1}\left(-\sum_{i=1}^{N}\mu_i\hat{\phi}^{1i}_1 x(t) + \sum_{i=1}^{N}\mu_i\hat{\phi}^{2i}_1 r(t) + \sum_{k=2}^{m}\hat{\phi}^0_{1k}u_k(t)\right). \end{aligned} \tag{5.113}$$

With the controller (5.98) (which is implemented by (5.113)) and the parameter adaptive laws (5.104)–(5.111), we have the following results.

Theorem 5.4 *Consider the T–S fuzzy system (5.88), the adaptive controller (5.98) and the parameter adaptive laws (5.104)–(5.111) guarantee that all the closed-loop signals bounded and* $\lim_{t\to\infty} e(t) = 0$.

Proof Consider the following Lyapunov function candidate

$$V = e^T Pe + \mathrm{tr}\left(\frac{\tilde{\Phi}_0^T S\tilde{\Phi}_0}{\gamma_0}\right) + \sum_{i=1}^N \mathrm{tr}\left(\frac{\tilde{\Phi}_{1i}^T S\tilde{\Phi}_{1i}}{\gamma_{3i}}\right) + \sum_{i=1}^N \mathrm{tr}\left(\frac{\tilde{\Phi}_{2i}^T S\tilde{\Phi}_{2i}}{\gamma_{4i}}\right) + \mathrm{tr}\left(\frac{\tilde{D}^T S\tilde{D}}{\gamma_D}\right). \tag{5.114}$$

The derivative of V along the error dynamics (5.103) is obtained as

$$\begin{aligned}
\dot{V} &= \dot{e}^T Pe + e^T P\dot{e} + 2\mathrm{tr}\left(\frac{\dot{\tilde{\Phi}}_0^T S\tilde{\Phi}_0}{\gamma_0}\right) + 2\sum_{i=1}^N \mathrm{tr}\left(\frac{\dot{\tilde{\Phi}}_{1i}^T S\tilde{\Phi}_{1i}}{\gamma_{3i}}\right) + 2\sum_{i=1}^N \mathrm{tr}\left(\frac{\dot{\tilde{\Phi}}_{2i}^T S\tilde{\Phi}_{2i}}{\gamma_{4i}}\right) \\
&\quad + 2\mathrm{tr}\left(\frac{\dot{\tilde{D}}^T S\tilde{D}}{\gamma_D}\right) \\
&= \sum_{i=1}^N \mu_i e^T (A_{mi}^T P + PA_{mi})e - 2\sum_{i=1}^N \mu_i e^T PS\tilde{\Phi}_{1i}x + 2\sum_{i=1}^N \mu_i e^T PS\tilde{\Phi}_{2i}r - 2e^T PS\tilde{D}u \\
&\quad + 2e^T PS\tilde{\Phi}_0 u + 2\mathrm{tr}\left(\frac{\dot{\tilde{\Phi}}_0^T S\tilde{\Phi}_0}{\gamma_0}\right) + 2\sum_{i=1}^N \mathrm{tr}\left(\frac{\dot{\tilde{\Phi}}_{1i}^T S\tilde{\Phi}_{1i}}{\gamma_{3i}}\right) + 2\sum_{i=1}^N \mathrm{tr}\left(\frac{\dot{\tilde{\Phi}}_{2i}^T S\tilde{\Phi}_{2i}}{\gamma_{4i}}\right) \\
&\quad + 2\mathrm{tr}\left(\frac{\dot{\tilde{D}}^T S\tilde{D}}{\gamma_D}\right) \\
&= \sum_{i=1}^N \mu_i e^T (A_{mi}^T P + PA_{mi})e - 2\sum_{i=1}^N \mathrm{tr}\left(\mu_i x e^T PS\tilde{\Phi}_{1i}\right) + 2\sum_{i=1}^N \mathrm{tr}\left(\mu_i r e^T PS\tilde{\Phi}_{2i}\right) \\
&\quad - 2\mathrm{tr}\left(ue^T PS\tilde{D}\right) + 2\mathrm{tr}\left(ue^T PS\tilde{\Phi}_0\right) + 2\mathrm{tr}\left(\frac{\dot{\tilde{\Phi}}_0^T S\tilde{\Phi}_0}{\gamma_0}\right) + 2\sum_{i=1}^N \mathrm{tr}\left(\frac{\dot{\tilde{\Phi}}_{1i}^T S\tilde{\Phi}_{1i}}{\gamma_{3i}}\right) \\
&\quad + 2\sum_{i=1}^N \mathrm{tr}\left(\frac{\dot{\tilde{\Phi}}_{2i}^T S\tilde{\Phi}_{2i}}{\gamma_{4i}}\right) + 2\mathrm{tr}\left(\frac{\dot{\tilde{D}}^T S\tilde{D}}{\gamma_D}\right).
\end{aligned} \tag{5.115}$$

Using (5.44) and (5.104)–(5.111) in (5.115), we have

$$\dot{V} \le -e^T(t)Qe(t) \le 0 \tag{5.116}$$

which means the tracking error $e(t)$ and the parameter estimates $\hat{\Phi}_0$, $\hat{\Phi}_{1i}$, $\hat{\Phi}_{2i}$ and $\hat{D}$ are all bounded. Since $e(t) = x(t) - x_m(t)$ and $x_m(t) \in L_\infty$, we have $x(t) \in L_\infty$.

Then with (5.113), we have $u \in L_\infty$. From (5.103), we have $\dot{e} \in L_\infty$. Hence, all the closed-loop signals are bounded. From (5.116), we also have

$$\int_0^\infty e^T(\tau)Qe(\tau)d\tau \le -\int_0^\infty \dot{V}d\tau = V(0) - V(\infty) < \infty, \qquad (5.117)$$

which means $e(t) \in L_2$. Then with Barbălat's lemma, we have the desired convergence property: $\lim_{t\to\infty} e(t) = 0$. ∇

Special Case 2: $B_i, i = 1, 2, \ldots, N$, are upper or lower triangular matrices. In this case, Assumption 5.5 can be replaced by the following assumption which requires less knowledge on B_i.

Assumption 5.8 There exist known upper and lower bounds on the diagonal elements of B_i such that for $\underline{b}^i_{jj} \le b^i_{jj} \le \bar{b}^i_{jj}$, $j = 1, 2, \ldots, m$, $\sum_{i=1}^N \mu_i(\xi)b^i_{jj} \ne 0$ for $\xi \in \Omega_\xi$ where Ω_ξ is a compact set representing the operating region of the fuzzy system (5.40).

For the special case 2, the adaptive controller is designed the same as (5.79), that is,

$$u(t) = \left(\sum_{i=1}^N \mu_i \hat{B}_i\right)^{-1} \left(-\sum_{i=1}^N \mu_i \hat{A}_i x(t) + \sum_{i=1}^N \mu_i A_{mi} x(t) + \sum_{i=1}^N \mu_i B_{mi} r(t)\right),$$

and the parameter adaptive laws for updating $\hat{A}_i$ and $\hat{B}_i$ also have the same forms as (5.82)–(5.83):

$$\dot{\hat{A}}_i(t) = \gamma_{1i}\mu_i e(t)x(t)^T \qquad (5.118)$$

$$\dot{\hat{B}}_i(t) = P_{\hat{B}_i}\left[\gamma_{2i}\mu_i e(t)u(t)^T\right], \qquad (5.119)$$

where the projection operator $P_{\hat{B}_i}[\cdot]$ is to make $\hat{B}_i = (\hat{b}^i_{jk})_{m\times m}$ keep the same structure as $B_i = (b^i_{jk})_{m\times m}$, i.e., lower triangular or upper triangular.

To facilitate the derivation of the parameter projection algorithm, define

$$G_i(t) = \gamma_{2i}\mu_i e(t)u(t)^T = (g^i_{jk}(t))_{m\times m}, \qquad (5.120)$$

where $g^i_{jk}(t)$ represents the jkth element in $G_i(t)$.

If B_i is a lower triangular matrix, i.e., $b^i_{jk} = 0$ for $j < k$, the projection operator $P_{\hat{B}_i}[\cdot]$ works in the following way:

$$
\begin{aligned}
&\hat{b}^i_{jk} = 0, \quad \forall j < k \\
&\dot{\hat{b}}^i_{jk} = g^i_{jk}(t), \quad \forall j > k \\
&\dot{\hat{b}}^i_{jj} = \begin{cases} g^i_{jj}(t) & \text{if } \hat{b}^i_{jj} \in [\underline{b}^i_{jj}, \overline{b}^i_{jj}] \\ & \text{or } \hat{b}^i_{jj} = \underline{b}^i_{jj} \text{ and } g^i_{jj}(t) \geq 0 \\ & \text{or } \hat{b}^i_{jj} = \overline{b}^i_{jj} \text{ and } g^i_{jj}(t) \leq 0 \\ 0 & \text{otherwise.} \end{cases}
\end{aligned}
\tag{5.121}
$$

If B_i is an upper triangular matrix, i.e., $b^i_{jk} = 0$ for $j > k$, the projection operator $P_{\hat{B}_i}[\cdot]$ works in the following way:

$$
\begin{aligned}
&\hat{b}^i_{jk} = 0, \quad \forall j > k \\
&\dot{\hat{b}}^i_{jk} = g^i_{jk}(t), \quad \forall j < k \\
&\dot{\hat{b}}^i_{jj} = \begin{cases} g^i_{jj}(t) & \text{if } \hat{b}^i_{jj} \in [\underline{b}^i_{jj}, \overline{b}^i_{jj}] \\ & \text{or } \hat{b}^i_{jj} = \underline{b}^i_{jj} \text{ and } g^i_{jj}(t) \geq 0 \\ & \text{or } \hat{b}^i_{jj} = \overline{b}^i_{jj} \text{ and } g^i_{jj}(t) \leq 0 \\ 0 & \text{otherwise.} \end{cases}
\end{aligned}
\tag{5.122}
$$

Based on Assumption 5.8 and using the parameter projection algorithm (5.121) or (5.122), it can be guaranteed that $\sum_{i=1}^{N} \mu_i \hat{B}_i$ is nonsingular. The closed-loop stability and tracking performance analysis can be carried out by following the way similar to that in proving Theorem 5.3 and is omitted here.

5.5 Simulation Study

In this section, we present simulation study on the state tracking control of a magnetic suspension system.

5.5.1 Simulation System

First, we construct a T–S fuzzy model to approximate the simplified magnetic suspension system. Then a fuzzy adaptive controller is designed based on the T–S fuzzy model. Finally, the controller is applied to the T–S fuzzy model and the original plant to illustrate the desired system performance under the proposed fuzzy adaptive scheme. Consider the following magnetic suspension system (Tanaka and Abuelenin 2009):

$$\begin{aligned}\dot{x}_1 &= x_2\\ \dot{x}_2 &= g - \frac{C}{4M}\left(\frac{x_3}{x_1}\right)^2\\ \dot{x}_3 &= \frac{x_3x_2}{x_2} - \frac{2Rx_1x_3}{C} + \frac{2x_1}{C}u,\end{aligned} \tag{5.123}$$

where x_1 and x_2 represent the position and velocity of the floater of the magnetic suspension system with respect to the stator, x_3 represents the current, u is the input voltage, M is the mass of the floater, C and R are the parameters of the magnetic suspension system, and g is the acceleration due to gravity. In the simulation, $R = 5$, $C = 7.2\pi \times 10^{-6}$, $M = 3\,\text{kg}$, $g = 9.8\,\text{m/s}^2$

The dynamic equations are approximated by the following two-rules T–S fuzzy model:

$$\begin{aligned}&\text{IF } x_1(t) \text{ is } F_1^1 \text{ THEN } \dot{x}(t) = A_1x(t) + B_1u(t),\\ &\text{IF } x_1(t) \text{ is } F_1^2 \text{ THEN } \dot{x}(t) = A_2x(t) + B_2u(t),\end{aligned} \tag{5.124}$$

where $x = [x_1, x_2, x_3]^T$, and F_1^1 and F_1^2 are two fuzzy sets representing "about 0.004m" and "about 0.008m". Membership functions for fuzzy sets F_1^1 and F_1^2 are shown in Fig. 5.2. The system matrices A_1, A_2, B_1 and B_2 are as follows:

$$\begin{aligned}A_1 &= \begin{pmatrix} 0 & 1 & 0\\ 4900.1 & 0 & -2.1\\ 0 & 2280.1 & -1768.4\end{pmatrix}, \quad B_1 = \begin{pmatrix} 0\\ 0\\ 353.6777\end{pmatrix}\\ A_2 &= \begin{pmatrix} 0 & 1 & 0\\ 2450.0 & 0 & -1.1\\ 0 & 2280.1 & -3536.8\end{pmatrix}, \quad B_2 = \begin{pmatrix} 0\\ 0\\ 707.3553\end{pmatrix}.\end{aligned} \tag{5.125}$$

The T–S reference model is chosen as

$$\begin{aligned}&\text{IF } x_1(t) \text{ is } F_1^1 \text{ THEN } \dot{x}_m(t) = A_{m1}x(t) + B_{m1}r(t),\\ &\text{IF } x_1(t) \text{ is } F_1^2 \text{ THEN } \dot{x}_m(t) = A_{m2}x(t) + B_{m2}r(t),\end{aligned} \tag{5.126}$$

$$\begin{aligned}A_{m1} &= \begin{pmatrix} 0 & 1 & 0\\ 4900.1 & 0 & -2.1\\ 1.64\times 10^5 & 2838.5 & -70\end{pmatrix}, \quad B_{m1} = \begin{pmatrix} 0\\ 0\\ 35.37\end{pmatrix}\\ A_{m2} &= \begin{pmatrix} 0 & 1 & 0\\ 2450.0 & 0 & -1.1\\ 1.69\times 10^5 & 3396.9 & -70\end{pmatrix}, \quad B_{m2} = \begin{pmatrix} 0\\ 0\\ 70.736\end{pmatrix}.\end{aligned} \tag{5.127}$$

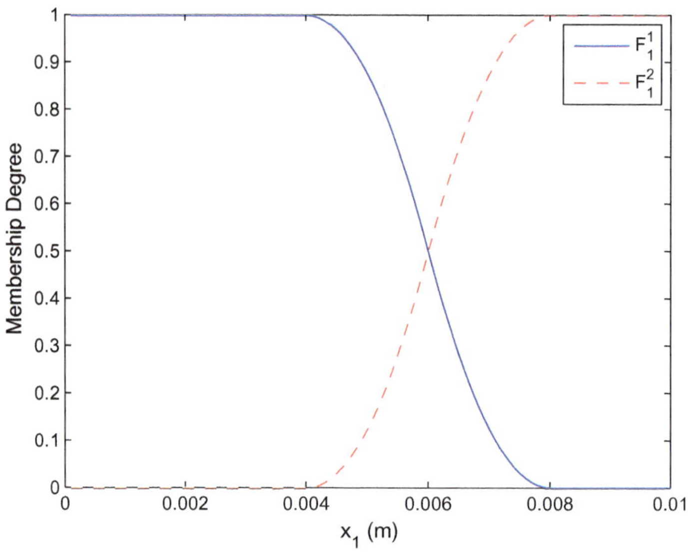

Fig. 5.2 Membership functions for x_1

Using LMI toolbox in MATLAB, the following common P can be found to make the condition (C.1) fulfilled:

$$P = \begin{pmatrix} 0.0001 & 0 & 0.1035 \\ 0 & 0.0046 & -0.1113 \\ 0.1035 & -0.1113 & 244.1140 \end{pmatrix} \tag{5.128}$$

which ensures the T–S reference model is globally asymptotically stable.

5.5.2 Simulation Results

The reference model has the input $r(t) = -0.4$. The initial states of the reference model and T–S fuzzy model are $x_m(0) = [0.002, 0, 1]^T$, $x(0) = [0.004, 0, 1]^T$. The initial values of parameters are $\hat{\Theta}_{11}(0) = [-464, -1.57, -4.8]$, $\hat{\Theta}_{21}(0) = 0.1$, $\hat{\Theta}_{12}(0) = [-235, -3, -4.9]$, $\hat{\Theta}_{22}(0) = 0.08$. The parameter learning rates $\Gamma_{11} = 2 \times 10^{-3}$, $\Gamma_{12} = 5 \times 10^{-7}$, $\Gamma_{21} = 10^{-8}$, $\Gamma_{22} = 10^{-6}$.

The control law is designed based on the T–S fuzzy system and applied to both the T–S fuzzy system and the original plant. The state tracking responses are shown in Figs. 5.3, 5.4 and 5.5. It can be observed that the T–S fuzzy system states track the reference model states asymptotically under the proposed adaptive fuzzy controller. The states of the original plant also can track the reference model states with desirable performance. A small position tracking error can be seen in Fig. 5.3 which is caused by

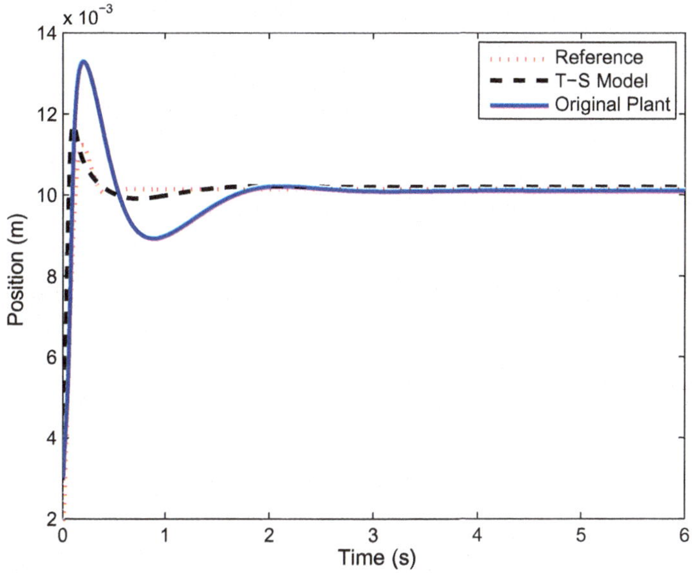

Fig. 5.3 State tracking response of x_1

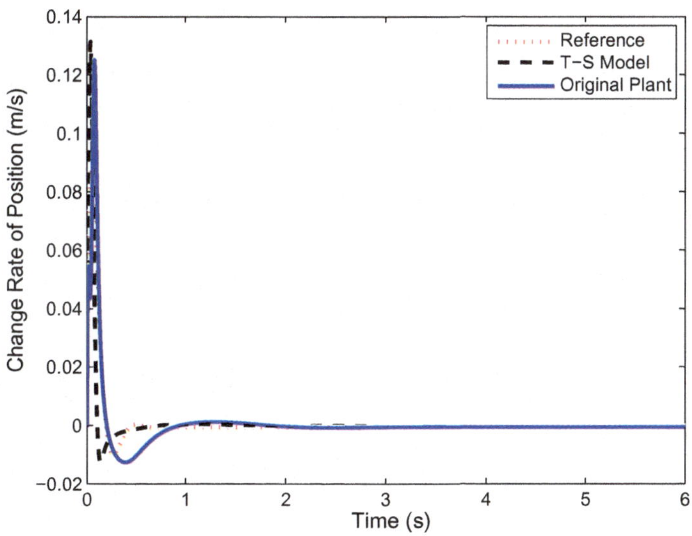

Fig. 5.4 State tracking response of x_2

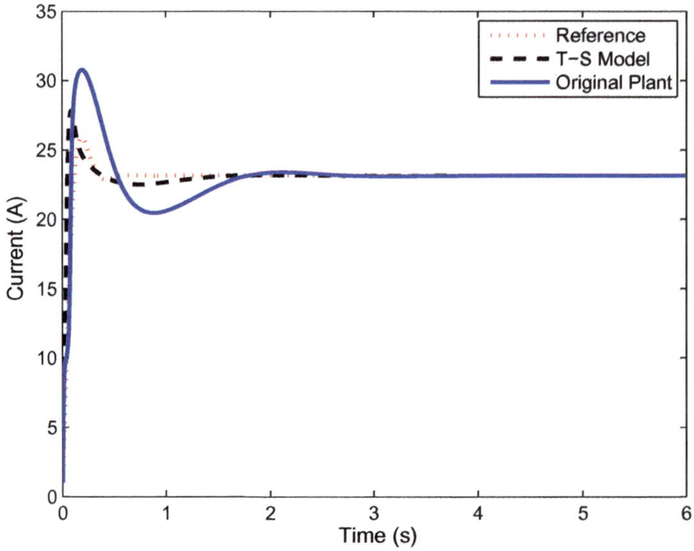

Fig. 5.5 State tracking response of x_3

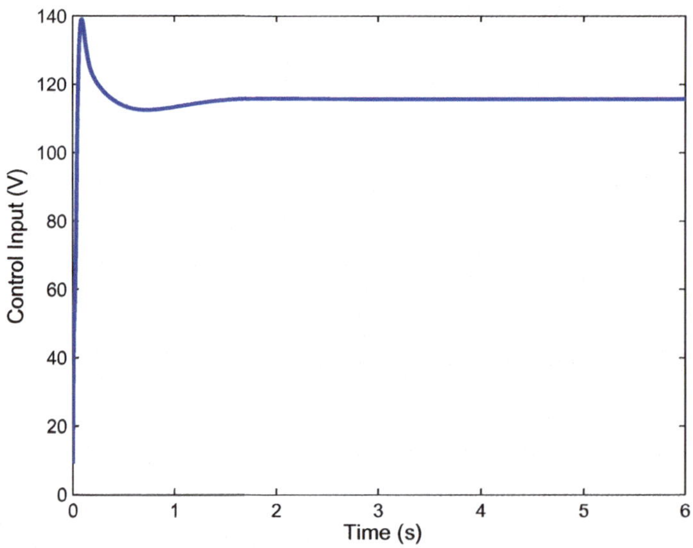

Fig. 5.6 Control input

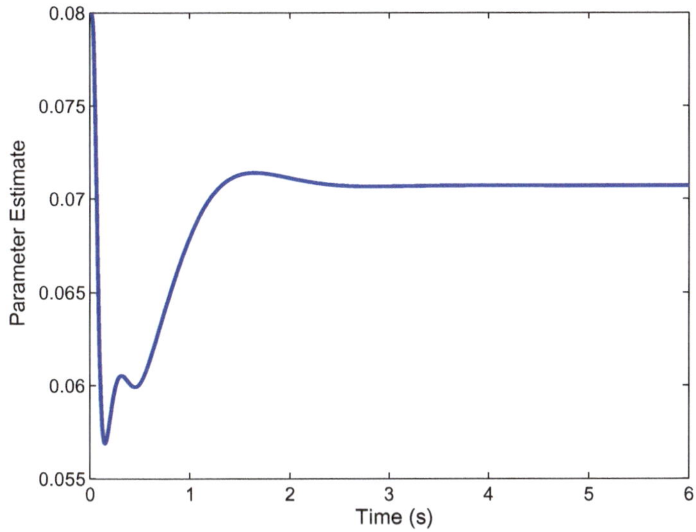

Fig. 5.7 Parameter adaptation of $\hat{\Theta}_{22}$

the modeling error which exists when a T–S fuzzy system is employed to approximate a nonlinear plant. The modeling error can be reduced through introducing more fuzzy rules and choosing more suitable membership functions. The adaptive control input is shown in Fig. 5.6 and the parameter adaptation of $\hat{\Theta}_{22}$ is shown in Fig. 5.7. Both the control input and the parameter estimates are bounded.

5.6 Summary

In this chapter, we have conducted detailed studies on adaptive state tracking control of T–S fuzzy systems and found there are several key factors that affect the adaptive state tracking control designs: the forms of system matrices A_i and B_i, the types of reference models and the way to construct the overall controller from local controllers.

For T–S fuzzy systems in canonical form, we proposed a state tracking controller with which feasible plant-model matching conditions can be derived for tracking the states of a stable linear reference model. The overall controller is constructed using a new weighted-average method that is different from conventional fuzzy weighted-average method, which can decouple the relationship between local controller parameters and local system parameters for different rules, leading to easy-to-fulfill plant-model matching conditions for state tracking control. For multi-input T–S fuzzy systems in general form, we have shown that by choosing a stable fuzzy system as the reference model, the restrict plant-model matching conditions for state tracking can be relaxed. As a key role for parameter estimation, an estimation model is designed based on which the estimation model state error dynamics is derived. Then an appropriate adaptive law for updating the parameters is designed and analyzed using Lyapunov theory.

For multi-input T–S fuzzy systems which have the same number of inputs as states, a state feedback controller is designed with feasible plant-model matching conditions for state tracking. The control singularity problem which may be caused due to the inversion of $\sum_{i=1}^{N} \mu_i \hat{B}_i$ in adaptive state tracking control is discussed. Although this problem is generally difficult to solve, for some special forms of B_i, solutions have been provided by using matrix decomposition and parameter projection.

The results in this chapter can provide a technical foundation of stability and tracking properties for adaptive state tracking control of T–S fuzzy systems. Further studies can be carried out regarding to the robustness issue of adaptive state tracking of T–S fuzzy systems and how to solve the control singularity problem for B_i in a general form.

References

Ioannou PA, Sun J (1996) Robust adaptive control. Prentice-Hall, Englewood Cliffs

Tanaka K, Abuelenin SM (2009) Design and simulation of a hybrid controller for a multi-input multi-output magnetic suspension system. In Proceedings of IEEE international conference on fuzzy system (FUZZ-IEEE'09), Jeju Island, Korea

Tanaka K, Sugeno M (1992) Stability analysis and design of fuzzy control systems. Fuzzy Sets Syst 45:135–156

Tao G (2003) Adaptive control design and analysis. Wiley, New York

Chapter 6
Adaptive T–S Fuzzy Output Tracking Control Using State Feedback

In Chap. 5, we have addressed the adaptive state tracking control problem for T–S fuzzy systems, which shows that under some restrictive structural conditions on the system matrices (A_i, B_i), the state tracking control objective can be achieved. In this chapter, we consider the output tracking control problem for discrete-time T–S fuzzy systems, which does not need any restrictive structural conditions on the system matrices (A_i, B_i, C_i), except for the knowledge of the T–S fuzzy system relative degree and a causality condition which will be introduced in this chapter.

In this chapter, we will develop a solution framework for adaptive state-feedback output tracking (SFOT) control of general discrete-time state-space T–S fuzzy systems with a relative degree ρ $(1 \leq \rho \leq n)$, including (i) the derivation of a normal form of a global noncanonical form state-space T–S fuzzy system model, which has an explicit relative degree structure and a specific input–output signal causality relationship desired for a feedback control design, (ii) the development of an adaptive control scheme for such a general T–S fuzzy system, and (iii) the simulation evaluation of the adaptive control system performance on an aircraft flight control system model.

The chapter is organized as follows. Section 6.1 formulates the problem addressed in this paper. In Sect. 6.2, a normal form of a global T–S fuzzy system is constructed with an explicit relative degree structure and causality. Section 6.3 develops an adaptive state-feedback output tracking scheme for the relative degree $\rho = 1$ and proves it has desired stability and tracking performance properties. Section 6.4 further considers the general case with $\rho > 1$ and solves the adaptive SFOT problem. In Sect. 6.5, a T–S fuzzy system model is constructed based on the linearized local models of a transport airplane and simulation results demonstrate the developed concepts and verify the desired performance of the new type of adaptive fuzzy control systems.

R. Qi et al., *Fuzzy System Identification and Adaptive Control*, Communications and Control Engineering,
https://doi.org/10.1007/978-3-030-19882-4_6

6.1 Problem Statement

In this section, we describe the adaptive control problem of noncanonical form state-space T–S fuzzy systems with general relative degrees.

In Chap. 2, we have shown that a discrete-time nonlinear

$$\begin{aligned} x(t+1) &= f(x(t), u(t)), \\ y(t) &= h(x(t)) \end{aligned} \tag{6.1}$$

can be approximated by a discrete-time T–S fuzzy system described by the following rule:

$$\begin{aligned} &\text{IF } \ \xi_1(t) \ \text{ is } \ F_1^i \ \text{ and } \ \ldots \ \text{ and } \ \xi_L(t) \ \text{ is } \ F_L^i \\ &\text{THEN } \ x(t+1) = A_i x(t) + B_i u(t), \\ &\qquad y(t) = C_i x(t) \end{aligned} \tag{6.2}$$

with t being the discrete-time sequence, where $x(t) \in R^n$ is the state vector, $u(t) \in R$ is the input, $y(t) \in R$ is the output, $\xi_1(t), \xi_2(t), \ldots, \xi_L(t)$ are some measurable system signals which are used to determine the operating area of the $i$the subsystem, F_j^i is a fuzzy set associated with which there is a membership function $F_j^i(\xi_j(t))$ to indicate the degree of membership of $\xi_j(t)$ in F_j^i, $A_i \in R^{n\times n}$, $B_i \in R^n$, and $C_i \in R^{1\times n}$ are system matrices, $i = 1, 2, \ldots, N$, $j = 1, 2, \ldots, L$.

The system matrices (A_i, B_i, C_i) in the local system (6.2) are in a noncanonical form, unlike most adaptive T–S fuzzy control scheme in literature which deal with special classes of systems whose matrices are in a canonical form. Furthermore, while from the linear local system models, a global T–S fuzzy system model can be constructed, the *relative degree* of a global T–S fuzzy system with general parameter matrices, which is crucial for adaptive control of such systems, was seldom addressed in literature.

Using the standard technique of *singleton fuzzification, product inference* and *weighted average*, we can obtain the following global T–S fuzzy model from (6.2):

$$\begin{aligned} x(t+1) &= \sum_{i=1}^{N} \mu_i(\xi(t))(A_i x(t) + B_i u(t)), \\ y(t) &= \sum_{i=1}^{N} \mu_i(\xi(t)) C_i x(t), \end{aligned} \tag{6.3}$$

where $\mu_i(\xi(t))$ is the normalized firing strength:

$$\mu_i(\xi(t)) = \frac{\lambda_i(\xi(t))}{\sum_{i=1}^{N} \lambda_i(\xi(t))}, \ \lambda_i(\xi(t)) = \prod_{j=1}^{L} F_j^i(\xi_j(t)),$$

$$\mu_i(\xi(t)) \geq 0, \ \sum_{i=1}^{N} \mu_i(\xi(t)) = 1. \tag{6.4}$$

Control objective. The control objective is to design an adaptive state-feedback controller to generate the control input $u(t)$ for the fuzzy system (6) with noncanonical (A_i, B_i, C_i) and parameter uncertainties to ensure all the signals in the closed-loop system are bounded and the output $y(t)$ asymptotically tracks a given bounded reference $y_m(t)$.

In Chap. 2, it has been shown that state tracking control design requires certain restrictive structural conditions on the system matrices $(A_i,\ B_i,\ C_i)$. In reality, it is usually unnecessary to make all the states track specific trajectories. For most practical problems, an output tracking control law can be designed with the system relative degree information, without the need of any restrictive structural conditions on the system matrices $(A_i,\ B_i,\ C_i)$. However, the problem on the relative degree of a T–S fuzzy system was seldom discussed in literature, especially for T–S fuzzy systems in a noncanonical form. Moreover, the causality of the global T–S fuzzy model needs to be addressed for state feedback output tracking control design. In the following section, we will show how to derive a global T–S fuzzy system with an explicit relative degree structure and causality, which is critical for state-feedback output tracking control design. We will also develop a solution to the adaptive state-feedback output tracking problem.

6.2 Modeling of T–S Fuzzy Systems with Relative Degree and Causality

For nonlinear systems, feedback linearization is a popular control design methodology. To employ feedback linearization techniques for T–S fuzzy system control design, it is important to derive the relative degree of the T–S fuzzy system and clarify its minimum-phase property. In this section,[1] we shall study the relative degree and minimum-phase property of noncanonical form state-space T–S fuzzy systems.

6.2.1 Relative Degree of a Dynamic System

In this subsection, we introduce the relative degree concept for a dynamic system. For a SISO continuous-time system with input $u \in R$, state $x \in R^n$ and output $y \in R$:

---

[1] Parts of Sect. 6.2 are reprinted from Qi et al. (2013), Copyright 2012, with permission from Elsevier.

$$\dot{x} = f(x, u), \quad y = h(x), \tag{6.5}$$

its relative degree is an integer ρ such that the ρth time derivative $y^{(\rho)}(t)$ of $y(t)$ depends on the input u explicitly, while all lower order time derivatives $y^{(i)}(t)$ of $y(t), i = 1, 2, \ldots, \rho - 1$, do not depend on the input u explicitly. The system relative degree thus depends on the functions f and h and its detailed theory has been developed (Isidori 1995). The specification of the relative degree of a nonlinear system is based on a feedback linearization procedure, and it is also crucial for a feedback linearization control design, which is popular and powerful for control of nonlinear systems.

As an example, consider a linear time-invariant (LTI) system: $\dot{x} = Ax + Bu, \ y = Cx$ with a transfer function $G(s) = C(sI - A)^{-1}B = \frac{N(s)}{D(s)}$. Its relative degree is the difference between the degree of the denominator polynomial $D(s)$ and the degree of the numerator polynomial $N(s)$, that is, it is the difference between the number of system poles and the number of system zeros. This definition can be expressed in a different form: the system has a relative degree ρ if $CA^iB = 0$ for $i = 0, 1, \ldots, \rho - 2$ and $CA^{\rho-1}B \neq 0$. For example, for a relative degree two system, we have $CB = 0$ and $CAB \neq 0$, and thus $\dot{y} = C\dot{x} = CAx + CBu = CAx$ which does not contain u explicitly, and $\ddot{y} = CA\dot{x} = CA^2x + CABu$ which contains u explicitly.

Such a relative degree concept can also be developed for a discrete-time system:

$$x(t+1) = f(x(t), u(t)), \quad y(t) = h(x(t)), \tag{6.6}$$

whose relative degree is the integer ρ such that $y(t+\rho)$ contains $u(t)$ explicitly, while $y(t+i)$, $i = 1, 2, \ldots, \rho - 1$, do not contain $u(t)$ explicitly. For a LTI system: $x(t+1) = Ax(t) + Bu(t), \ y(t) = Cx(t)$, its relative degree is the relative order of the system transfer function $G(z) = C(zI - A)^{-1}B$, and it can be defined as the integer ρ such that $CA^iB = 0$ for $i = 0, 1, \ldots, \rho - 2$ and $CA^{\rho-1}B \neq 0$. For example, for a relative degree $\rho = 3$ system, we have $CB = CAB = 0$ and $CA^2B \neq 0$, and hence $y(t+1) = Cx(t+1) = CAx(t) + CBu(t) = CAx(t)$ and $y(t+2) = CAx(t+1) = CA^2x(t) + CABu(t) = CA^2x(t)$ (both do not depend on $u(t)$ explicitly), and $y(t+3) = CA^2x(t+1) = CA^3x(t) + CA^2Bu(t)$ which depends on $u(t)$ explicitly.

For the global fuzzy system (6.3), its relative degree can be defined in a similar way: the system has relative degree ρ if $y(t+\rho)$ contains $u(t)$ explicitly, while $y(t+i)$, $i = 1, 2, \ldots, \rho - 1$, do not contain $u(t)$ explicitly. The detailed relative degree conditions and feedback linearization procedure will be given in the next subsection.

6.2.2 Relative Degree of T–S Fuzzy System

Considering the global fuzzy system (6.3), if its relative degree is one ($\rho = 1$), we have

$$\begin{aligned} y(t+1) &= \sum_{i=1}^{N} \mu_i(\xi(t+1))C_i x(t+1) \\ &= \sum_{i=1}^{N} \mu_i(\xi(t+1))C_i \sum_{i=1}^{N} \mu_i(\xi(t))(A_i x(t) + B_i u(t)) \\ &= \sum_{i=1}^{N} \mu_i(\xi(t+1))C_i \sum_{i=1}^{N} \mu_i(\xi(t))A_i x(t) \\ &\quad + \sum_{i=1}^{N} \mu_i(\xi(t+1))C_i \sum_{i=1}^{N} \mu_i(\xi(t))B_i u(t), \end{aligned} \tag{6.7}$$

which requires $\mu_i(\xi(t+1))$ available at time t. If $\rho = 2$, which means

$$\sum_{i=1}^{N} \mu_i(\xi(t+1))C_i \sum_{i=1}^{N} \mu_i(\xi(t))B_i = 0 \tag{6.8}$$

in (6.7), we have

$$\begin{aligned} y(t+2) &= \sum_{i=1}^{N} \mu_i(\xi(t+2))C_i \sum_{i=1}^{N} \mu_i(\xi(t+1))A_i x(t+1) \qquad (6.9) \\ &= \sum_{i=1}^{N} \mu_i(\xi(t+2))C_i \sum_{i=1}^{N} \mu_i(\xi(t+1))A_i \sum_{i=1}^{N} \mu_i(\xi(t))(A_i x(t) + B_i u(t)), \end{aligned}$$

which requires $\mu_i(\xi(t+2))$ and $\mu_i(\xi(t+1))$ are available at time t.

Following the similar procedure, we can derive that for a general relative degree ρ, it is required $\mu_i(\xi(t+\rho))$, $\mu_i(\xi(t+\rho-1))$, ..., and $\mu_i(\xi(t+1))$ are available at time t to connect $y(t+\rho)$ explicitly with $u(t)$.

Since the premise variables ξ_j, $j = 1, 2, \ldots, L$, are usually related to the state vector x, i.e.,

$$\xi = [\xi_1, \xi_2, \ldots, \xi_L]^T = [\xi_1(x), \xi_2(x), \ldots, \xi_L(x)]^T, \tag{6.10}$$

the values of the vector ξ at time t should only depend on the values of x at/before time $t - \rho$, i.e.,

$$\begin{aligned}\xi(t) &= [\xi_1(t), \xi_2(t), \ldots, \xi_L(t)]^T \\ &= [\xi_1(\underline{x}(t-\rho)), \xi_2(\underline{x}(t-\rho)), \ldots, \xi_L(\underline{x}(t-\rho))]^T,\end{aligned} \tag{6.11}$$

where $\underline{x}(t-\rho) = [x(t-\rho), x(t-\rho-1), \ldots, x(t-\rho-N_d)]^T$, and N_d is determined by how many past steps the values of x are used in $\xi_i(t) = \xi_i(\underline{x}(t-\rho))$, $i = 1, 2, \ldots, L$. Note that, $N_d = 0$ for the case when the premise variables ξ_j are related to the state variable x statically, as commonly seen in the literature, and a choice of $N_d \geq 1$ can provide more flexibility in modeling a dynamic system using T–S fuzzy functions. With the definition in (6.11), $\mu_i(\xi(t+\rho))$, $\mu_i(\xi(t+\rho-1))$, ..., and $\mu_i(\xi(t+1))$ are available at time t for a fuzzy system with a relative degree ρ.

Therefore, in order to obtain a global T–S fuzzy system with an explicit relative degree ρ and causality for state-feedback output tracking control design, the premise variable vector $\xi(t)$ should only include the values of states at/before time $t-\rho$ as shown in (6.11). The firing strength of the ith rule in (6.3), $\mu_i(\xi(t))$, is calculated as

$$\mu_i(\xi(t)) = \mu_i(\xi_1(\underline{x}(t-\rho)), \xi_2(\underline{x}(t-\rho)), \ldots, \xi_L(\underline{x}(t-\rho))). \tag{6.12}$$

To show the relative degree information clearly and to simplify the notation, here we replace $\mu(\xi(t))$ by $\mu(t-\rho))$ in (6.12), that is,

$$\mu_i(t-\rho) \triangleq \mu_i(\xi(t)) = \mu_i(\xi_1(\underline{x}(t-\rho)), \xi_2(\underline{x}(t-\rho)), \ldots, \xi_L(\underline{x}(t-\rho))). \tag{6.13}$$

As discussed and analyzed above, we know that the global T–S fuzzy system (6.3) cannot be directly used for state-feedback output tracking control design since it does not have an explicit relative degree structure and the choice of $\xi(t)$ may cause non-causality problem when deriving the normal form of the global T–S fuzzy system. To ensure the causality for a T–S fuzzy system with a relative degree ρ, $\xi(t)$ can only include states/outputs values at/before time $t-\rho$. One possible choice of $\xi(t)$ is given in (6.11). Based on such a choice and with the definitions in (6.12) and (6.13), a global state-space form T–S fuzzy system is derived as

$$\begin{aligned}x(t+1) &= \sum_{i=1}^{N} \mu_i(t-\rho)(A_i x(t) + B_i u(t)), \\ y(t) &= \sum_{i=1}^{N} \mu_i(t-\rho) C_i x(t),\end{aligned} \tag{6.14}$$

which can be further formulated into a normal form with an explicit relative degree structure, as will be shown in the following.

Relative degree conditions. Before deriving the normal form of the global T–S fuzzy system, we have specified the relative degree conditions for the global fuzzy system (6.14) as the assumption:

Assumption 6.1 The global fuzzy system (6.14) satisfies the general relative degree conditions:

$$C_i A_{j_1} \cdots A_{j_k} B_l = 0, \ i, j_1, \ldots, j_k, l = 1, 2, \ldots, N, \tag{6.15}$$

for $k = 0, 1, \ldots, \rho - 2$, and for all $t \geq 0$,

$$\sum_{i=1}^{N} \mu_i(t) C_i \sum_{j_1=1}^{N} \mu_{j_1}(t-1) A_{j_1} \cdots \sum_{j_{\rho-1}=1}^{N} \mu_{j_{\rho-1}}(t-\rho+1) A_{j_{\rho-1}} \sum_{l=1}^{N} \mu_l(t-\rho) B_l \neq 0. \tag{6.16}$$

The condition (6.15) in Assumption 6.1 implies that when we derive the expressions for $y(t+1)$, $y(t+2)$, ..., $y(t+\rho-1)$, all the terms multiplying the input $u(t)$ are zeros. And the condition (6.16) ensures that $u(t)$ appears explicitly in the expression of $y(t+\rho)$, thus ensures the system has a relative degree ρ. Introducing new state variables: $z_1(t) = y(t)$, $z_2(t) = y(t+1)$, ..., and $z_\rho(t) = y(t+\rho-1)$, we can derive the normal form of the global fuzzy system (6.14) as

$$\begin{aligned} z_j(t+1) &= z_{j+1}(t), \ j = 1, 2, \ldots, \rho - 1, \\ z_\rho(t+1) &= R(t)x(t) + G(t)u(t), \end{aligned} \tag{6.17}$$

where

$$\begin{aligned} R(t) &= \sum_{i=1}^{N} \mu_i(t) C_i \sum_{j_1=1}^{N} \mu_{j_1}(t-1) A_{j_1} \cdots \\ &\quad \sum_{j_{\rho-1}=1}^{N} \mu_{j_{\rho-1}}(t-\rho+1) A_{j_{\rho-1}} \sum_{j_\rho=1}^{N} \mu_{j_\rho}(t-\rho) A_{j_\rho}, \end{aligned} \tag{6.18}$$

$$\begin{aligned} G(t) &= \sum_{i=1}^{N} \mu_i(t) C_i \sum_{j_1=1}^{N} \mu_{j_1}(t-1) A_{j_1} \cdots \\ &\quad \sum_{j_{\rho-1}=1}^{N} \mu_{j_{\rho-1}}(t-\rho+1) A_{j_{\rho-1}} \sum_{j_\rho=1}^{N} \mu_{j_\rho}(t-\rho) B_{j_\rho}. \end{aligned} \tag{6.19}$$

From this derivation, we also have formed a set of parametrizations for the signals $z_i(t)$, $i = 2, 3, \ldots, \rho$, which are useful for adaptive control design with estimates of those signals which are not available when the system parameters are unknown. The system model (6.17) has the explicit relative degree ρ (that is, $z_\rho(t+1) = y(t+\rho)$ is related to $u(t)$) and causality (that is, $y(t+\rho) = z_\rho(t+1)$ depends on $\mu_i(t)$ (not

$\mu_i(t+j)$ with $j > 0$) and $x(t)$ and $u(t)$. In particular, the coefficients $G(t)$ and $R(t)$ depend on $\mu_i(t-j)$ ($j \geq 0$) available at time t, which is crucial for control design.

Thus far, we have developed a new global T–S fuzzy system model (6.14). The developed global fuzzy system model has several key features: (i) generality, (ii) relative degree, (iii) causality, (iv) relaxed matching conditions for adaptive control, and (iv) time-varying system feature.

We summarize the development of the T–S fuzzy system model for a set of general local system models as:

Proposition 6.1 *Following a standard fuzzy modeling procedure, a nonlinear dynamic system (6.1), via the local fuzzy system models (6.2) under the relative degree conditions in Assumption 6.1, can be approximated by a global fuzzy system model (6.14) which can be further transformed into the normal form (6.17) with an explicit relative degree structure and involves only causal membership functions.*

For the adaptive control problem to be solved in the next section, we make the following assumptions:

Assumption 6.2 The coefficient $G(t)$ satisfies: $G(t) \neq 0$ for all $t \geq 0$.

Assumption 6.3 The fuzzy system (6.14) is minimum phase.

A regular input–output linear time-invariant system

$$A(z^{-1})[y](t) = z^{-d}\bar{B}(z^{-1})[u](t) \tag{6.20}$$

with $A(z^{-1}) = 1 + a_1 z^{-1} + a_2 z^{-2} \cdots + a_n z^{-n}$ and $\bar{B}(z^{-1}) = b_0 + b_1 z^{-1} + \cdots + b_{n-d} z^{-n+d}$, is minimum phase if all zero of $\bar{B}(z^{-1})$ are in $|z| < 1$. For the equivalent system expression

$$\begin{aligned} & y(t+n) + a_1 y(t+n-1) + \cdots + a_n y(t) \\ = \; & b_0 u(t+n-d) + b_1 u(t+n-d-1) + \cdots + b_{n-d} u(t-d), \end{aligned} \tag{6.21}$$

this condition implies that

$$|u(t-d)| \leq c_1 |y(t)| + c_2 \sum_{\tau=0}^{t-1} \lambda^{t-\tau-1} |y(\tau)|, \; t \geq d, \tag{6.22}$$

for some constants $c_1 > 0$, $c_2 > 0$ and $\lambda \in (0, 1)$. This minimum-phase condition can also be used for an input–output T–S fuzzy system model (Qi et al. 2012).

In our study, the fuzzy system (6.14) is a nonlinear state-space model, and its minimum-phase property needs to be defined through the zero dynamics subsystem of (6.14), leading to a condition similar to that in (6.22).

As in the literature, the minimum-phase condition may be specified based on a zero dynamics subsystem: for some state variables $z^{(2)}(t) = [z_{\rho+1}(t), \ldots, z_n(t)]^T \in R^{n-\rho}$ and a vector function $Q(\cdot, \cdot; t) \in R^{(n-\rho)}$, the global zero dynamics subsystem

$$z^{(2)}(t+1) = Q(z^{(1)}(t), z^{(2)}(t); t) \tag{6.23}$$

with $z^{(1)}(t) = [z_1(t), \ldots, z_\rho(t)]^T \in R^\rho$, is bounded-input bounded-state (BIBS) stable (with $z^{(1)}(t)$ being the input), that is,

$$\|z^{(2)}(t)\| \leq k_q \max_{0 \leq \tau \leq t-1} \|z^{(1)}(\tau)\| + \epsilon_q(t), \tag{6.24}$$

for some $k_q > 0$ and some exponentially decaying signal $\epsilon_q(t)$ related to the initial conditions of $z^{(2)}(t)$, where $\|\cdot\|$ is any chosen vector norm.

Remark 6.1 We note that the construction of such a global zero dynamics subsystem for the global system models (6.14) and (6.17) is still an open issue under investigation. The complementary state variables $z^{(2)}(t) = [z_{\rho+1}(t), \ldots, z_n(t)]^T \in R^{n-\rho}$ are functions of the original state variables $x(t)$: $z^{(2)}(t) = T_2(x(t); t)$ for some matrix function $T_2(\cdot; t) \in R^{(n-\rho)\times n}$ such that the state transformation $z(t) = [(z^{(1)}(t))^T, (z^{(2)}(t))^T]^T = T(x(t); t)$ is an invertible and one-to-one mapping (diffeomorphism) so that the boundedness of $z(t)$ implies that of $x(t)$ and vice versa (that is, the minimum-phase condition means that the boundedness of $z^{(1)}(t)$, to be ensured by a feedback control law, implies the boundedness of $x(t)$ (Isidori 1995)).

Finally, we note that a global normal form T–S fuzzy model, similar to that in Proposition 6.1, can be derived for the system (6.14) with $x(t+1) = \sum_{i=1}^{N} \mu_i(t-\rho+1)(A_i x(t) + B_i u(t))$.

In this section, we have presented the analysis for state-space T–S fuzzy systems with a general relative degree ρ. In the next section, we will start with the case of $\rho = 1$ to show the basic adaptive control design ideas and then address the general case with $\rho \geq 1$ to develop the complete design procedure and stability theory.

6.3 Designs for T–S Fuzzy Systems with Relative Degree $\rho = 1$

In this section,[2] we consider the adaptive control design for the T–S fuzzy system with relative degree $\rho = 1$.

For the special case when $\rho = 1$, the model (6.14) reduces to

$$x(t+1) = \sum_{i=1}^{N} \mu_i(t-1) A_i x(t) + \sum_{i=1}^{N} \mu_i(t-1) B_i u(t),$$
$$y(t) = \sum_{i=1}^{N} \mu_i(t-1) C_i x(t), \tag{6.25}$$

[2]Parts of Sect. 6.3 are reprinted from Qi et al. (2013), Copyright 2012, with permission from Elsevier.

which is similar to many T–S system models commonly seen in the literature but with the difference that $\mu_i(t-1)$ is presented in the output equation, to specify the relative degree $\rho = 1$.

This formulation is necessary for an output tracking design which is applicable to systems for which a state tracking design is impossible due to some restrictive state-space plant and reference model matching conditions as presented in Chap. 5. For output tracking, we show that the plant model matching is always possible.

6.3.1 Nominal Controller

For the control design, the model (6.25) can be written as

$$\begin{aligned} y(t+1) &= \sum_{i=1}^{N} \mu_i(t) C_i x(t+1) \\ &= R(t)x(t) + G(t)u(t), \end{aligned} \tag{6.26}$$

where $R(t) = \sum_{i=1}^{N} \mu_i(t) C_i \sum_{j=1}^{N} \mu_j(t-1) A_j \in R^{1\times n}$ is a row vector gain and $G(t) = \sum_{i=1}^{N} \mu_i(t) C_i \sum_{j=1}^{N} \mu_j(t-1) B_j \in R$ is a scalar gain.

For a given reference output signal $y_m(t)$ to be tracked by the output signal $y(t)$, to generate a control signal for (6.26), we can use the deadbeat nominal control law:

$$u(t) = \frac{1}{G(t)}(-R(t)x(t) + y_m(t+1)), \tag{6.27}$$

which can bring the output $y(t+1)$ to $y_m(t+1)$ in one step (that is, $y(t+1) = y_m(t+1)$, $t \geq 0$), to ensure the tracking of $y_m(t)$ by $y(t)$ and thus the boundedness of $y(t)$. With (6.26) and (6.27), we have the boundedness of $y(t)$, and from the definition of $z^{(1)}(t)$ we have the boundedness of $z^{(1)}(t)$. Then, from the BIBS stability property of (6.23), that is, (6.24), we have the boundedness of $z(t)$, and from the diffeomorphism of $z(t) = T(x(t); t)$, we have the boundedness of $x(t)$ and that of $u(t)$, so that all closed-loop system signals are bounded.

Remark 6.2 In discrete-time control theory, the dead beat control problem consists of finding what input signal to be applied to a system in order to bring the output to the steady state in the smallest number of time steps (Dorf and Bishop 2005).

Remark 6.3 The nominal control law can also be design as

$$\begin{aligned} u(t) = \frac{1}{G(t)} & (-R(t)x(t) + y_m(t+1) - k_1 e(t) - k_2 e(t-1) \\ & - \cdots - k_q e(t-q+1)\big), \end{aligned} \tag{6.28}$$

where $e(t) = y(t) - y_m(t)$ and the parameters $k_1, k_2, \ldots, k_q$ are selected such that the zeros of the polynomial $1 + k_1 z^{-1} + \cdots + k_q z^{-q}$ lie strictly inside the unit circle. With (6.28), $y(t+1)$ will approach $y_m(t+1)$ after some transient process determined by the values of $k_1, k_2, \ldots, k_q$.

A nominal control law provides the basic controller structure which can be parametrized for parameter adaptation when the system parameters are unknown.

6.3.2 Adaptive Control Design

For simplicity of expression and without loss of generality, we consider the case of $N = 2$ in developing a parametrized model for adaptive control. With some manipulations on (6.25), we obtain the parametrized model:

$$y(t+1) = k_1^T w_1(t) + k_2^T w_2(t), \tag{6.29}$$

where

$$k_1 = [C_1 A_1, C_1 A_2, C_2 A_1, C_2 A_2]^T, \tag{6.30}$$

$$k_2 = [C_1 B_1, C_1 B_2, C_2 B_1, C_2 B_2]^T, \tag{6.31}$$

$$\begin{aligned} w_1(t) = [&\mu_1(t)\mu_1(t-1)x^T(t), \mu_1(t)\mu_2(t-1)x^T(t), \\ &\mu_1(t)\mu_2(t-1)x^T(t), \mu_2(t)\mu_2(t-1)x^T(t)]^T, \end{aligned} \tag{6.32}$$

$$\begin{aligned} w_2(t) = [&\mu_1(t)\mu_1(t-1)u(t), \mu_1(t)\mu_2(t-1)u(t), \\ &\mu_1(t)\mu_2(t-1)u(t), \mu_2(t)\mu_2(t-1)u(t)]^T. \end{aligned} \tag{6.33}$$

The model (6.29) can be further formulated as

$$y(t+1) = \theta^T \phi(t), \tag{6.34}$$

where $\theta = [k_1^T, k_2^T]^T$ and $\phi(t) = [w_1^T(t), w_2^T(t)]^T$.

Parameter adaptive law. The formulation (6.34), with the parameter vector θ unknown and vector signal $\phi(t)$ known, is a standard regression form with a linearized parametrization for which many parameter adaptation laws can be adopted to estimate the unknown parameters in θ. As a choice, the following gradient type adaptive law is employed to obtain the estimate $\hat{\theta}$ of θ:

$$\hat{\theta}(t) = \hat{\theta}(t-1) + \frac{\gamma(t)\phi(t-1)\varepsilon(t)}{c + \phi^T(t-1)\phi(t-1)}, \tag{6.35}$$

where $\gamma(t) \in (\gamma_0, 2 - \gamma_0)$ is an adaptation gain for some constant $\gamma_0 \in (0, 1)$, $c > 0$ is a small design parameter, and

$$\varepsilon(t) = y(t) - \hat{\theta}^T(t-1)\phi(t-1). \tag{6.36}$$

We note that for adaptive control, the second and third components of k_1 and k_2 in (6.30) and (6.31) can also be combined as one parameter to be adaptively estimated, as their corresponding regressor components are the same.

Adaptive control law. With $\hat{k}_1(t)$ and $\hat{k}_2(t)$ being the estimates of k_1 and k_2 at time t, the adaptive version of the nominal (deadbeat) control law (6.27) can be implemented as

$$u(t) = \frac{1}{\hat{G}(t)}(-\hat{k}_1^T(t)w_1(t) + y_m(t+1)), \tag{6.37}$$

where

$$\hat{G}(t) = \hat{k}_2^T(t)[\mu_1(t)\mu_1(t-1), \mu_1(t)\mu_2(t-1), \mu_1(t)\mu_2(t-1), \mu_2(t)\mu_2(t-1)]^T. \tag{6.38}$$

For this adaptive control law, parameter projection can be used for the parameter estimation algorithm (6.35) to ensure that $|\hat{G}(t)| > g_0$ for some constant $g_0 > 0$ (Tao 2003).

6.4 Designs for T–S Fuzzy Systems with Relative Degree $\rho \geq 1$

In this section,[3] we develop an adaptive control scheme based on the global normal form system (6.17), apply it to the global T–S fuzzy system (6.14), and evaluate its performance. Based on (6.17) and (6.14), our goal is to derive and study a desired baseline adaptive control design for control of nonlinear systems using the T–S fuzzy system approach.

6.4.1 Nominal Control Law

A nominal control law is designed for the case when the parameters of the fuzzy system (6.14) are known. Recall from Sect. 6.2.2 that the fuzzy system (6.14) with relative degree $1 < \rho < n$ can be transformed into

$$\begin{aligned} z_i(t+1) &= z_{i+1}(t), \quad i = 1, \ldots, \rho - 1, \\ z_\rho(t+1) &= R(t)x(t) + G(t)u(t) \end{aligned} \tag{6.39}$$

with $z^{(1)}(t) = [z_1(t), z_2(t), \ldots, z_\rho(t)]^T = [y(t), y(t+2), \ldots, y(t+\rho-1)]^T$, plus a possible zero dynamics subsystem: $z^{(2)}(t+1) = Q(z^{(1)}(t), z^{(2)}(t); t)$ (see

[3] Parts of Sect. 6.4 are reprinted from Qi et al. (2013), Copyright 2012, with permission from Elsevier.

(6.23)), where $R(t)$ and $G(t)$ are given in (6.18) and (6.19) respectively, and are known when the system parameters are known.

Nominal control law. For a given reference output signal $y_m(t)$ to be tracked by the output signal $y(t)$, to generate a control signal for (6.39), we can use the deadbeat nominal control law:

$$u(t) = \frac{1}{G(t)}(-R(t)x(t) + y_m(t+\rho)), \tag{6.40}$$

which can bring the output $y(t+\rho)$ to $y_m(t+\rho)$ in one step (that is, $y(t+\rho) = y_m(t+\rho)$, $t \geq 0$), to ensure the tracking of $y_m(t)$ by $y(t)$ and thus the boundedness of $y(t)$ (and that of $z^{(1)}(t)$), and, under Assumptions 6.2 and 6.3, the boundedness of $z^{(2)}(t)$ and $x(t)$, and finally, that of $u(t)$, so that all closed-loop system signals are bounded.

6.4.2 Adaptive Control Design

The model (6.39) can be written as

$$y(t+\rho) = R(t)x(t) + G(t)u(t). \tag{6.41}$$

Following the similar procedure as presented for the special case $\rho = 1$, we can formulate (6.41) into the linearly parametrized form:

$$y(t+\rho) = k_1^T \omega_1(t) + k_2^T \omega_2(t), \tag{6.42}$$

where $k_1 \in R^l, l = n \times N^{\rho+1}$, is a vector consisting of $N^{\rho+1}$ basic vectors $(C_i A_{j_1} \cdots A_{j_\rho})^T \in R^n$, and $k_2 \in R^{N^{\rho+1}}$ consisting of $N^{\rho+1}$ elements $C_i A_{j_1} \cdots A_{j_\rho - 1} B_{j_\rho}$, $i, j_1, \ldots, j_\rho \in \{1, 2, \ldots, N\}$. The two regressors $\omega_1(t)$ and $\omega_2(t)$ have the same dimensions as k_1 and k_2, respectively, with $\omega_1(t)$ consisting of $N^{\rho+1}$ basic vectors $\mu_i(t)\mu_{j_1}(t-1)\mu_{j_2}(t-2)\cdots\mu_{j_\rho}(t-\rho)x(t)$ and $\omega_2(t)$ of $N^{\rho+1}$ elements $\mu_i(t)\mu_{j_1}(t-1)\mu_{j_2}(t-2)\cdots\mu_{j_\rho}(t-\rho)u(t)$.

With $\theta = [k_1^T, k_2^T]^T$ and $\phi(t) = [\omega_1^T(t), \omega_2^T(t)]^T$, the model (6.42) can be further formulated as

$$y(t+\rho) = \theta^T \phi(t). \tag{6.43}$$

Parameter adaptive law. Based on (6.43), the following gradient type adaptive law is employed to obtain the estimate $\hat{\theta}$ of θ:

$$\hat{\theta}(t) = \hat{\theta}(t-1) + \frac{\gamma(t)\phi(t-\rho)\varepsilon(t)}{m^2(t)}, \tag{6.44}$$

where $\gamma(t) \in (\gamma_0, 2-\gamma_0)$ is an adaptation gain for some constant $\gamma_0 \in (0, 1)$ and

$$m(t) = \sqrt{c + \phi^T(t-\rho)\phi(t-\rho)}, \tag{6.45}$$

$$\varepsilon(t) = y(t) - \hat{\theta}^T(t-1)\phi(t-\rho) \tag{6.46}$$

with $c > 0$ being a small design parameter.

Adaptive control law. With the estimated parameters, we can have the estimates: $\hat{G}(t)$ and $\hat{R}(t)$, so that the nominal control law (6.40) can be implemented in the following adaptive way:

$$u(t) = \frac{1}{\hat{G}(t)}(-\hat{R}(t)x(t) + y_m(t+\rho)). \tag{6.47}$$

6.4.3 *Stability and Tracking Properties*

For this parameter estimation algorithm(6.44), we have the following properties.

Lemma 6.1 *The parameter adaptive law (6.44), when applied to the parametrized fuzzy system (6.43), ensures that*

(i) $\|\hat{\theta}(t) - \theta\| \leq \|\hat{\theta}(t-1) - \theta\| \leq \|\hat{\theta}(0) - \theta\|$, *for the* l^2*-vector norm* $\|\cdot\|$;
(ii) $\frac{\varepsilon(t)}{m(t)} \in L^2$;
(iii) $\lim_{t\to\infty} \frac{\varepsilon(t)}{m(t)} = 0$;
(iv) $\|\hat{\theta}(t) - \hat{\theta}(t-t_1)\| \in L^2$; *and*
(vi) $\lim_{t\to\infty} \|\hat{\theta}(t) - \hat{\theta}(t-t_1)\| = 0$, *for any finite* $t_1 > 0$.

Proof Defining the parameter error $\tilde{\theta} = \hat{\theta} - \theta$ and with (6.44), we have

$$\tilde{\theta}(t) = \tilde{\theta}(t-1) + \frac{\gamma(t)\phi(t-\rho)\varepsilon(t)}{m^2(t)}. \tag{6.48}$$

Considering the following positive definite function:

$$V(\tilde{\theta}) = \tilde{\theta}^T\tilde{\theta},$$

the time increment of $V(\tilde{\theta})$ along (6.48) is

$$\begin{aligned} V(\tilde{\theta}(t)) - V(\tilde{\theta}(t-1)) &= -\gamma(t)\left(2 - \frac{\gamma(t)\phi^T(t-\rho)\phi(t-\rho)}{m^2(t)}\right)\frac{\varepsilon^2(t)}{m^2(t)} \\ &\leq -\gamma(t)(2-\gamma(t))\frac{\varepsilon^2(t)}{m^2(t)} \leq -\gamma_0^2\frac{\varepsilon^2(t)}{m^2(t)}, \end{aligned} \tag{6.49}$$

for $\gamma(t) \in (\gamma_0, 2-\gamma_0)$, which implies $\|\hat{\theta}(t)-\theta\| \leq \|\hat{\theta}(t-1)-\theta\| \leq \|\hat{\theta}(0)-\theta\|$ and $\hat{\theta}(t) \in L^\infty$.

It follows from (6.49) and the boundedness of $V(\tilde{\theta})$ that

$$\gamma_0^2 \sum_{\tau=0}^{t} \frac{\varepsilon^2(\tau)}{m^2(\tau)} = V(\tilde{\theta}(0)) - V(\tilde{\theta}(t)) \leq V(\tilde{\theta}(0)), \tag{6.50}$$

that is, $\frac{\varepsilon(t)}{m(t)} \in L^2$.

From (6.44), we have

$$\hat{\theta}(t) = \hat{\theta}(t-t_1) + \sum_{\tau=0}^{t_1-1} \frac{\gamma(t-\tau)\phi(t-\rho-\tau)\varepsilon(t-\tau)}{m^2(t-\tau)}, \tag{6.51}$$

and with $\frac{\varepsilon(t)}{m(t)} \in L^2$, we obtain $\hat{\theta}(t)-\hat{\theta}(t-t_1) \in L^2$, $\lim_{t\to\infty} \|\hat{\theta}(t)-\hat{\theta}(t-t_1)\| = 0$. ∇

Based on the desired properties of the parameter adaptation law (6.44), the following closed-loop stability and asymptotic tracking results can be proved.

We first present a desired property for $\phi(t)$.

Lemma 6.2 *Under Assumptions 6.2 and 6.3, the regressor $\phi(t)$ defined in (6.43) satisfies*

$$\|\phi(t)\| \leq \rho_1 + \rho_2 \max_{0\leq\tau\leq t} |e(\tau+\rho)|, \tag{6.52}$$

for some positive constants ρ_1 and ρ_2.

Proof From the definition of $\phi(t)$ in (6.43) whose components $\omega_1(t)$ and $\omega_2(t)$ are the products of μ_i at various time instants and $x(t)$, and products of μ_i at various time instants and $u(t)$, and the property of μ_i in (6.4), we have

$$\|\phi(t)\| \leq \kappa_0 \|[x^T(t), u(t)]\|, \tag{6.53}$$

for some constant $\kappa_0 > 0$.

With $y(t) = e(t) + y_m(t)$, the system Eq. (6.41) can be formulated as

$$e(t+\rho) + y_m(t+\rho) = R(t)x(t) + G(t)u(t). \tag{6.54}$$

With Assumption 6.2, we obtain

$$|u(t)| \leq \kappa_1 |e(t+\rho)| + \kappa_2 \|x(t)\| + \kappa_3, \tag{6.55}$$

where κ_i, $i = 1, 2, 3$, are some positive constants.

Since $z(t) = [z^{(1)T}(t), z^{(2)T}(t)]^T = T(x(t); t)$ is an invertible and one-to-one mapping (diffeomorphism) so that the boundedness of $z(t)$ implies that of $x(t)$ and vice versa. With the minimum-phase property assumed in Assumption 6.3, the boundedness of $z^{(1)}(t)$ implies that of $x(t)$ and vice versa. Hence, from (6.24), ignoring the effect of the exponentially decaying $\epsilon_q(t)$, we have

$$\|x(t)\| \leq \kappa_4 \|z^{(1)}(t)\| + \kappa_4 \max_{0 \leq \tau \leq t-1} \|z^{(1)}(\tau)\|, \tag{6.56}$$

for some constant $\kappa_4 > 0$.

From (6.53), (6.55) and (6.56), we have

$$\|\phi(t)\| \leq \kappa_5 |e(t+\rho)| + \kappa_6 \|z^{(1)}(t)\| + \kappa_6 \max_{0 \leq \tau \leq t-1} \|z^{(1)}(\tau)\| + \kappa_7, \tag{6.57}$$

where κ_i, $i = 5, 6, 7$, are some positive constants.

As defined previously: $z^{(1)}(t) = [y(t), y(t+1), \ldots, y(t+\rho-1)]^T$, and again with $y(t) = e(t) + y_m(t)$, we have

$$\begin{aligned} z^{(1)}(t) = {} & [e(t), e(t+1), \ldots, e(t+\rho-1)]^T \\ & + [y_m(t), y_m(t+1), \ldots, y_m(t+\rho-1)]^T. \end{aligned} \tag{6.58}$$

Then, we have

$$\|z^{(1)}(t)\| \leq \kappa_8 \max_{0 \leq \tau \leq t-1} |e(\tau+\rho)| + \kappa_9, \tag{6.59}$$

where κ_i, $i = 8, 9$, are some positive constants.

Finally, using (6.57) and (6.59), we obtain

$$\|\phi(t)\| \leq \rho_1 + \rho_2 \max_{0 \leq \tau \leq t} |e(\tau+\rho)|, \tag{6.60}$$

where ρ_1 and ρ_2 are some positive constants. ∇

We now show the desired closed-loop system properties.

Theorem 1 *The adaptive controller (6.47) with the adaptive law (6.44), applied to the system (6.14) under Assumptions 6.2 and 6.3, guarantees that all closed-loop system signals are bounded and* $\lim_{t\to\infty}(y(t) - y_m(t)) = 0$.

Proof From (6.43) and (6.46), we have

$$\varepsilon(t) = -(\hat{\theta}(t-1) - \theta)^T \phi(t-\rho), \tag{6.61}$$

and from (6.43) and (6.47), we have

$$\begin{aligned} y(t+\rho) &= R(t)x(t) + \frac{G(t)}{\hat{G(t)}}(-\hat{R}(t)x(t) + y_m(t+\rho)) \\ &= R(t)x(t) + \frac{G(t) - \hat{G}(t) + \hat{G}(t)}{\hat{G(t)}}(-\hat{R}(t)x(t) + y_m(t+\rho)) \end{aligned}$$

$$
\begin{aligned}
&= R(t)x(t) - \hat{R}(t)x(t) + y_m(t+\rho) \\
&+ \frac{G(t) - \hat{G}(t)}{\hat{G(t)}}(-\hat{R}(t)x(t) + y_m(t+\rho)).
\end{aligned} \tag{6.62}
$$

With $e(t) = y(t) - y_m(t)$ and (6.47) again, we obtain

$$
\begin{aligned}
e(t+\rho) &= R(t)x(t) - \hat{R}(t)x(t) + (G(t) - \hat{G}(t))u(t) \\
&= y(t+\rho) - \hat{y}(t+\rho) \\
&= \theta^T \phi(t) - \hat{\theta}^T(t)\phi(t) \\
&= -\tilde{\theta}^T(t)\phi(t),
\end{aligned} \tag{6.63}
$$

where $\tilde{\theta}(t) = \hat{\theta}(t) - \theta$.

With (6.63), we have $e(t) = -\tilde{\theta}^T(t-\rho)\phi(t-\rho)$, which can be expressed as

$$
\begin{aligned}
e(t) = \bar{\varepsilon}(t)\sqrt{c + \phi^T(t-\rho)\phi(t-\rho)} - (\hat{\theta}(t-\rho) - \\
\hat{\theta}(t-1))^T \bar{\phi}(t-\rho)\sqrt{c + \phi^T(t-\rho)\phi(t-\rho)},
\end{aligned} \tag{6.64}
$$

where with Lemma 6.1, $\hat{\theta}(t-\rho) - \hat{\theta}(t-1) \in L^2 \cap L^\infty$,

$$
\begin{aligned}
\bar{\varepsilon}(t) &= \frac{\varepsilon(t)}{\sqrt{c + \phi^T(t-\rho)\phi(t-\rho)}} \in L^2 \cap L^\infty, \\
\bar{\phi}(t-\rho) &= \frac{\phi(t-\rho)}{\sqrt{c + \phi^T(t-\rho)\phi(t-\rho)}} \leq 1.
\end{aligned} \tag{6.65}
$$

Using the inequality: $\sqrt{c + \phi^T(t-\rho)\phi(t-\rho)} \leq \sqrt{c} + \|\phi^T(t-\rho)\|$, we express $e(t)$ from (6.64) as

$$
|e(t)| \leq c_1 + |\bar{\varepsilon}(t)| \, \|\phi^T(t-\rho)\| + \|\hat{\theta}(t-\rho) - \hat{\theta}(t-1)\| \, \|\phi^T(t-\rho)\|,
$$

for some constant $c_1 > 0$. Using Lemma 6.2: $\|\phi(t-\rho)\| \leq \rho_1 + \rho_2 \max_{0\leq\tau\leq t} |e(\tau)|$, we obtain

$$
\begin{aligned}
|e(t)| \leq c_2 + c_3|\bar{\varepsilon}(t)| \max_{0\leq\tau\leq t} |e(\tau)| \\
+ c_4\|\hat{\theta}(t-\rho) - \hat{\theta}(t-1)\| \max_{0\leq\tau\leq t} |e(\tau)|,
\end{aligned} \tag{6.66}
$$

for some constants $c_i > 0$, $i = 2, 3, 4$.

From Lemma 6.1, we have that $\lim_{t\to\infty} \bar{\varepsilon}(t) = 0$ and $\lim_{t\to\infty} \|\hat{\theta}(t) - \hat{\theta}(t-\rho)\| = 0$, and with these results, it follows from (6.66) that $e(t)$ is bounded, which implies $y(t)$ is bounded, and in turn from the system's minimum-phase property that $u(t)$ is bounded. Hence, all signals in the closed-loop system are bounded, based on which, from (6.64) in which $\bar{\varepsilon}(t) \in L^2$ and $\hat{\theta}(t-\rho) - \hat{\theta}(t-1) \in L^2$, we have that $e(t) \in L^2$ so that $\lim_{t\to\infty} e(t) = 0$. ∇

Remark 6.4 In practice, there may exist modeling errors between original nonlinear systems and T–S fuzzy systems. To deal with modeling errors, a robust adaptive control law is commonly used in the adaptive control literature. Various methods have been developed to modify the adaptive law such as dead-zone modification and σ-modification in Tao (2003) to ensure the robustness of adaptive control systems, that is, the closed-loop system signal boundedness and a tracking error bounded by the modeling error in an average sense. Adaptive control system robustness can be achieved for any bounded external disturbances and for general state variable-dependent uncertainties, which satisfy some smallness properties (this is in general the case for many adaptive control systems). This conclusion is also applicable to the adaptive fuzzy control design of this paper.

Design procedure for adaptive state-feedback output tracking control. We now summarize the adaptive state-feedback output tracking control design procedure for the general case of systems with $N > 0$ and $0 < \rho < n$, as follows:

Step 1: determine the relative degree ρ of the global fuzzy system (6.14) based on the conditions in Assumption 6.1;

Step 2: derive the normal form (6.17) of the system (6.14);

Step 3: parametrize the normal form (6.17) with $z_\rho(t+1) = y(t+\rho)$ into the forms (6.42) and (6.43): $y(t+\rho) = \theta^T\phi(t)$;

Step 4: estimate the parameters $y(t+\rho) = \theta^T\phi(t)$ by an adaptive law (6.44);

Step 5: implement an adaptive control law (6.47) with the parameter estimates.

6.5 Simulation Study

In this section,[4] we verify the desired performance of the adaptive control scheme for a global T–S fuzzy system model, designed based its normal form, by simulation results. A T–S fuzzy model constructed from the linearized longitudinal dynamics models of a transport airplane is used as the controlled plant to test the adaptive control algorithm.

6.5.1 Simulation System

The linearized longitudinal dynamics model has the form

$$\begin{aligned} \dot{x}(t) &= A_i x(t) + B_i u(t), \\ y(t) &= C_i x(t), \end{aligned} \tag{6.67}$$

[4] Parts of Sect. 6.5 are reprinted from Qi et al. (2013), Copyright 2012, with permission from Elsevier.

where $x(t) = [U_b, W_b, Q_b, \theta]^T$ (U_b: forward velocity (x-axis), W_b: vertical velocity (z-axis), Q_b: pitch rate (y-axis), θ: Euler pitch angle (y-axis)), $u(t) = \delta_e$ is the deflection angle of the elevator, and A_i, B_i, and C_i, $i = 1, \ldots, N$, represent system matrices at different operating points for linearization.

In the case of climb with $U_{b0} = 454$ ft/s, $W_{b0} = 27.2$ ft/s, $Q_{b0} = 0$ rad/s, and $\theta_0 = 0.147$ rad, the linearized dynamics matrices for the model (6.67) are (A_1, B_1). In the case of descent with $U_{b0} = 454$ ft/s, $W_{b0} = 28.1$ ft/s, $Q_{b0} = 0$ rad/s and $\theta_0 = 0.0443$ rad, the linearized dynamics matrices for the model (6.67) are (A_2, B_2). The output matrices are $C_1 = C_2 = C = [0, 0, 0, 1]$.

Choosing the sampling time $T = 0.1$ s, we obtain the corresponding discrete-time state-space model of the above continuous-time model, with the matrices (A_{d1}, B_{d1}):

$$\begin{bmatrix} 0.999 & 0.00994 & -2.53 & -3.18 \\ -0.00750 & 0.901 & 41.2 & -0.439 \\ 3.51 \times 10^{-5} & -7.04 \times 10^{-4} & 0.891 & 1.75 \times 10^{-4} \\ 1.74 \times 10^{-6} & -3.64 \times 10^{-5} & 0.0947 & 1 \end{bmatrix}$$

$$\begin{bmatrix} 0.0124 & -0.207 & -0.00700 & -0.000357 \end{bmatrix}^T,$$

and (A_{d2}, B_{d2}):

$$\begin{bmatrix} 0.999 & 0.0102 & -2.61 & -3.21 \\ -0.00686 & 0.901 & 41.2 & -0.123 \\ 6.18 \times 10^{-5} & -7.02 \times 10^{-4} & 0.891 & -5.95 \times 10^{-5} \\ 3.10 \times 10^{-6} & -3.63 \times 10^{-5} & 0.0947 & 1 \end{bmatrix}$$

$$\begin{bmatrix} 0.0129 & -0.207 & -0.0070 & -0.000357 \end{bmatrix}^T$$

with $C_{d1} = C_{d2} = C$.

A T–S fuzzy model constructed from linearized models. From the above two local linearized dynamics models, we can construct a T–S fuzzy model with two rules. Choosing the pitch angle θ as the premise variable ξ_1 and assuming $\theta \in [0, 0.3]$ rad, which covers the operating points for the two linearized model, we obtain:

$$\begin{aligned} \text{IF } \xi_1 \text{ is } F_1^i, \text{ THEN } x(t+1) &= A_{di}x(t) + B_{di}u(t) \\ y(t) &= C_{di}x(t), \end{aligned} \tag{6.68}$$

where the membership function F_j^i characterizing the membership degree of $\xi_j(t)$ in the fuzzy set F_j^i is chosen as Gaussian function: $F_j^i(\xi_j(t)) = \exp\{-\frac{(\xi_j(t)-c_j^i)^2}{\sigma_j^{i2}}\}$, $j = 1, i = 1, 2$, with c_j^i and σ_j^i being the center and radius of the chosen Gaussian function and $[c_1^1, \sigma_1^1, c_1^2, \sigma_1^2] = [0.147, 0.1, 0.0443, 0.1]$.

Since $C_i B_j \neq 0, i, j = 1, 2$, we have the relative degree $\rho = 1$. The global T–S fuzzy model is obtained as (6.25).

Following the procedures in Sect. 4.2, we obtain the parametrized model in the form of (6.29) with

$$\begin{aligned} k_1 &= [C_{d1}A_{d1},\ C_{d1}A_{d2},\ C_{d2}A_{d1},\ C_{d2}A_{d2}]^T \in R^{16}, \\ k_2 &= [C_{d1}B_{d1},\ C_{d1}B_{d2},\ C_{d2}B_{d1}C_{d2}B_{d2}]^T \in R^4. \end{aligned} \tag{6.69}$$

6.5.2 Simulation Results

In the simulation, we set the initial values of $\hat{k}_1$ and $\hat{k}_2$ as: $\hat{k}_1(0) = 80\%k_1$, $\hat{k}_2(0) = 80\%k_2$, and the initial states $x(0) = [22.7, 1.5, 0, 0.0025]^T$. The parameters for the adaptive law (6.35)–(6.36) are chosen as : $\gamma(t) = 0.2$ and $c = 0.01$.

The adaptive tracking response and control signal for a constant reference signal $y_m(t) = 0.15$ are shown in Fig. 6.1. We can see that with the adaptive controller, the output converges to the constant reference signal quickly and smoothly. The adaptations of $\hat{k}_1$ and $\hat{k}_2$ are demonstrated in Figs. 6.2 and 6.3. $\hat{k}_1 \in R^{16}$ and $\hat{k}_2 \in R^4$ are the estimates of k_1 and k_2 in (6.69), respectively. Due to the space limit, only eight parameter estimates in $\hat{k}_1$ are shown in Fig. 6.2, which have effectively demonstrated the adaptation of $\hat{k}_1$. From Fig. 6.2, it can be observed that the parameter estimates converge to constant values after some initial transient responses. If the simulation is run for longer time, the parameter estimates in Fig. 6.3 will also approach constant values. The parameter adaptation process is in accordance with Lemma 6.1.

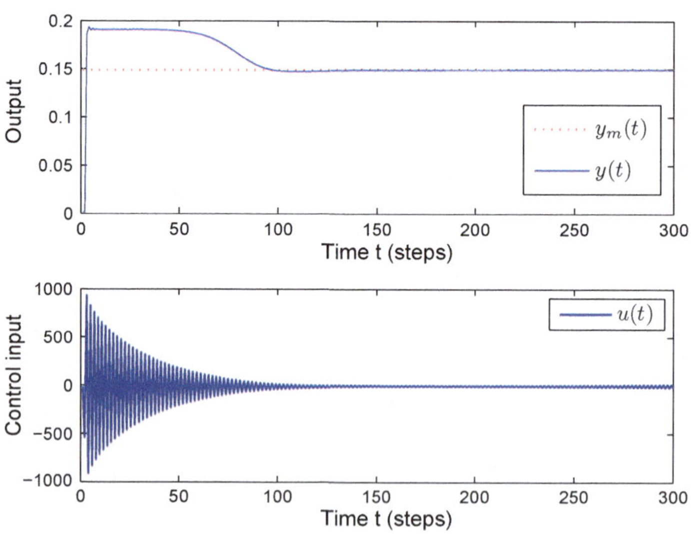

Fig. 6.1 Adaptive tracking response and control signal. Reprinted from Qi et al. (2013), Copyright 2012, with permission from Elsevier

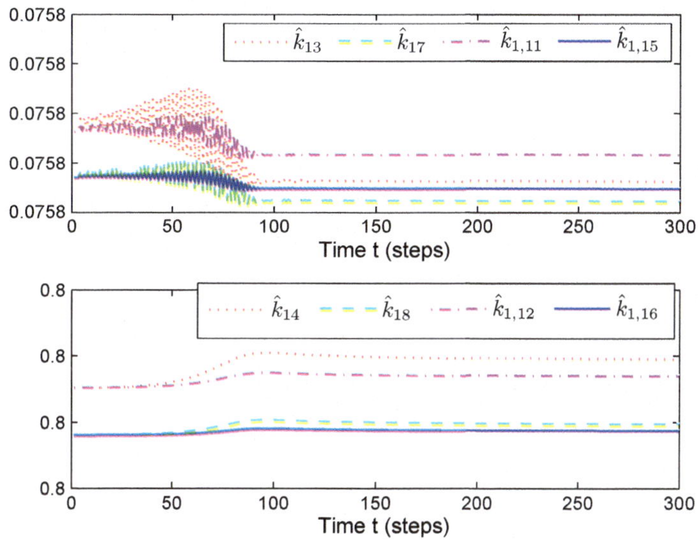

Fig. 6.2 Adaptation of parameters of $\hat{k}_1$. Reprinted from Qi et al. (2013), Copyright 2012, with permission from Elsevier

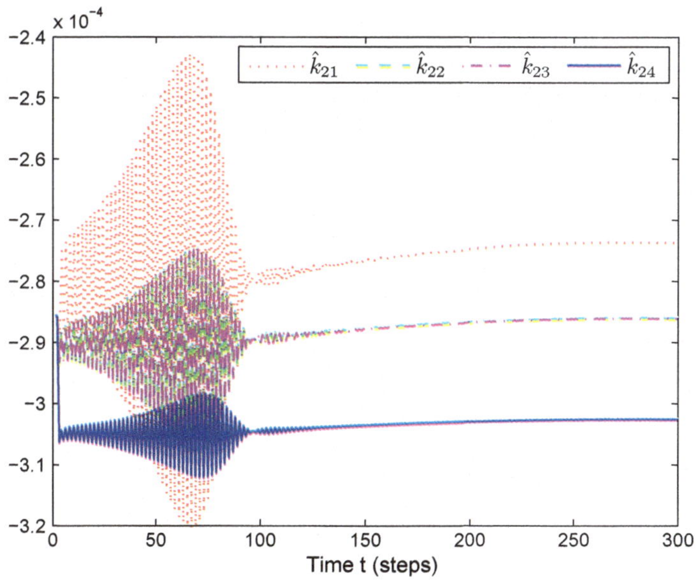

Fig. 6.3 Adaptation of parameters of $\hat{k}_2$. Reprinted from Qi et al. (2013), Copyright 2012, with permission from Elsevier

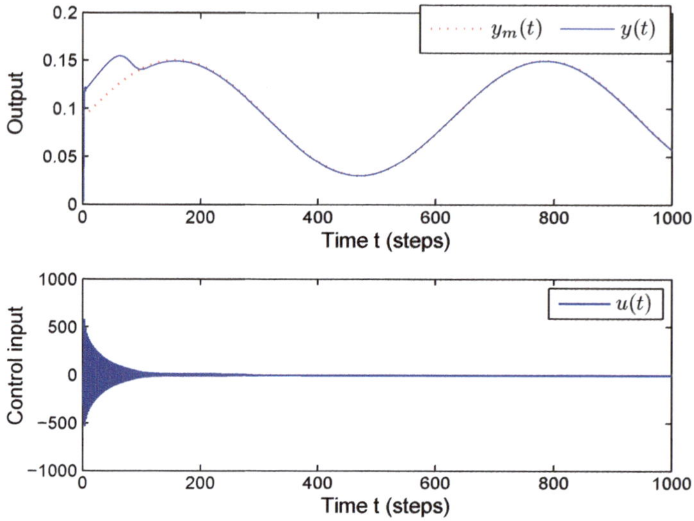

Fig. 6.4 Adaptive tracking response and control signal. Reprinted from Qi et al. (2013), Copyright 2012, with permission from Elsevier

For a time-varying reference signal $y_m(t) = 0.09 + 0.06\sin(0.01t)$, which covers both operating regions of the two linearized local models, the adaptive tracking results are presented in Fig. 6.4. The adaptations of $\hat{k}_1$ and $\hat{k}_2$ are shown in Figs. 6.5 and 6.6. It

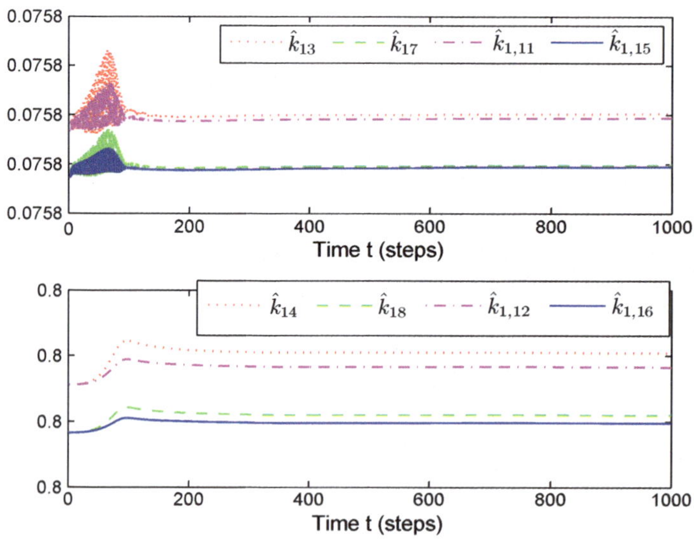

Fig. 6.5 Adaptation of parameters of $\hat{k}_1$. Reprinted from Qi et al. (2013), Copyright 2012, with permission from Elsevier

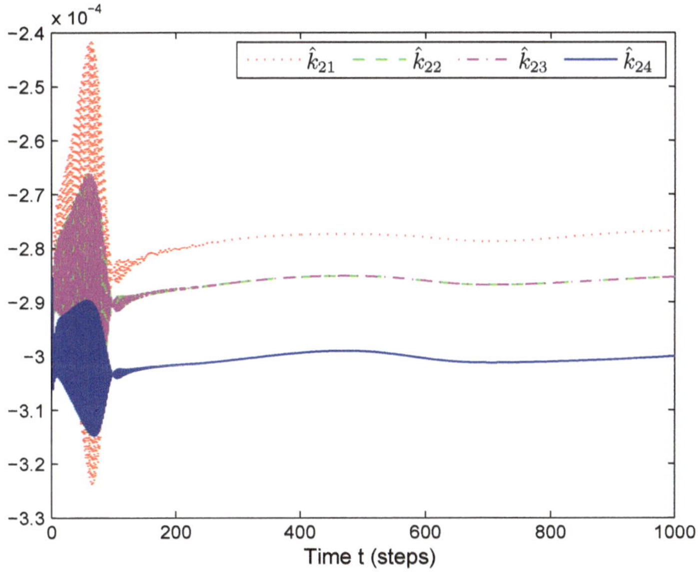

Fig. 6.6 Adaptation of parameters of $\hat{k}_2$. Reprinted from Qi et al. (2013), Copyright 2012, with permission from Elsevier

can be seen that for a time-varying reference, those parameter estimates also converge to certain constant values under the adaptive law (the parameter estimates of $\hat{k}_2$ will show such convergence for longer simulation time).

The simulation results verify the effectiveness of the desired closed-loop system performance. As shown from the simulation results, at the beginning of the simulation, there is a large transient response in both the control and tracking error signals, due to the initial parameter errors. As time elapses, the tracking error decreases with the adaptation of parameters. While this simulation study is for verifying the developed adaptive T–S control theory, the improvement of system transient responses for adaptive fuzzy control systems is an important issue to be further studied.

6.6 Summary

In this chapter, we have formulated a solution framework for adaptive control of T–S fuzzy systems with their local state-space models in noncanonical form. A normal form has been derived for a global T–S fuzzy model, which has an explicit relative degree structure and a proper causality property. Such a global T–S fuzzy model is an approximation of a discrete-time nonlinear system with a general relative degree, based on which an adaptive state-feedback control scheme has been developed with desired stability and output tracking performance. The key design conditions for the proposed adaptive output tracking control scheme are much relaxed as compared with that for a state tracking design. Simulation results on a T–S fuzzy system of a transport airplane example have also verified the desired performance of the developed adaptive fuzzy control systems.

References

Dorf RC, Bishop RH (2005) Modern control systems. Prentice Hall, Upper Saddle River

Isidori A (1995) Nonlinear control systems, 3rd edn. Springer, London

Qi R, Tao G, Jiang B, Tan C (2012) Adaptive control schemes for discrete-time T-S fuzzy systems with unknown parameters and actuator failures. IEEE Trans Fuzzy Syst 20:471–486

Qi R, Tao G, Tan C, Yao X (2013) Adaptive control of discrete-time state-space T–S fuzzy systems with general relative degree. Fuzzy Sets Syst 217:22–40

Tao G (2003) Adaptive control design and analysis. Wiley, New York

Chapter 7
Adaptive T–S Fuzzy Control Using Output Feedback: SISO Cases

In Chaps. 5 and 6, we have presented adaptive state feedback control designs for state-space T–S fuzzy systems with unknown parameters to achieve state tracking and output tracking, respectively. However, in reality, many systems may have states that are unmeasurable. For this situation, output feedback designs are more practical than state feedback. In this chapter, we consider adaptive output feedback control for discrete-time single-input single-output (SISO) T–S fuzzy systems described by input–output models. A key advantage of using an input–output approach over a state-space approach is that desired system tracking performance can be characterized and achieved based on much relaxed design conditions for adaptive control, in the presence of system parameter uncertainties. As presented in Chap. 5, for non-canonical state-space T–S fuzzy systems, some restrictive matching conditions were assumed for state tracking designs.

For tracking control design using input–output T–S fuzzy models, several approaches have been proposed in the literature. An input–output T–S fuzzy model structure with single delay was developed by Ying (1999), based on which a feedback linearization controller was designed to achieve perfect tracking and a necessary and sufficient condition was derived to check the stability of the controller. Using the same system and controller structure and considering parameter uncertainties, an adaptive feedback linearization control scheme for T–S fuzzy systems was rigorously developed in Feng (2010). Other studies within the similar framework can be found in Shi (2008), Tseng (2006), and Qi and Brdys (2008). In those studies, the delay from the input to the output was assumed to be one.

In reality, plants may have multiple input–output delays so that the T–S fuzzy systems to approximate such plants may have multiple delays as well. To enable feedback linearization design for such systems, usually a prediction model is required. T–S fuzzy systems have the feature that their overall models are nonlinear models but local models in each rule are linear models, which motivates two ways to develop their prediction models: one is to derive local prediction models from local linear

R. Qi et al., *Fuzzy System Identification and Adaptive Control*, Communications and Control Engineering,
https://doi.org/10.1007/978-3-030-19882-4_7

models first and then obtain the global prediction model by fuzzily blending all the local prediction models; the other is to derive the global prediction model directly through the overall nonlinear system model. The former approach was developed by Qi et al. (2012a) while the latter was resolved by Qi et al. (2012b).

This chapter is organized as follows. In Sect. 7.1, we formulate the output feedback and output tracking adaptive control problems for SISO nonlinear systems via T–S fuzzy modeling and prediction. Next, in Sect. 7.2, we develop a d-step prediction model for a T–S fuzzy system with multiple input–output delays through local predictions, design an adaptive output feedback control law, and establish the basic stability and tracking properties as the foundations for this adaptive fuzzy control method. Then, in Sect. 7.3, a d-step prediction model is developed through global T–S fuzzy prediction, based on which an adaptive output feedback control scheme is proposed with rigorous stability and tracking performance analysis.

7.1 Problem Statement

Consider a SISO nonlinear system in its discrete-time input–output form

$$y(t) = f(y(t-1), \ldots, y(t-n), u(t-d), \ldots, u(t-n)), \tag{7.1}$$

where $f(\cdot, \ldots, \cdot)$ is some nonlinear function, $y(\cdot)$ is the system output signal, $u(\cdot)$ is the system input signal, $t = 0, 1, 2, \ldots,$ is the discrete-time time variable, n is the system order, and d is the number of system input–output delays. We are interested in the general case with $d \geq 1$ (especially, $d > 1$).

For the feedback control design of (7.1) with d-step delays, it is crucial to derive a prediction model of the form

$$\begin{aligned} y(t+d) = \ & f_d(y(t), y(t-1), \ldots, y(t-n+1), \\ & u(t), u(t-1), \ldots, u(t-n+1)), \end{aligned} \tag{7.2}$$

in terms of the current and past system input $u(\tau)$ and output $y(\tau)$, $\tau \leq t$, for some function $f_d(\cdot, \ldots, \cdot)$. Then a nominal control law can be chosen for $u(t)$ to satisfy $y_m(t+d) = f_d(y(t), y(t-1), \ldots, y(t-n+1), u(t), u(t-1), \ldots, u(t-n+1))$ where $y_m(t+d)$ is a bounded reference signal.

In this chapter, we consider the system (7.1) in the form of a global T–S system model for approximation and control of nonlinear dynamic systems. Such a global T–S fuzzy system model can be derived based on a set of general local system models of the form

$$\begin{aligned} & y(t) + a_{i,1}y(t-1) + \cdots + a_{i,n}y(t-n) \\ & = b_{i,0}u(t-d) + b_{i,1}u(t-d-1) + \cdots + b_{i,n-d}u(t-n) \end{aligned} \tag{7.3}$$

for the system (7.1) operating at different points.

To be more precise, let $\xi_j(t)$, $j = 1, 2, \ldots, L$, be the premise variables for fuzzy rules which are chosen from $y(t-1), \ldots, y(t-n)$ and $u(t-d), \ldots, u(t-n)$ for specific systems and applications. When the values of $\xi_j(t)$ are set to be in some given intervals, they define certain regions for the variables $y(t-1), \ldots, y(t-n)$ and $u(t-d), \ldots, u(t-n)$. With the variables $y(t-1), \ldots, y(t-n)$ and $u(t-d), \ldots, u(t-n)$ being in each of such regions, the system function $f(y(t-1), \ldots, y(t-n), u(t-d), \ldots, u(t-n))$ can be approximated by a linear system model.

This motivates the following discrete-time input–output T–S fuzzy system model:

$$\begin{aligned} R^i :\ & \text{IF } \xi_1(t) \text{ is } F_1^i \text{ and } \xi_2(t) \text{ is } F_2^i \text{ and } \ \ldots \ \text{ and } \xi_L(t) \text{ is } F_L^i \\ & \text{THEN } y(t) + a_{i,1} y(t-1) + \cdots + a_{i,n} y(t-n) = b_{i,0} u(t-d) \\ & \qquad + b_{i,1} u(t-d-1) + \cdots + b_{i,\,n-d} u(t-n),\ b_{i,0} \neq 0, \end{aligned} \tag{7.4}$$

where R^i, $i = 1, 2, \ldots, N$, denotes the ith fuzzy rule defining the ith subsystem, N is the number of fuzzy rules, $a_{i,1}, \ldots, a_{i,n}, b_{i,0}, \ldots, b_{i,\,n-d}$ are coefficients of the ith subsystem, $d > 0$ denotes the system delay, and "$\xi_j(t)$ is F_j^i" is a part of the ith fuzzy rule, with the premise variables $\xi(t) = [\xi_1(t), \xi_2(t), \ldots, \xi_L(t)]^T$ being some measurable system signals or their functions and F_j^i being typically an interval of real numbers, called a fuzzy set associated with which there is a membership function $F_j^i(\xi_j(t))$ to indicate the membership degree of $\xi_j(t)$ in F_j^i.

In this chapter, we shall obtain the global T–S fuzzy prediction model (7.2) for (7.1) through two approaches:

Approach I: Obtain the prediction models for local linear subsystems in (7.4) first and then construct the overall nonlinear prediction model by fuzzily blending all the local prediction models.

Approach II: Obtain the overall nonlinear fuzzy model from (7.4) first and then derive the nonlinear prediction model directly from it.

These two approaches represent two different ways to use T–S fuzzy system modeling theory to approximate a nonlinear dynamic system (7.1).

In the following sections, we will show how to derive these two different prediction models by Approach I and Approach II. Different approaches lead to different parameterized models, based on which adaptive controllers are designed to ensure closed-loop signal boundedness and asymptotic tracking of a bounded reference signal $y_m(t)$ by the system output $y(t)$ when there exist unknown parameters.

7.2 Approach I: Design Based on Linear Prediction Models

Each equation in (7.4) defines a local linear model for the original nonlinear system (7.1). To derive a global fuzzy system model for (7.1), we will first derive a d-step prediction model for each local linear model and then use them to construct a global fuzzy (nonlinear) system model.[1]

7.2.1 T–S Fuzzy Prediction Model via Linear Prediction

Introducing the polynomials in z^{-1}: $A_i(z^{-1}) = 1 + a_{i,1}z^{-1} + \cdots + a_{i,n}z^{-n}$ and $\bar{B}_i(z^{-1}) = b_{i,0} + b_{i,1}z^{-1} + \cdots + b_{i,n-d}z^{-n+d}, i = 1, 2, \ldots, N$, with z^{-1} and z being the time-delay and time-advance operators: $z^{-1}[x](t) = x(t-1)$ and $z[x](t) = x(t+1)$, we express the local linear system (7.4) as

$$A_i(z^{-1})[y](t) = z^{-d}\bar{B}_i(z^{-1})[u](t), \tag{7.5}$$

where, as a notation, for a polynomial $P(z^{-1}) = p_0 + p_1z^{-1} + \cdots + p_{n_p}z^{-n_p}$ and a signal $x(t) \in R$, we define $P(z^{-1})[x](t) = p_0x(t) + p_1x(t-1) + \cdots + p_{n_p}x(t-n_p)$. Following the procedure proposed by Goodwin and Sin (1984), and solving the polynomial equation

$$1 = F_i(z^{-1})A_i(z^{-1}) + z^{-d}G_i(z^{-1}), \tag{7.6}$$

we obtain the polynomials

$$\begin{aligned} F_i(z^{-1}) &= 1 + f_{i,1}z^{-1} + \cdots + f_{i,d-1}z^{-d+1} \\ G_i(z^{-1}) &= g_{i,0} + g_{i,1}z^{-1} + \cdots + g_{i,n-1}z^{-n+1} \end{aligned} \tag{7.7}$$

with $f_{i,k} = -\sum_{j=0}^{k-1} f_{i,j}a_{i,k-j}, k = 1, 2, \ldots, d-1$, and $g_{i,k} = -\sum_{j=1}^{d-1} f_{i,j}a_{i,k+d-j}$, $k = 0, 1, \ldots, n-1$, for $a_{i,n+1} = a_{i,n+2} = \cdots = a_{i,n+d-1} = 0$. Then, operating both sides of (7.6) on $y(t)$, we obtain

$$y(t+d) = z^d F_i(z^{-1})A_i(z^{-1})[y](t) + G_i(z^{-1})[y](t). \tag{7.8}$$

Substituting (7.5) into (7.8) yields the local d-step prediction equation

$$y(t+d) = F_i(z^{-1})\bar{B}_i(z^{-1})[u](t) + G_i(z^{-1})[y](t), \tag{7.9}$$

which can be written as

$$y(t+d) = \alpha_i(z^{-1})[y](t) + \beta_i(z^{-1})[u](t), \tag{7.10}$$

where $\alpha_i(z^{-1}) = G_i(z^{-1})$ and $\beta_i(z^{-1}) = F_i(z^{-1})\bar{B}_i(z^{-1})$:

$$\alpha_i(z^{-1}) = \alpha_{i,0} + \alpha_{i,1}z^{-1} + \cdots + \alpha_{i,n-1}z^{-n+1} \tag{7.11}$$

$$\beta_i(z^{-1}) = \beta_{i,0} + \beta_{i,1}z^{-1} + \cdots + \beta_{i,n-1}z^{-n+1} \tag{7.12}$$

with $\beta_{i,0} = b_{i,0} \neq 0$.

Based on the above derivation, we have the following proposition.

Proposition 7.1 *The local linear system (7.4) can be equivalently represented by the local d-step prediction model (7.10).*

Different from what has been done by Goodwin and Sin (1984), in our study, we will now use the local models in (7.10) to form a global d-step prediction fuzzy system model. To proceed, based on (7.10), using the standard technique of *singleton fuzzification*, *product inference*, and *weighted average*, we obtain the following global d-step prediction nonlinear fuzzy dynamic system model (Qi et al. 2012a).

Proposition 7.2 *Following a standard fuzzy modeling procedure, a nonlinear dynamic system (7.1), via the local fuzzy system model (7.4), can be approximated by a global d-step SISO prediction fuzzy system model:*

$$y(t+d) = \sum_{i=1}^{N} \mu_i \alpha_i(z^{-1})[y](t) + \sum_{i=1}^{N} \mu_i \beta_i(z^{-1})[u](t), \tag{7.13}$$

where μ_i is the normalized membership function satisfying

$$\begin{aligned} \mu_i(\xi) &= \frac{\lambda_i(\xi)}{\sum_{i=1}^{N} \lambda_i(\xi)}, \quad \lambda_i(\xi) = \prod_{j=1}^{L} F_j^i(\xi_j) \\ \mu_i(\xi) &\geq 0, \quad \sum_{i=1}^{N} \mu_i(\xi) = 1. \end{aligned} \tag{7.14}$$

The fuzzy system model (7.13) is an approximate model for the original nonlinear system model (7.1). It has been shown that T–S fuzzy systems with linear rule consequents can uniformly approximate any continuous function on a compact domain to any degree of accuracy (Tanaka and Wang 2001). Given a specific approximation accuracy, a sufficient condition on calculating the minimal number of fuzzy sets and fuzzy rules is also provided, which is independent of the shape of membership functions, fuzzy logic operators, and the defuzzifier type so that the computed number may be very large. In practice, a smaller number of fuzzy sets and rules may be found to meet the desired approximation accuracy depending on the function to be approximated and the type of membership functions chosen.

Remark 7.1 Choosing a suitable number of fuzzy rules and proper membership functions plays a critical role in constructing T–S fuzzy models. In the literature, various methods have been developed to determine the number of fuzzy rules and the position and distribution of membership functions (Barada and Singh 1998; Angelov and Filev 2004). Usually, if the T–S fuzzy model is constructed from linearizing a nonlinear plant at N different operating points represented by $\bar{\xi}^i = [\bar{\xi}_1^i, \bar{\xi}_2^i, \ldots, \bar{\xi}_L^i]$, $i = 1, 2, \ldots, N$, there will be N rules and the ith membership function for the premise variable ξ_j, $j = 1, 2, \ldots, L$, can be designed as a triangular or Gaussian function with its maximum placed at the corresponding ith operating point $\bar{\xi}_j^i$. The neighboring membership functions for ξ_j should overlap each other with some degree and the combination of all the membership functions should cover the whole universe of discourse of ξ_j, that is, there is at least one positive membership degree for any allowed value of ξ_j. If the T–S fuzzy model is identified from input–output data, the number of fuzzy rules and the centers of membership functions can be obtained by fuzzy clustering methods such as fuzzy c-means and subtractive clustering algorithms (Barada and Singh 1998).

Remark 7.2 Another important technical issue is the robustness of feedback control system properties with respect to modeling errors in a system model used for control design. When applied to a realistic system, the system models (7.13) and even (7.17) are subject to certain modeling error $\Delta(y(\cdot), u(\cdot); t)$, especially for the fuzzy system model (7.13) which involves approximations to a regular nonlinear system. For example, in the presence of modeling errors, the system fuzzy model (7.13) may become

$$\begin{aligned} y(t+d) &= \sum_{i=1}^{N} \mu_i \alpha_i(z^{-1})[y](t) + \sum_{i=1}^{N} \mu_i \beta_i(z^{-1})[u](t) \\ &+\Delta(y(t), y(t-1), \ldots, u(t-1), u(t-2), \ldots; t). \end{aligned} \tag{7.15}$$

It is well-known that for a regular system model with modeling errors Δ, the robustness issue can be dealt with by robust adaptive control laws (Ioannou and Sun 1996; Tao 2003). We note that for adaptive control of the fuzzy system model (7.13), the robustness issue can be addressed, in a similar way to that for regular (non-fuzzy) systems, by using robust adaptive control designs (e.g., such a design is given in Feng 2010). More specifically, for $\Delta(y(t), y(t-1), \ldots, u(t-1), u(t-2), \ldots, t) \in L^2$, a normal adaptive control law is robust with respect to such Δ; for $\Delta(y(t), y(t-1), \ldots, u(t-1), u(t-2), \ldots, t) \in L^\infty$, an adaptive law with an additional modification term is able to ensure adaptive control system robustness; and for $\Delta(y(t), y(t-1), \ldots, u(t-1), u(t-2), \ldots, t)$ whose boundedness cannot be guaranteed a priori, an adaptive law with normalization and robustification can ensure closed-loop system boundedness stability provided the normalized $\Delta(y(t), y(t-1), \ldots, u(t-1), u(t-2), \ldots, t)$ is small (Tao 2003). □

Control objective. The control objective is to find a control signal $u(t)$ for the system (7.13) with unknown parameters $\alpha_{i,0}, \alpha_{i,1}, \ldots, \alpha_{i,n}, \beta_{i,0}, \beta_{i,1}, \ldots, \beta_{i,n-d}$, $i = 1, \ldots, N$, to ensure closed-loop signal boundedness and asymptotic tracking of a bounded reference output $y_m(t)$ by the system output $y(t)$, under the assumptions:

Assumption 7.1 The fuzzy system (7.13) is minimum phase.

Assumption 7.2 $\sum_{i=1}^{N} \mu_i(\xi(t))\beta_{i,0} \neq 0$, for all $t \geq 0$.

The exact definition of a minimum phase T–S fuzzy system (7.13) is to be defined later on. The uncertainty of the parameters $\alpha_{i,0}, \alpha_{i,1}, \ldots, \alpha_{i,n}, \beta_{i,0}, \beta_{i,1}, \ldots, \beta_{i,n-d}, i = 1, \ldots, N$, is equivalent to that of the parameters $a_{i,1}, \ldots, a_{i,n}, b_{i,0}, b_{i,1}, \ldots, b_{i,n-d}$ of the local fuzzy systems (7.4). From (7.12), we have $\beta_{i,0} = b_{i,0} \neq 0, i = 1, 2, \ldots, N$. Since $b_{i,0}$, $i = 1, 2, \ldots, N$, are the (high-frequency) gains of the linear subsystems which approximate the original nonlinear system, their signs can be practically assumed to be known. Without loss of generality, we assume $\beta_{i,0} = b_{i,0} > 0$, $i = 1, 2, \ldots, N$. Under this practical assumption, due to the properties of μ_i in (7.14): $\mu_i(\xi) \geq 0$ and $\sum_{i=1}^{N} \mu_i(\xi) = 1$, we actually do have Assumption 7.2 satisfied.

However, for adaptive control, we will need to have the following condition satisfied:

$$\sum_{i=1}^{N} \mu_i(\xi(t))\hat{\beta}_{i,0} \neq 0, \ \forall t \geq 0, \tag{7.16}$$

where $\hat{\beta}_{i,0}, i = 1, 2, \ldots, N$, are the estimates of $\beta_{i,0}$. This condition can be ensured by using parameter projection (Ioannou and Sun 1996) on the parameter estimates $\hat{\beta}_{i,0}, i = 1, 2, \ldots, N$, using the knowledge of the positive upper and lower bounds of $\beta_{i,0}, i = 1, 2, \ldots, N$.

Minimum phase condition of a regular LTI system. A regular linear time-invariant (LTI) system

$$A(z^{-1})[y](t) = z^{-d}\bar{B}(z^{-1})[u](t) \tag{7.17}$$

with $A(z^{-1}) = 1 + a_1 z^{-1} + a_2 z^{-2} \cdots + a_n z^{-n}$ and $\bar{B}(z^{-1}) = b_0 + b_1 z^{-1} + \cdots + b_{n-d} z^{-n+d}$ is minimum phase if all zeros of $\bar{B}(z^{-1})$ are in $|z| < 1$. For the equivalent system expression

$$\begin{aligned} &(z^n + a_1 z^{n-1} + \cdots + a_n)[y](t) \\ &= (b_0 z^n + b_1 z^{n-1} + \cdots + b_{n-d} z^d)[u](t-d), \end{aligned} \tag{7.18}$$

this condition implies that

$$|u(t-d)| \leq c_1|y(t)| + c_2 \sum_{\tau=0}^{t-1} \lambda^{t-\tau-1}|y(\tau)|, \ t \geq d, \tag{7.19}$$

for some constants $c_1 > 0$, $c_2 > 0$, and $\lambda \in (0, 1)$.

Minimum phase SISO fuzzy system. Based on this analogue, we propose the following minimum phase definition for the global SISO fuzzy system model (7.13).

Definition 7.1 The fuzzy system (7.13) is minimum phase if the condition (7.19) is satisfied.

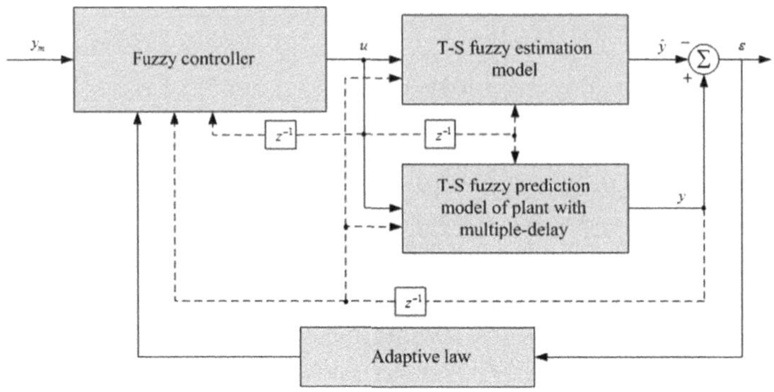

Fig. 7.1 Adaptive control structure for T–S fuzzy systems with general delay. © [2012] IEEE. Reprinted, with permission, from Qi et al. (2012a)

Unlike the case with a regular (LTI) system (7.17) whose minimum phase property can be checked using the knowledge of the zeros of $\bar{B}(z^{-1})$, the fuzzy system (7.13) is nonlinear and time-varying in nature and its zeros cannot be simply defined (they are only partially related to the zeros of each $\bar{B}_i(z^{-1})$ but largely related to $\mu_i(t)$ and their combined effect). Moreover, after fuzzy modeling, the explicit role of the polynomials $\bar{B}_i(z^{-1})$ of the original subsystems (7.4) in contributing to the minimum phase property of the fuzzy system (7.13) becomes less clear.

In the following section, we design a fuzzy adaptive controller for a minimum phase fuzzy system which can guarantee boundedness of all the closed-loop signals and the convergence of the tracking error based on the d-step prediction fuzzy system model derived in Sect. 7.2.1 and the minimum phase fuzzy system definition given in Definition 7.1.

The layout of the approach is presented in Fig. 7.1. After the above derivation the T–S prediction model (which captures the system order and delay information and gives a reasonable approximation of a nonlinear dynamic system), a parameterized model of the T–S system will be derived, an adaptive law will be developed to update the estimates of the unknown system parameters, and an adaptive controller will be constructed based on the estimated T–S model.

7.2.2 Nominal Controller

If the system parameters are all known, the control problem can be solved by the following nominal prediction-based control scheme .

The global T–S system (7.13), based on the local fuzzy system (7.4), can be written as

$$y(t+d) = \sum_{i=1}^{N} \mu_i \alpha_i(z^{-1})[y](t) + \sum_{i=1}^{N} \mu_i \bar{\beta}_i(z^{-1})[u](t) + \sum_{i=1}^{N} \mu_i \beta_{i,0} u(t) \quad (7.20)$$

where $\bar{\beta}_i(z^{-1}) = \beta_{i,1} z^{-1} + \cdots + \beta_{i,n-1} z^{-n+1}$. If those local system parameters $a_{i,1}, \ldots, a_{i,n}, b_{i,0}, \ldots, b_{i,\,n-d}$, $i = 1, 2, \ldots, N$, are known, one can calculate the parameters $\alpha_{i,0}, \alpha_{i,1}, \ldots, \alpha_{i,\,n-1}, \beta_{i,0}, \beta_{i,1}, \ldots, \beta_{i,\,n-1}$ of $\alpha_i(z^{-1})$, and $\beta_i(z^{-1})$, $i = 1, 2, \ldots, N$, from (7.10). Then, for a given bounded reference output signal $y_m(t)$ to be tracked by the system output $y(t)$, the fuzzy control law for the global system (7.13) can be obtained as

$$\begin{aligned} u(t) = \frac{1}{\sum_{i=1}^{N} \mu_i \beta_{i,0}} \Bigg(& -\sum_{i=1}^{N} \mu_i \alpha_i(z^{-1})[y](t) \\ & -\sum_{i=1}^{N} \mu_i \bar{\beta}_i(z^{-1})[u](t) + y_m(t+d) \Bigg) \end{aligned} \quad (7.21)$$

which, when applied to the system (7.13), brings $y(t+d)$ to $y_m(t+d)$ in one step and leads the closed-loop system to

$$y(t) = y_m(t), \quad t > d, \quad (7.22)$$

$$\sum_{i=1}^{N} \mu_i \beta_i(z^{-1})[u](t) = y_m(t+d) - \sum_{i=1}^{N} \mu_i \alpha_i(z^{-1})[y_m](t), \quad t \geq d+n. \quad (7.23)$$

Hence, $y(t)$ is bounded, and with Assumption 7.1, $u(t)$ is bounded.

This nominal control law provides the basic controller structure which can be parameterized for parameter adaptation when the system parameters are unknown.

7.2.3 Adaptive Control Scheme

To estimate the system parameters, we need to develop a parameterized model. With the knowledge of n and d, the fuzzy system (7.13) can be expressed as

$$y(t+d) = \theta^T \phi(t), \quad (7.24)$$

where, for $i = 1, 2, \ldots, N$,

$$\begin{aligned} \phi(t) &= [\phi_1^T(t), \phi_2^T(t), \ldots, \phi_N^T(t)]^T \\ \theta &= [\theta_1^T, \theta_2^T, \ldots, \theta_N^T]^T \\ \phi_i(t) &= [\mu_i y(t), \mu_i y(t-1), \ldots, \mu_i y(t-n+1), \end{aligned} \quad (7.25)$$

$$\mu_i u(t), \mu_i u(t-1), \ldots, \mu_i u(t-n+1)]^T$$
$$\theta_i = [\alpha_{i,0}, \alpha_{i,1}, \ldots, \alpha_{i,n-1}, \beta_{i,0}, \beta_{i,1}, \ldots, \beta_{i,n-1}]^T.$$

The expression (7.24), with θ unknown and $\phi(t)$ known, is a regression form with a linear parametrization for which many parameter adaptation algorithms can be adopted to estimate these unknown parameters in θ.

Parameter adaptive law. As a choice, the following parameter adaptive law is employed to obtain the estimate $\hat{\theta}(t)$ of θ:

$$\hat{\theta}(t) = \hat{\theta}(t-1) + \frac{\gamma(t)\phi(t-d)\varepsilon(t)}{c + \phi^T(t-d)\phi(t-d)}, \tag{7.26}$$

where $\gamma(t) \in (\gamma_0, 2-\gamma_0)$ is an adaptation gain for some constant $\gamma_0 \in (0, 1)$, $c > 0$ is a small design parameter, and the estimation error is

$$\varepsilon(t) = y(t) - \hat{\theta}^T(t-1)\phi(t-d) \tag{7.27}$$

with an initial estimate $\hat{\theta}(0)$ (chosen to be as close as possible to θ).

For the parameter adaptive law (7.26), the following results have been established (Qi et al. 2012a).

Lemma 7.1 *The parameter adaptive law (7.26), for the fuzzy system (7.24), has the following properties:*

(i) $\|\hat{\theta}(t) - \theta\| \le \|\hat{\theta}(t-1) - \theta\| \le \|\hat{\theta}(0) - \theta\|$, *for the l^2 (Euclidean) norm* $\|\cdot\|$;
(ii) $\frac{\varepsilon(t)}{\sqrt{c+\phi^T(t-d)\phi(t-d)}} \in L^2$;
(iii) $\lim_{t\to\infty} \frac{\varepsilon(t)}{\sqrt{c+\phi^T(t-d)\phi(t-d)}} = 0$;
(iv) $\|\hat{\theta}(t) - \hat{\theta}(t-t_1)\| \in L^2$; *and*
(vi) $\lim_{t\to\infty} \|\hat{\theta}(t) - \hat{\theta}(t-t_1)\| = 0$, *for any finite* $t_1 > 0$.

Similar to that in Goodwin and Sin (1984), the proof of this lemma follows from the linear error equation $\varepsilon(t) = -(\hat{\theta}(t-1) - \theta)^T\phi(t-d)$ (as from (7.24) and (7.27)) and the adaptive law (7.26), which lead to the squared error norm equation

$$\begin{aligned} &\|\hat{\theta}(t) - \theta\|^2 - \|\hat{\theta}(t-1) - \theta\|^2 \\ &= -\gamma(t)\left[2 - \frac{\gamma(t)\phi^T(t-d)\phi(t-d)}{c + \phi^T(t-d)\phi(t-d)}\right] \\ &\quad \cdot \frac{\varepsilon^2(t)}{c + \phi^T(t-d)\phi(t-d)}, \end{aligned} \tag{7.28}$$

where

$$\gamma(t)\left[2-\frac{\gamma(t)\phi^T(t-d)\phi(t-d)}{c+\phi^T(t-d)\phi(t-d)}\right]=\gamma(t)\left[2-\gamma(t)+\frac{\gamma(t)c}{c+\phi^T(t-d)\phi(t-d)}\right]$$
$$\geq \gamma_0(2-\gamma_0)>0. \tag{7.29}$$

Hence, we have the desired inequality

$$\|\hat{\theta}(t)-\theta\|^2-\|\hat{\theta}(t-1)-\theta\|^2$$
$$\leq -\gamma_0(2-\gamma_0)\frac{\varepsilon^2(t)}{c+\phi^T(t-d)\phi(t-d)} \tag{7.30}$$

from which the lemma's properties can be readily derived.

This adaptive law generates online estimate $\hat{\theta}(t)$ of the unknown parameter θ, with desired stability and L_2 properties, to be used for implementing an adaptive control law.

Adaptive control law. For the system (7.4) with unknown parameters, one may choose the following adaptive fuzzy control law based on the certainty equivalence principle:

$$R^i: \text{ IF } \xi_1 \text{ is } F_1^i \text{ and } \xi_2 \text{ is } F_2^i \text{ and} \ldots \text{and } \xi_L \text{ is } F_L^i$$
$$\text{THEN } u(t)=\frac{1}{\sum_{i=1}^N \mu_i\hat{\beta}_{i,0}}\Big(-\hat{\alpha}_i(z^{-1})[y](t)-\widehat{\bar{\beta}}_i(z^{-1})[u](t)$$
$$+y_m(t+d)\Big), \tag{7.31}$$

where $\hat{\alpha}_i(z^{-1})=\hat{\alpha}_{i,0}+\hat{\alpha}_{i,1}z^{-1}+\cdots+\hat{\alpha}_{i,n-1}z^{-n+1}$ and $\widehat{\bar{\beta}}_i(z^{-1})=\hat{\beta}_{i,1}z^{-1}+\cdots+\hat{\beta}_{i,n-1}z^{-n+1}$ are the estimates of $\alpha_i(z^{-1})$ and $\bar{\beta}_i(z^{-1})$ in (7.11) and (7.12), with parameter estimates $\hat{\alpha}_{i,j}$ and $\hat{\beta}_{i,j}$. We then choose the global fuzzy control law as

$$u(t)=\frac{1}{\sum_{i=1}^N \mu_i\hat{\beta}_{i,0}}\left(-\sum_{i=1}^N \mu_i\hat{\alpha}_i(z^{-1})[y](t)\right.$$
$$\left.-\sum_{i=1}^N \mu_i\widehat{\bar{\beta}}_i(z^{-1})[u](t)+y_m(t+d)\right). \tag{7.32}$$

For this adaptive control law, parameter projection (Tao 2003) may be used for the parameter adaptive law (7.26) to ensure that $\sum_{i=1}^N \mu_i\hat{\beta}_{i,0}>\beta_0$ for some constant $\beta_0>0$, and all $t\geq 0$ (see the discussion on Assumption 7.2 in Sect. 7.2.1).

Stability Analysis. We now show that the adaptive control system has desired stability and tracking properties. Substituting (7.32) into (7.20), we obtain the closed-loop system as

$$
\begin{aligned}
y(t+d) = & \sum_{i=1}^{N} \mu_i \alpha_i(z^{-1})[y](t) - \sum_{i=1}^{N} \mu_i \hat{\alpha}_i(z^{-1})[y](t) \\
& + \sum_{i=1}^{N} \mu_i \bar{\beta}_i(z^{-1})[u](t) - \sum_{i=1}^{N} \mu_i \widehat{\bar{\beta}}_i(z^{-1})[u](t) \\
& + \left(\sum_{i=1}^{N} \mu_i \beta_{i,0} - \sum_{i=1}^{N} \mu_i \hat{\beta}_{i,0} \right) u(t) + y_m(t+d) \\
= & \; \theta^T \phi(t) - \hat{\theta}^T(t)\phi(t) + y_m(t+d).
\end{aligned} \tag{7.33}
$$

Defining the output tracking error $e(t) = y(t) - y_m(t)$ and $\tilde{\theta}(t) = \hat{\theta}(t) - \theta$, we obtain

$$
e(t+d) = -\tilde{\theta}^T(t)\phi(t). \tag{7.34}
$$

First, an upper bounding property for the regressor $\phi(t)$ is established in the following lemma (Qi et al. 2012a).

Lemma 7.2 *Under Assumption 7.1, the regressor $\phi(t)$ defined in (7.25) satisfies*

$$
\|\phi(t-d)\| \le \rho_1 + \rho_2 \max_{0 \le \tau \le t} |e(\tau)| \tag{7.35}
$$

for some positive constants ρ_1 and ρ_2.

Proof From the definition of $\phi(t)$ in (7.25) and μ_i in (7.14), we have

$$
\|\phi(t)\| \le \kappa_1 \|\psi(t)\|, \tag{7.36}
$$

for some constant $\kappa_1 > 0$, where $\psi(t) = [y(t), y(t-1), \ldots, y(t-n+1), u(t), u(t-1), \ldots, u(t-n+1)]^T$, satisfying

$$
\|\psi(t)\| \le \kappa_2 \max_{t-n+1 \le \tau \le t} |e(\tau)| + \kappa_3 \max_{t-n+1 \le \tau \le t} |u(\tau)| + \kappa_4, \tag{7.37}
$$

with κ_i, $i = 1, 2, 3$, being some positive constants.

With Assumption 7.1 and Definition 7.1, we obtain from (7.13) that

$$
\begin{aligned}
|u(t)| & \le \kappa_5 |e(t+d)| + \kappa_6 \sum_{\tau=0}^{t-1} \lambda^{t-\tau-1} |e(\tau+d)| + \kappa_7 \\
& \le \kappa_8 \max_{0 \le \tau \le t} |e(\tau+d)| + \kappa_7,
\end{aligned} \tag{7.38}
$$

where κ_i, $i = 5, 6, 7, 8$, are some positive constants.

Finally, using (7.36)–(7.38), we obtain

$$
\|\phi(t)\| \le \rho_1 + \rho_2 \max_{0 \le \tau \le t} |e(\tau+d)|, \tag{7.39}
$$

where ρ_1 and ρ_2 are some positive constants. ∇

Based on the results of Lemmas 7.1 and 7.2, we have the desired closed-loop system properties (Qi et al. 2012a).

Theorem 7.1 *All signals in the closed-loop system, with the plant* (7.13) *satisfying Assumptions 7.1 and 7.2, the controller (7.32) and the adaptive law (7.26), are bounded, and* $\lim_{t\to\infty}(y(t) - y_m(t)) = 0$.

Proof From (7.24) and (7.27), we have $\varepsilon(t) = -(\hat{\theta}(t-1) - \theta)^T \phi(t-d)$ and from (7.95), we have

$$\begin{aligned} e(t) &= \bar{\varepsilon}(t)\sqrt{c + \phi^T(t-d)\phi(t-d)} \\ &\quad -(\hat{\theta}(t-d) - \hat{\theta}(t-1))^T \bar{\phi}(t-d)\sqrt{c + \phi^T(t-d)\phi(t-d)}, \end{aligned} \tag{7.40}$$

where $\bar{\varepsilon}(t) = \frac{\varepsilon(t)}{\sqrt{c+\phi^T(t-d)\phi(t-d)}}$ and $\bar{\phi}(t-d) = \frac{\phi(t-d)}{\sqrt{c+\phi^T(t-d)\phi(t-d)}}$, with $\bar{\varepsilon}(t) \in L_2 \cap L_\infty$, $\hat{\theta}(t-d) - \hat{\theta}(t-1) \in L_2 \cap L_\infty$, and $\|\bar{\phi}(t-d)\| \le 1$.

With the inequality: $\sqrt{c + \phi^T(t-d)\phi(t-d)} \le \sqrt{c} + \|\phi^T(t-d)\|$, we express $e(t)$ in (7.40) as

$$|e(t)| \le c_1 + |\bar{\varepsilon}(t)|\, \|\phi^T(t-d)\| + \|\hat{\theta}(t-d) - \hat{\theta}(t-1)\|\, \|\phi^T(t-d)\|, \tag{7.41}$$

for some constant $c_1 > 0$. Using Lemma 7.2: $\|\phi(t-d)\| \le \rho_1 + \rho_2 \max_{0\le\tau\le t} |e(\tau)|$, we obtain

$$\begin{aligned} |e(t)| &\le c_2 + c_3|\bar{\varepsilon}(t)| \max_{0\le\tau\le t} |e(\tau)| \\ &\quad + c_4\|\hat{\theta}(t-d) - \hat{\theta}(t-1)\| \max_{0\le\tau\le t} |e(\tau)|, \end{aligned} \tag{7.42}$$

for some constants $c_i > 0$, $i = 2, 3, 4$.

From Lemma 7.1, we have that $\lim_{t\to\infty} \bar{\varepsilon}(t) = 0$ and $\lim_{t\to\infty} \|\hat{\theta}(t) - \hat{\theta}(t-d)\| = 0$, and with these results, it follows from (7.42) that $e(t)$ is bounded, which implies $y(t)$ is bounded, and in turn from the system's minimum phase property that $u(t)$ is bounded. Hence, all signals in the closed-loop system are bounded, based on which, from (7.40) in which $\bar{\varepsilon}(t) \in L_2$ and $\hat{\theta}(t-d) - \hat{\theta}(t-1) \in L_2$, we have that $e(t) \in L_2$ so that $\lim_{t\to\infty} e(t) = 0$. ∇

Remark 7.3 The closed-loop stability and asymptotic output tracking properties given in Theorem 1 are the basic adaptive system properties only for the case when there are no modeling and approximation errors in the system (7.13). In the presence of such errors as described in Remark 7.2, the adaptive law (7.26) needs to be modified with a robustifying signal such as dead-zone, σ-modification, and parameter projection (Ioannou and Sun 1996; Tao 2003), for robust parameter adaptation. While the closed-loop signal boundedness can be ensured globally or locally for small normalized modeling and approximation errors, the tracking error $e(t) = y(t) - y_m(t)$

can be shown to be bounded by normalized modeling and approximation errors in a mean sense. □

7.2.4 Simulation Study

In this section, we show the simulation results from a flexible-joint manipulator system, with details of control design and system responses.

Simulation system. Consider a flexible-joint manipulator system described by the following equations (Ghorbel et al. 1989):

$$\begin{aligned} I\ddot{q}_1 + Mgl\sin(q_1) + k(q_1 - q_2) &= 0 \\ J\ddot{q}_2 + B\dot{q}_2 - k(q_1 - q_2) &= u, \end{aligned} \tag{7.43}$$

where q_1 and q_2 are the angle positions (in radius) of the link and the rotor, M denotes the mass, I and J are the link and rotor inertia, respectively, k is the elasticity constant of the joint spring, M and l represent the mass and the position of the center of the gravity of the link, B is the rotor friction constant, and u is the force (in Newtons) applied to the link. For simulation, the parameters are set as $Mgl = 0.8$ N · m, $I = 0.031$ kg · m^2, $J = 0.004$ kg · m^2, $k = 31$ N · m/rad, and $B = 0.007$ N · m · s/rad.

The nonlinear system (7.43) can be written into the following state-space form:

$$\begin{cases} \dot{x}_1 = x_2 \\ \dot{x}_2 = -\frac{Mgl}{I}\sin(x_1) - \frac{k}{I}(x_1 - x_3) \\ \dot{x}_3 = x_4 \\ \dot{x}_4 = -\frac{B}{J}x_3 + \frac{k}{J}(x_1 - x_3) + \frac{1}{J}u \\ y = x_1 \end{cases} \tag{7.44}$$

by selecting the state vector: $x = [x_1, x_2, x_3, x_4]^T = [q_1, \dot{q}_1, q_2, \dot{q}_2]^T$ and the output: $y = q_1$.

In this example, a T–S fuzzy model is constructed by linearizing the nonlinear plant (7.43) at five equilibrium points for $x_1 \in (-\pi/3, \pi/3)$ and $x_3 \in (-\pi/3, \pi/3)$. The ith rule of the T–S fuzzy model has the following form:

$$\begin{aligned} R^i:\ &\text{IF}\ \ x_1\ \ \text{is}\ \ F_1^i\ \ \text{and}\ \ x_3\ \ \text{is}\ \ F_2^i, \\ &\text{THEN}\ \ \dot{x} = A_i x + B_i u, \end{aligned} \tag{7.45}$$

where $i = 1, 2, \ldots, 5$, A_i and B_i are the local state-space matrices corresponding to the ith linearized model, and F_1^i and F_2^i are fuzzy sets for the two premise variables x_1 and x_3 in the ith rule. The membership functions characterizing the fuzzy sets F_1^i and F_2^i are selected as Gaussian functions, which have their maximum values at their corresponding linearizing points and decrease toward their neighbor linearizing points, as shown in Fig. 7.2.

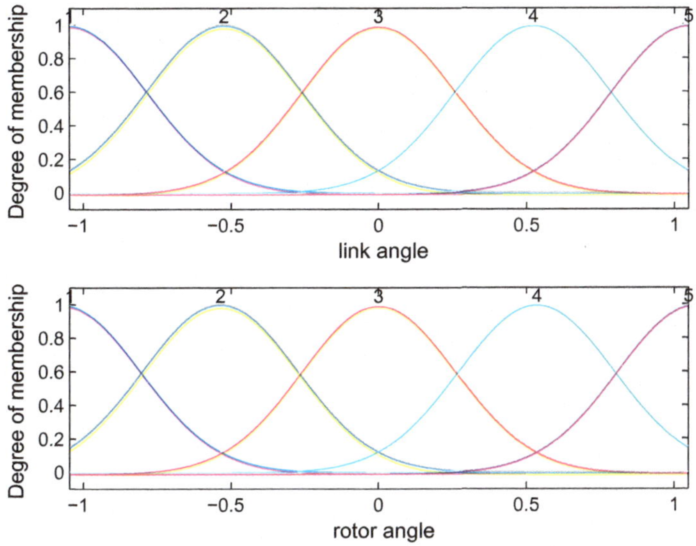

Fig. 7.2 Membership functions for link angle x_1 and rotor angle x_2. © [2012] IEEE. Reprinted, with permission, from Qi et al. (2012a)

Selecting the sampling time as $T = 0.6\,\mathrm{s}$ and using a zero-order-hold discretization method, a discrete-time T–S fuzzy model can be obtained from (7.45) with its ith rule as

$$\begin{aligned} R^i :\ & \text{IF}\ \ x_1(t)\ \ \text{is}\ \ F_1^i\ \ \text{and}\ \ x_3(t)\ \ \text{is}\ \ F_2^i, \\ & \text{THEN}\ \ x(t+1) = A_{di}x(t) + B_{di}u(t), \end{aligned} \tag{7.46}$$

where (A_{di}, B_{di}), $i = 1, 2, \ldots, 5$, can be found in Qi et al. (2012a).

The state-space model (7.46) can be then expressed in the input–output T–S fuzzy model:

$$\begin{aligned} R^i :\ & \text{IF}\ \ x_1(t)\ \ \text{is}\ \ F_1^i\ \ \text{and}\ \ x_3(t)\ \ \text{is}\ \ F_2^i, \\ & \text{THEN}\ y(t) + a_{i,1}y(t-1) + \cdots + a_{i,4}y(t-4) \\ & = b_{i,0}u(t-1) + \cdots + b_{i,3}u(t-4) \end{aligned}$$

with output $y(t) = x_1(t)$. While this model with $d = 1$ itself can be simulated with our adaptive control scheme, in out study, we introduce one additional delay at the input to model the finite computation time delay of digital control and actuator delay of operation (Franklin et al. 1998), to obtain the following discrete-time T–S fuzzy model with delay $d = 2$:

$$
\begin{aligned}
R^i: &\ \text{IF}\ \ x_1(t)\ \ \text{is}\ \ F_1^i\ \ \text{and}\ \ x_3(t)\ \ \text{is}\ \ F_2^i,\\
&\ \text{THEN}\ y(t) + a_{i,1}y(t-1) + \cdots + a_{i,4}y(t-4)\\
&\ = b_{i,0}u(t-2) + \cdots + b_{i,3}u(t-5).
\end{aligned}
$$

With Proposition 7.2, we obtain the following global 2-step prediction fuzzy system model:

$$
y(t+2) = \sum_{i=1}^{5} \mu_i \alpha_i(z^{-1})[y](t) + \sum_{i=1}^{5} \mu_i \beta_i(z^{-1})[u](t),
$$

where

$$
\begin{aligned}
\alpha_i(z^{-1}) &= \alpha_{i,0} + \alpha_{i,1}z^{-1} + \alpha_{i,2}z^{-2} + \alpha_{i,3}z^{-3}\\
&= a_{i,1}^2 - a_{i,2} + (a_{i,1}a_{i,2} - a_{i,3})z^{-1}\\
&\quad + (a_{i,1}a_{i,3} - a_{i,4})z^{-2} + a_{i,1}a_{i,4}z^{-3}\\
\beta_i(z^{-1}) &= \beta_{i,0} + \beta_{i,1}z^{-1} + \beta_{i,2}z^{-2} + \beta_{i,3}z^{-3} + \beta_{i,4}z^{-4}\\
&= b_{i,0} + (b_{i,1} - a_{i,1}b_{i,0})z^{-1} + (b_{i,2} - a_{i,1}b_{i,1})z^{-2}\\
&\quad + (b_{i,3} - a_{i,1}b_{i,2})z^{-3} - a_{i,1}b_{i,3}z^{-4}.
\end{aligned}
$$

Simulation results. For simulation, we chose $\gamma(t) = 1$, $c = 0.01$, $y_m(t) = 0.5\sin(0.2t)$, and the initial parameter values $\hat{\theta}(0) = 65\% \times \theta$. The system out-

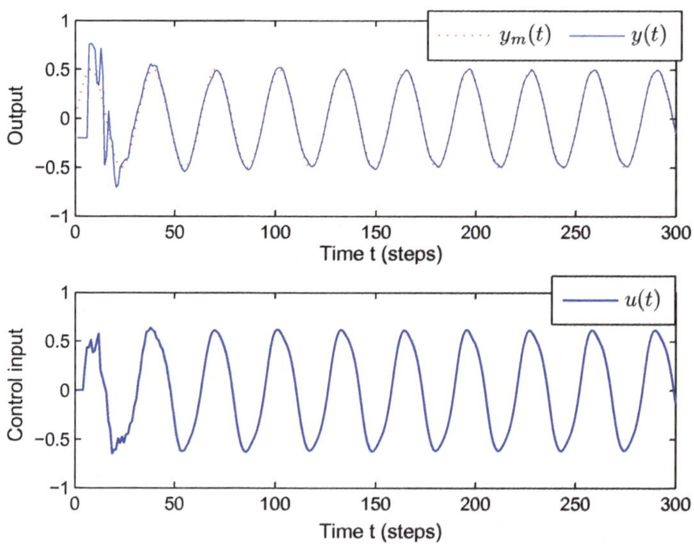

Fig. 7.3 Adaptive system response and control signal. © [2012] IEEE. Reprinted, with permission, from Qi et al. (2012a)

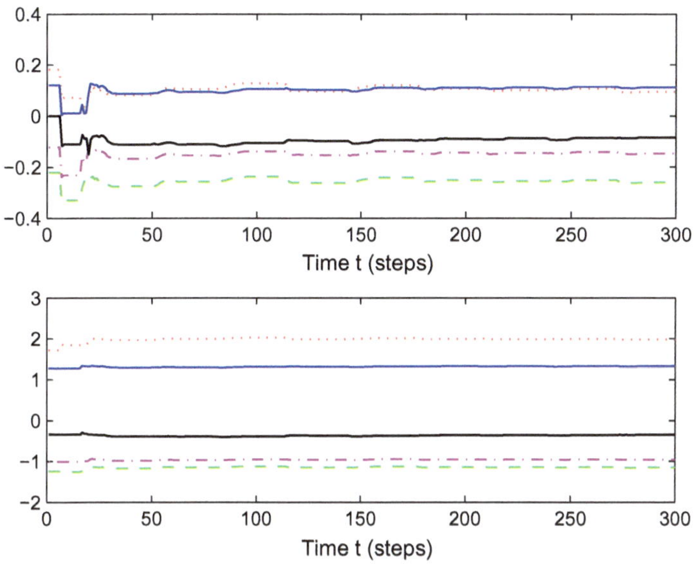

Fig. 7.4 Parameter adaptation of Rule 2. © [2012] IEEE. Reprinted, with permission, from Qi et al. (2012a)

put $y(t)$ and the reference $y_m(t)$, and the control signal $u(t)$ are shown in Fig. 7.3. It can be observed from Figs. 7.3 and 7.4 that with the designed adaptive fuzzy control law, the system output tracks the reference accurately after some initial transient and the parameters tend to converge to some constant values, which verifies Lemma 7.1 and Theorem 7.1.

7.3 Approach II: Design Based on Nonlinear Prediction Model

In this section,[2] we derive a prediction model of the global discrete-time input–output T–S fuzzy system with multi-delays and employ it for adaptive control design in the presence of parameter uncertainties.

Introduce the polynomials in z^{-1}:

$$\bar{A}_i(z^{-1}) = -a_{i1}z^{-1} - \cdots - a_{in}z^{-n}, \tag{7.47}$$

$$\bar{B}_i(z^{-1}) = b_{i0} + b_{i1}z^{-1} + \cdots + b_{i\,n-d}z^{-n+d}. \tag{7.48}$$

[2]Parts of Sect. 7.3 are reproduced from Qi et al. (2012b) by permission of John Wiley & Sons Ltd.

We express the local linear system (7.4) as

$$y(t) = \bar{A}_i(z^{-1})[y](t) + z^{-d}\bar{B}_i(z^{-1})[u](t). \tag{7.49}$$

Applying the z^d operator on both sides of (7.49) yields

$$y(t+d) = \bar{A}_i(z^{-1})[y](t+d) + \bar{B}_i(z^{-1})[u](t). \tag{7.50}$$

We then use the local models in (7.50), based on a standard fuzzy system modeling technique of *singleton fuzzification*, *product inference*, and *weighted average*, to form a global T–S fuzzy system model as

$$y(t+d) = \sum_{i=1}^{N} \mu_i(\xi(t))\bar{A}_i(z^{-1})[y](t+d) + \sum_{i=1}^{N} \mu_i(\xi(t))\bar{B}_i(z^{-1})[u](t), \tag{7.51}$$

where μ_i is defined in (7.14).

Control objective. The adaptive control objective is to find a control signal $u(t)$ for the system (7.51) with unknown parameters, to ensure closed-loop signal boundedness and asymptotic tracking of a bounded reference output $y_m(t)$ by the system output $y(t)$, under the following assumptions:

Assumption 7.3 The fuzzy system (7.51) is minimum phase.

Assumption 7.4 $\sum_{i=1}^{N} \mu_i(\xi(t))b_{i0} \neq 0$, for all $t \geq 0$.

In this section, we conduct a study for a general global discrete-time T–S fuzzy system, to derive a prediction model for the global T–S fuzzy system (7.51) and use it for the design of an adaptive predictor for T–S fuzzy systems.

Prediction is to forecast the future values of the system output signal y using the current and past values of y and the system input signal u. A predictor for (7.51) with delay d is an equation for $y(t+d)$ in terms of $u(\tau)$ and $y(\tau)$ for $\tau \leq t$. An adaptive predictor uses the adaptive estimates of the unknown system parameters for prediction.

In the following sections, we use $\mu_i(t)$ to denote $\mu_i(\xi(t))$, $i = 1, 2, \ldots, N$, for the simplicity of expression.

7.3.1 *T–S Fuzzy Prediction Model via Nonlinear Prediction*

Since the model (7.51) contains $y(t+d-1)$, $y(t+d-2), \ldots, y(t+1)$, $y(t)$, $y(t-1), \ldots, y(t+d-n)$, of which $y(t+d-1)$, $y(t+d-2), \ldots, y(t+1)$ are not available at time t. We use (7.51) backward to derive their expressions in terms of available signals. By using $\mu(t)$ to denote $\mu(\xi(t))$ for simplicity, we can rewrite (7.51) as

$$y(t+d) = \sum_{i=1}^{N} \mu_i(t)\bar{A}_i(z^{-1})[y](t+d) + \sum_{i=1}^{N} \mu_i(t)\bar{B}_i(z^{-1})[u](t), \tag{7.52}$$

in which $\mu_i(t)\bar{A}_i(z^{-1})[y](t+d)$ contains $\mu_i(t)$, $y(t+d-1)$, $y(t+d-2), \ldots,$ $y(t+1)$, $y(t)$, $y(t-1), \ldots,$ $y(t+d-n)$, and $\mu_i(t)\bar{B}_i(z^{-1})[u](t)$ contains $\mu_i(t)$, $u(t)$, $u(t-1), \ldots,$ $u(t+d-n)$. Replacing t with $t-1$ in (7.52), we can express $y(t+d-1)$ as

$$\begin{aligned} y(t+d-1) = &\sum_{i=1}^{N} \mu_i(t-1)\bar{A}_i(z^{-1})[y](t+d-1) \\ &+ \sum_{i=1}^{N} \mu_i(t-1)\bar{B}_i(z^{-1})[u](t-1), \end{aligned} \tag{7.53}$$

in which $\mu_i(t-1)\bar{A}_i(z^{-1})[y](t+d-1)$ contains $\mu_i(t-1)$, $y(t+d-2)$, $y(t+d-3), \ldots,$ $y(t+1)$, $y(t)$, $y(t-1), \ldots,$ $y(t+d-n-1)$, and $\mu_i(t-1)\bar{B}_i(z^{-1})$ $[u](t-1)$ contains $\mu_i(t-1)$, $u(t-1)$, $u(t-2), \ldots,$ $u(t+d-n-1)$. Replacing t with $t-2$ in (7.52), we express $y(t+d-2)$ as

$$\begin{aligned} y(t+d-2) = &\sum_{i=1}^{N} \mu_i(t-2)\bar{A}_i(z^{-1})[y](t+d-2) \\ &+ \sum_{i=1}^{N} \mu_i(t-2)\bar{B}_i(z^{-1})[u](t-2), \end{aligned} \tag{7.54}$$

in which $\mu_i(t-2)\bar{A}_i(z^{-1})[y](t+d-2)$ contains $\mu_i(t-2)$, $y(t+d-3)$, $y(t+d-4), \ldots, y(t+1), y(t), y(t-1), \ldots, y(t+d-n-2)$, and $\mu_i(t-2)\bar{B}_i(z^{-1})[u]$ $(t-2)$ contains $\mu_i(t-2)$, $u(t-2)$, $u(t-3), \ldots,$ $u(t+d-n-2)$. Similarly, for $j = 3, 4, \ldots, d-2$, replacing t with $t-j$ in (7.52), we express $y(t+d-j)$ as

$$\begin{aligned} y(t+d-j) = &\sum_{i=1}^{N} \mu_i(t-j)\bar{A}_i(z^{-1})[y](t+d-j) \\ &+ \sum_{i=1}^{N} \mu_i(t-j)\bar{B}_i(z^{-1})[u](t-j), \end{aligned} \tag{7.55}$$

in which $\mu_i(t-j)\bar{A}_i(z^{-1})[y](t+d-j)$ contains $\mu_i(t-j)$, $y(t+d-j-1)$, $y(t+d-j-2), \ldots,$ $y(t+1)$, $y(t)$, $y(t-1), \ldots,$ $y(t+d-n-j)$, and $\mu_i(t-j)\bar{B}_i(z^{-1})[u](t-j)$ contains $\mu_i(t-j)$, $u(t-j)$, $u(t-j-1), \ldots, u(t+d-n-j)$. Finally, replacing t with $t-d+1$ in (7.52), we express $y(t+1)$ as

$$\begin{aligned} y(t+1) = & \sum_{i=1}^{N} \mu_i(t-d+1)\bar{A}_i(z^{-1})[y](t+1) \\ & + \sum_{i=1}^{N} \mu_i(t-d+1)\bar{B}_i(z^{-1})[u](t-d+1), \end{aligned} \tag{7.56}$$

in which $\mu_i(t-d+1)\bar{A}_i(z^{-1})[y](t+1)$ contains $\mu_i(t-d+1), y(t), y(t-1), \ldots, y(t-n+1)$, and $\mu_i(t-d+1)\bar{B}_i(z^{-1})[u](t-j)$ contains $\mu_i(t-d+1)$, $u(t-d+1), u(t-d), \ldots, u(t-n+1)$.

We now replace $y(t+d-1)$, $y(t+d-2), \ldots, y(t+1)$ in (7.51) by their expressions in (7.53), (7.54), ..., (7.56), to obtain an expression of $y(t+d)$ in terms of $\{\mu_i(t), \mu_i(t-1), \ldots, \mu_i(t-d+1)\}$, $\{y(t), y(t-1), \ldots, y(t+d-n), y(t+d-n-1), \ldots, y(t-n+1)\}$, $\{u(t), u(t-1), \ldots, u(t+d-n), u(t+d-n-1), \ldots, u(t-n+1)\}$, and the parameters a_{ij} and b_{ik}, $i = 1, 2, \ldots, N$, $j = 1, 2, \ldots, n$, $k = 0, 1, \ldots, n-d$. Such an expression can be denoted as

$$y(t+d) = f_y(\mu_i(\cdot), y(\cdot)) + f_u(\mu_i(\cdot), u(\cdot)) \tag{7.57}$$

for some functions

$$f_y(\mu_i(\cdot), y(\cdot)) = f_y(\mu_i(t), \ldots, \mu_i(t-d+1), y(t), \ldots, y(t-n+1)), \tag{7.58}$$

$$f_u(\mu_i(\cdot), u(\cdot)) = f_u(\mu_i(t), \ldots, \mu_i(t-d+1), u(t), \ldots, u(t-n+1)), \tag{7.59}$$

which are linear in the products of μ_i and y, or products of μ_i and u; in particular,

$$\begin{aligned} & f_u(\mu_i(t), \ldots, \mu_i(t-d+1), u(t), \ldots, u(t-n+1)) \\ = & \textstyle\sum_{i=1}^{N} \mu_i(t) b_{i0} u(t) + f_{u1}(\mu_i(t), \ldots, \mu_i(t-d+1), u(t-1), \ldots, u(t-n+1)) \end{aligned} \tag{7.60}$$

for some function $f_{u1}(\mu_i(t), \ldots, \mu_i(t-d+1), u(t-1), \ldots, u(t-n+1))$ independent of $u(t)$.

Thus, we have reached a general prediction equation (7.57) for the global T–S fuzzy system (7.51), which expresses $y(t+d)$ in terms of signals independent of $y(t+d-1), \ldots, y(t+1)$.

Remark 7.4 We can also choose a stable polynomial $P_m(z) = z^d + a_1 z^{d-1} + \cdots + a_{d-1} z + a_d$ and define $\bar{y}(t+d) = P_m(z)[y](t)$ as a signal to be predicted. Using the above procedure, we can derive the prediction equation for $\bar{y}(t+d)$:

$$\bar{y}(t+d) = \bar{f}_y(\mu_i(\cdot), y(\cdot)) + \bar{f}_u(\mu_i(\cdot), u(\cdot)), \tag{7.61}$$

for some functions $\bar{f}_y(\mu_i(\cdot), y(\cdot))$ and $\bar{f}_u(\mu_i(\cdot), u(\cdot))$ which have the same properties as that of $f_y(\mu_i(\cdot), y(\cdot))$ and $f_u(\mu_i(\cdot), u(\cdot))$ such as (7.60). □

7.3.2 Adaptive Predictor

To derive an adaptive version of the prediction equation, we parameterize (7.57) as

$$y(t+d) = \theta_y^T \phi_y(t) + \theta_{u0}^T \phi_{u0}(t) + \theta_{u1}^T \phi_{u1}(t), \tag{7.62}$$

where, in view of (7.60),

$$\theta_y^T \phi_y(t) = f_y(\mu_i(\cdot), y(\cdot)), \tag{7.63}$$

$$\theta_{u0}^T \phi_{u0}(t) = \sum_{i=1}^{N} \mu_i(t) b_{i0} u(t), \tag{7.64}$$

$$\theta_{u1}^T \phi_{u1}(t) = f_{u1}(\mu_i(\cdot), u(\cdot)), \tag{7.65}$$

for some parameter vectors $\theta_y \in R^{n_y}$, $\theta_{u0} \in R^N$, and $\theta_{u1} \in R^{n_{u1}}$, and vector signals $\phi_y(t)$, $\phi_{u0}(t)$, and $\phi_{u1}(t)$ whose components are the products of μ_i and y at various time instants, and products of μ_i and u, respectively, in particular,

$$\theta_{u0} = [b_{10}, b_{20}, \ldots, b_{N0}]^T \tag{7.66}$$
$$\phi_{u0} = [\mu_1(t)u(t), \mu_2(t)u(t), \ldots, \mu_N(t)u(t)]^T. \tag{7.67}$$

Remark 7.5 The vector functions $\phi_y(t)$ and $\phi_{u1}(t)$ in (7.62) (which may be of a high dimension and a complicated expression) are both linear functions of the products $\mu_i y$ and $\mu_i u$ of different combinations of μ_i, y, and u. Since $0 \le \mu_i \le 1$, the norms of the functions $\phi_y(t)$ and $\phi_{u1}(t)$ are bounded by the norms of their corresponding forms with $|y|$ and $|u|$ only. This is an important property useful for stability analysis. □

Define

$$\theta = [\theta_y^T, \theta_{u0}^T, \theta_{u1}^T]^T \in R^{n_\theta}, \tag{7.68}$$
$$\phi(t) = [\phi_y^T(t), \phi_{u0}^T(t), \phi_{u1}^T(t)]^T \in R^{n_\theta}, \tag{7.69}$$

where $n_\theta = n_y + N + n_{u1}$, then the Eq. (7.62) can be reformulated into

$$y(t+d) = \theta^T \phi(t), \tag{7.70}$$

or equivalently,

$$y(t) = \theta^T \phi(t-d). \tag{7.71}$$

Let $\hat{\theta}_y(t-1)$, $\hat{\theta}_{u0}(t-1)$, and $\hat{\theta}_{u1}(t-1)$ be the estimates of the unknown parameter vectors θ_y, θ_{u0}, and θ_{u1} at time $t-1$, and introduce the estimation error at time

t as

$$\begin{aligned}\varepsilon(t) &= y(t) - \hat{\theta}_y^T(t-1)\phi_y(t-d) - \hat{\theta}_{u0}^T(t-1)\phi_{u0}(t-d) - \hat{\theta}_{u1}^T(t-1)\phi_{u1}(t-d) \\ &= y(t) - \hat{\theta}^T(t-1)\phi(t-d),\end{aligned} \tag{7.72}$$

where

$$\hat{\theta}(t) = [\hat{\theta}_y^T(t), \hat{\theta}_{u0}^T(t), \hat{\theta}_{u1}^T(t)]^T \tag{7.73}$$

$$\phi(t-d) = [\phi_y^T(t-d), \phi_{u0}^T(t-d), \phi_{u1}^T(t-d)]^T. \tag{7.74}$$

Then, with (7.71), we have

$$\varepsilon(t) = -\tilde{\theta}^T(t-1)\phi(t-d), \tag{7.75}$$

where $\tilde{\theta}(t-1) = \hat{\theta}(t-1) - \theta$.

Parameter adaptive law. Based on the estimation error Eq. (7.72), the following adaptive law can be employed to update the parameter estimate $\hat{\theta}(t)$:

$$\hat{\theta}(t) = \hat{\theta}(t-1) + \frac{\gamma(t)\phi(t-d)\varepsilon(t)}{m^2(t)}, \tag{7.76}$$

where

$$m(t) = \sqrt{c + \phi^T(t-d)\phi(t-d)}, \tag{7.77}$$

and $\gamma(t) \in (\gamma_0, 2-\gamma_0)$ is an adaptation gain for some constant $\gamma_0 \in (0, 1)$, and $c > 0$ is a small design parameter.

To ensure $\sum_{i=1}^{N} \mu_i(\xi(t))\hat{b}_{i0}(t) > b_0$ to avoid the singularity problem in calculating the control signal, the parameter adaptive law (7.76) is modified by using parameter projection:

$$\hat{\theta}(t) = \hat{\theta}(t-1) + \frac{\gamma(t)\phi(t-d)\varepsilon(t)}{m^2(t)} + f(t), \tag{7.78}$$

where $f(t)$ is a projection function.

It is assumed that b_{i0} belongs to a known interval $[\underline{b}_{i0}, \bar{b}_{i0}]$, $i = 1, 2, \ldots, N$, and a base adaptation vector is defined as

$$g(t) = \frac{\gamma(t)\phi(t-d)\varepsilon(t)}{m^2(t)}. \tag{7.79}$$

Denoting the jth component of $\theta(t) \in R^{n_\theta}$, $f(t)$, $g(t)$ as $\theta_j(t)$, $f_j(t)$, $g_j(t)$, $j = 1, 2, \ldots, n_\theta$, with $b_{i0} \in [\underline{b}_{i0}, \bar{b}_{i0}]$, we have $\theta_j \in [\underline{b}_{i0}, \bar{b}_{i0}]$ for $j = n_y + i$. We choose the initial estimates for θ_j, $j = n_y + 1, n_y + 2, \ldots, n_y + N$ as

$$\hat{\theta}_j \in [\underline{b}_{i0}, \bar{b}_{i0}], \quad j = n_y + i, \quad i = 1, 2, \ldots, N, \tag{7.80}$$

and set the projection function components as

$$f_j(t) = 0, \quad \text{for } j \le n_y \text{ or } j > n_y + N \tag{7.81}$$

$$f_{n_y+i}(t) = \begin{cases} 0 & \text{if } \hat{\theta}_{n_y+i}(t) + g_{n_y+i}(t) \in [\underline{b}_{i0}, \bar{b}_{i0}] \\ \bar{b}_{i0} - \hat{\theta}_{n_y+i}(t) - g_{n_y+i}(t) & \text{if } \hat{\theta}_{n_y+i}(t) + g_{n_y+i}(t) > \bar{b}_{i0} \\ \underline{b}_{i0} - \hat{\theta}_{n_y+i}(t) - g_{n_y+i}(t) & \text{if } \hat{\theta}_{n_y+i}(t) + g_{n_y+i}(t) < \underline{b}_{i0}, \quad i = 1, 2, \ldots, N. \end{cases} \tag{7.82}$$

Stable adaptation. For this parameter estimation algorithm, we have the following properties (Qi et al. 2012b).

Lemma 7.3 *The parameter adaptive law (7.78), when applied to the fuzzy system (7.51), has the standard properties:*

(i) $\|\hat{\theta}(t) - \theta\| \le \|\hat{\theta}(t-1) - \theta\| \le \|\hat{\theta}(0) - \theta\|$;
(ii) $\frac{\varepsilon(t)}{m(t)} \in L^2$;
(iii) $\lim_{t\to\infty} \frac{\varepsilon(t)}{m(t)} = 0$;
(iv) $\|\hat{\theta}(t) - \hat{\theta}(t - t_1)\| \in L^2$; *and*
(vi) $\lim_{t\to\infty} \|\hat{\theta}(t) - \hat{\theta}(t - t_1)\| = 0, \forall t_1 > 0$.

Proof Defining the parameter error $\tilde{\theta} = \hat{\theta} - \theta$ and with (7.78), we have

$$\tilde{\theta}(t) = \tilde{\theta}(t-1) + \frac{\gamma(t)\phi(t-d)\varepsilon(t)}{m^2(t)} + f(t). \tag{7.83}$$

Consider the following positive definite function

$$V(\tilde{\theta}) = \tilde{\theta}^T\tilde{\theta},$$

then the time increment of $V(\tilde{\theta})$ along (7.83) is

$$\begin{aligned} V(\tilde{\theta}(t)) - V(\tilde{\theta}(t-1)) = & -\gamma(t)\left(2 - \frac{\gamma(t)\phi^T(t-d)\phi(t-d)}{m^2(t)}\right)\frac{\varepsilon^2(t)}{m^2(t)} \\ & +2f^T\left(\tilde{\theta}(t-1) + \frac{\gamma(t)\phi(t-d)\varepsilon(t)}{m^2(t)} + f(t)\right) \\ & -f^T(t)f(t). \end{aligned} \tag{7.84}$$

With the parameter projection functions (7.81) and (7.82), we have

$$\begin{aligned} & f^T\left(\tilde{\theta}(t-1) + \frac{\gamma(t)\phi(t-d)\varepsilon(t)}{m^2(t)} + f(t)\right) \\ & = \sum_{i=1}^{N} f_{n_y+i}(\hat{\theta}_{n_y+i} - \theta_{n_y+i} + g_{n_y+i} + f_{n_y+i}) \le 0, \end{aligned} \tag{7.85}$$

which, together with

$$\begin{aligned} -\gamma(t)\left(2-\frac{\gamma(t)\phi^T(t-d)\phi(t-d)}{m^2(t)}\right)\frac{\varepsilon^2(t)}{m^2(t)} &\le -\gamma(t)\,(2-\gamma(t))\,\frac{\varepsilon^2(t)}{m^2(t)} \\ &\le -\gamma_0^2\frac{\varepsilon^2(t)}{m^2(t)} \end{aligned} \tag{7.86}$$

for $\gamma(t) \in (\gamma_0, 2-\gamma_0)$, ensures

$$V(\tilde{\theta}(t)) - V(\tilde{\theta}(t-1)) \le -\gamma_0^2\frac{\varepsilon^2(t)}{m^2(t)} - f^T(t)f(t) \le 0, \tag{7.87}$$

which implies $\|\hat{\theta}(t)-\theta\| \le \|\hat{\theta}(t-1)-\theta\| \le \|\hat{\theta}(0)-\theta\|$ and $\hat{\theta}(t) \in L_\infty$.

It follows from (7.87) and the boundedness of $V(\tilde{\theta})$ that

$$\sum_{\tau=0}^{t}\left(\gamma_0^2\frac{\varepsilon^2(\tau)}{m^2(\tau)} + f^T(\tau)f(\tau)\right) = V(\tilde{\theta}(0)) - V(\tilde{\theta}(t)) \le V(\tilde{\theta}(0)), \tag{7.88}$$

that is, $\frac{\varepsilon(t)}{m(t)} \in L_2$ and $f(t) \in L_2$.

From (7.78), we have

$$\begin{aligned} \hat{\theta}(t) = \hat{\theta}(t-t_1) &+ \sum_{\tau=0}^{t_1-1}\frac{\gamma(t-\tau)\phi(t-d-\tau)\varepsilon(t-\tau)}{m^2(t-\tau)} \\ &+ \sum_{\tau=0}^{t_1-1} f^T(t-\tau)f(t-\tau), \end{aligned} \tag{7.89}$$

with $\frac{\varepsilon(t)}{m(t)} \in L_2$ and $f(t) \in L_2$, we obtain $\hat{\theta}(t)-\hat{\theta}(t-t_1) \in L_2$, $\lim_{t\to\infty}\|\hat{\theta}(t) - \hat{\theta}(t-t_1)\| = 0$. ∇

From Lemma 7.3, we can derive the following two properties: (i) $\varepsilon(t) \in L_2$ and $\lim_{t\to\infty}\varepsilon(t) = 0$ if $\phi(t-d) \in L_\infty$, and (ii) $\lim_{t\to\infty}\|\hat{\theta}(t+d-1)-\hat{\theta}(t)\| = 0$. The first property means that if all system signals are bounded, then $\lim_{t\to\infty}\varepsilon(t+d) = 0$, that is,

$$\begin{aligned} \hat{y}_0(t+d) = {} & \hat{\theta}_y^T(t+d-1)\phi_y(t) + \hat{\theta}_{u0}^T(t+d-1)\phi_{u0}(t) \\ & + \hat{\theta}_{u1}^T(t+d-1)\phi_{u1}(t) \end{aligned} \tag{7.90}$$

is an adaptive prediction of $y(t+d)$ such that $\lim_{t\to\infty}(\hat{y}_0(t+d)-y(t+d)) = 0$. The second property implies that if $\phi(t) \in L_\infty$, the signal

$$\hat{y}(t+d) = \hat{\theta}_y^T(t)\phi_y(t) + \hat{\theta}_{u0}^T(t)\phi_{u0}(t) + \hat{\theta}_{u1}^T(t)\phi_{u1}(t) \tag{7.91}$$

is also an adaptive prediction of $y(t+d)$ such that $\lim_{t\to\infty}(\hat{y}(t+d) - y(t+d)) = 0$. Note that the signal $\hat{y}(t+d)$ is available at time t, while $\hat{y}_0(t+d)$ is not for $d > 1$. Thus, we have developed an adaptive predictor (7.91) updated by (7.78), for the T–S fuzzy system (7.51).

This adaptive prediction theory also provides the basis for adaptive control of the T–S fuzzy system (7.51). In the adaptive prediction equation (7.91), we have

$$\hat{\theta}_{u0}^T(t)\phi_{u0}(t) = \left(\sum_{i=1}^{N} \mu_i(\xi(t))\hat{b}_{i0}(t)\right) u(t). \tag{7.92}$$

The control signal $u(t)$ at time t has direct influence on the prediction signal $\hat{y}(t+d)$. It would be desirable to choose $u(t)$ to make $\hat{y}(t+d) = y_m(t+d)$ for a given reference output signal $y_m(t)$ for the system output $y(t)$ to track. This is the main idea to be developed in the next section.

7.3.3 Adaptive Control Scheme

In this section, we design and analyze an adaptive control scheme for the T–S fuzzy system (7.51).

Adaptive control law. We then choose the adaptive fuzzy control law as

$$u(t) = \frac{1}{\sum_{i=1}^{N} \mu_i(\xi(t))\hat{b}_{i0}(t)} \left(y_m(t+d) - \hat{\theta}_y^T(t)\phi_y(t) - \hat{\theta}_{u1}^T(t)\phi_{u1}(t)\right). \tag{7.93}$$

For this adaptive control law, the parameter projection component (7.81) has been added in the parameter adaptation algorithm (7.78) to ensure that $\sum_{i=1}^{N} \mu_i(\xi(t))\hat{b}_{i0}(t) > b_0$ for some constant $b_0 > 0$.

Stability analysis. We now show that the adaptive control system has desired stability and tracking properties. Substituting (7.93) into (7.51), we obtain the closed-loop system as

$$y(t+d) = \theta^T\phi(t) - \hat{\theta}^T(t)\phi(t) + y_m(t+d). \tag{7.94}$$

With $e(t) = y(t) - y_m(t)$ and $\tilde{\theta}(t) = \hat{\theta}(t) - \theta$, we obtain

$$e(t+d) = -\tilde{\theta}^T(t)\phi(t). \tag{7.95}$$

We first present a desired property for $\phi(t)$ (Qi et al. 2012b).

Lemma 7.4 *Under Assumption 7.3, the regressor $\phi(t)$ defined in (7.69) satisfies*

$$\|\phi(t)\| \leq \rho_1 + \rho_2 \max_{0\leq\tau\leq t} |e(\tau+d)| \tag{7.96}$$

for some positive constants ρ_1 and ρ_2.

Proof From the definition of $\phi(t)$ in (7.69) whose components are the products of μ_i and y at various time instants, and products of μ_i and u, and the property of μ_i in (7.14), we have

$$\|\phi(t)\| \leq \kappa_1 \|\psi(t)\|, \tag{7.97}$$

for some constant $\kappa_1 > 0$, where

$$\psi(t) = [y(t), y(t-1), \ldots, y(t-n+1), u(t), u(t-1), \ldots, u(t-n+1)]^T. \tag{7.98}$$

With $y(t) = e(t) + y_m(t)$, $\psi(t)$ can be expressed as

$$\begin{aligned}\psi(t) = {} & [e(t), e(t-1), \ldots, e(t-n+1), u(t), u(t-1), \ldots, u(t-n+1)]^T \\ & + [y_m(t), y_m(t-1), \ldots, y_m(t-n+1), 0, \ldots, 0]^T.\end{aligned}$$

Then, we have

$$\|\psi(t)\| \leq \kappa_2 \max_{t-n+1 \leq \tau \leq t} |e(\tau)| + \kappa_3 \max_{t-n+1 \leq \tau \leq t} |u(\tau)| + \kappa_4, \tag{7.99}$$

where κ_i, $i = 2, 3, 4$, are some positive constants.

Again with $y(t) = e(t) + y_m(t)$, the system Eq. (7.51) can be formulated as

$$\begin{aligned}e(t+d) + y_m(t+d) = {} & \sum_{i=1}^{N} \mu_i \alpha_i(z^{-1})[e](t) + \sum_{i=1}^{N} \mu_i \alpha_i(z^{-1})[y_m](t) \\ & + \sum_{i=1}^{N} \mu_i \beta_i(z^{-1})[u](t).\end{aligned} \tag{7.100}$$

With Assumption 7.3 and Definition 7.1, we obtain

$$\begin{aligned}|u(t)| & \leq \kappa_5 |e(t+d)| + \kappa_6 \sum_{\tau=0,1,\ldots,t-1} \lambda^{t-\tau-1} |e(\tau+d)| + \kappa_7 \\ & \leq \kappa_8 \max_{0 \leq \tau \leq t} |e(\tau+d)| + \kappa_7,\end{aligned} \tag{7.101}$$

where κ_i, $i = 5, 6, 7, 8$, are some positive constants.

Finally, using (7.97), (7.99), and (7.101), we obtain

$$\|\phi(t)\| \leq \rho_1 + \rho_2 \max_{0 \leq \tau \leq t} |e(\tau+d)|, \tag{7.102}$$

where ρ_1 and ρ_2 are some positive constants. ∇

We now show the desired closed-loop system properties (Qi et al. 2012b).

Theorem 7.2 *All signals in the closed-loop system, with the plant (7.51) satisfying Assumptions 7.3 and 7.4, the controller (7.93) and the adaptive law (7.78), are bounded, and* $\lim_{t\to\infty}(y(t) - y_m(t)) = 0$.

Proof From (7.75), we have

$$\varepsilon(t) = -(\hat{\theta}(t-1) - \theta)^T \phi(t-d), \tag{7.103}$$

and from (7.95), we have $e(t) = -\tilde{\theta}^T(t-d)\phi(t-d)$, which can be expressed as

$$\begin{aligned} e(t) &= \bar{\varepsilon}(t)\sqrt{c + \phi^T(t-d)\phi(t-d)} - (\hat{\theta}(t-d) \\ &\quad -\hat{\theta}(t-1))^T \bar{\phi}(t-d)\sqrt{c + \phi^T(t-d)\phi(t-d)}, \end{aligned} \tag{7.104}$$

where $\hat{\theta}(t-d) - \hat{\theta}(t-1) \in L_2 \cap L_\infty$,

$$\begin{aligned} \bar{\varepsilon}(t) &= \frac{\varepsilon(t)}{\sqrt{c + \phi^T(t-d)\phi(t-d)}} \in L_2 \cap L_\infty, \\ \bar{\phi}(t-d) &= \frac{\phi(t-d)}{\sqrt{c + \phi^T(t-d)\phi(t-d)}} \le 1. \end{aligned} \tag{7.105}$$

Using the inequality: $\sqrt{c + \phi^T(t-d)\phi(t-d)} \le \sqrt{c} + \|\phi^T(t-d)\|$, we express $e(t)$ from (7.104) as

$$|e(t)| \le c_1 + |\bar{\varepsilon}(t)|\, \|\phi^T(t-d)\| + \|\hat{\theta}(t-d) - \hat{\theta}(t-1)\|\, \|\phi^T(t-d)\|,$$

for some constant $c_1 > 0$. Using Lemma 7.4: $\|\phi(t-d)\| \le \rho_1 + \rho_2 \max_{0\le\tau\le t} |e(\tau)|$, we obtain

$$\begin{aligned} |e(t)| \le\ & c_2 + c_3|\bar{\varepsilon}(t)| \max_{0\le\tau\le t} |e(\tau)| \\ & + c_4\|\hat{\theta}(t-d) - \hat{\theta}(t-1)\| \max_{0\le\tau\le t} |e(\tau)|, \end{aligned} \tag{7.106}$$

for some constants $c_i > 0$, $i = 2, 3, 4$.

From Lemma 7.3, we have that $\lim_{t\to\infty} \bar{\varepsilon}(t) = 0$ and $\lim_{t\to\infty} \|\hat{\theta}(t) - \hat{\theta}(t-d)\| = 0$, and with these results, it follows from (7.106) that $e(t)$ is bounded, which implies $y(t)$ is bounded, and in turn from the system's minimum phase property that $u(t)$ is bounded. Hence, all signals in the closed-loop system are bounded, based on which, from (7.104) in which $\bar{\varepsilon}(t) \in L_2$ and $\hat{\theta}(t-d) - \hat{\theta}(t-1) \in L_2$, we have that $e(t) \in L^2$ so that $\lim_{t\to\infty} e(t) = 0$. ∇

A system may have two kinds of uncertainties: parameterized uncertainties (the error-free case) and unparameterizable uncertainties (the modeling errors including the approximation errors in the fuzzy control case). The most important feature (advantage) of adaptive control is its ability to deal with parameterized uncertainties, which is the key component of all adaptive control designs making up the adaptive control theory seen in adaptive control literature. Thus, it is important to develop the adaptive control design and analyze its desired properties for the error-free case, like what is about Theorem 7.2. The modeling errors can be dealt with using robust adaptive control (Ioannou and Sun 1996). The closed-loop stability and asymptotic output tracking properties given in Theorem 7.2 are the basic adaptive system properties only for the case when there are no modeling and approximation errors in the system (7.51). In the presence of such errors as described in (7.15), the adaptive law (7.76) needs to be modified with robustifying signals such as dead-zone, σ-modification, and parameter projection (Goodwin and Sin 1984; Ioannou and Sun 1996; Tao 2003), for robust parameter adaptation. While the closed-loop signal boundedness can be ensured globally or locally for small normalized modeling and approximation errors, the tracking error $e(t) = y(t) - y_m(t)$ can be shown to be bounded by normalized modeling and approximation errors in a mean sense similar to that seen in the literature for robust adaptive of non-fuzzy systems, as an extension of the L_2 tracking property in Theorem 7.2 (see its proof at the end), for the no modeling/approximation error case. □

Remark 7.6 About the constants c_i in Theorem 7.2, only the constant c is used in the parameter adaptation in (38), which can be selected as a small positive number to ensure the boundedness of $\frac{\phi^T(t-d)\phi(t-d)}{m^2(t)}$. All the other constants c_i in the proof of Theorem 7.2 are not directly used in the implementation of the adaptive controller. They are just used to describe how the tracking error $e(t)$ is bounded. Therefore, it is enough for us to know the existence of such constants, but we do not need to select the exact values for them. □

7.3.4 Simulation Study

In this section, we present an illustrative example with simulation results to show the control design and evaluation, based on the mass–spring–damper mechanical system (Tanaka et al. 1996).

Simulation system. The mass–spring–damper mechanical system is described by the following equation:

$$M\ddot{x} + c_1\dot{x} + c_2 x = (1 + c_3\dot{x}^3)u, \tag{7.107}$$

where M denotes the mass, x is the displacement (in meters) of the mass, u is the force (in Newtons) applied to the spring, c_1 is the damping constant, c_2 is the spring constant, and c_3 is a constant related to the nonlinear term $\dot{x}^3$. For simulation, the parameters are set as $M = 1$ kg, $c_1 = 15$ N · s/m, $c_2 = 20$ N/m, and $c_3 = 0.13$ N/(m/s)^3.

Choose the output $y = x$. To use the method developed in Sect. 7.3, we need a discrete-time T–S fuzzy system to approximate the dynamics of the nonlinear plant

(7.107). Assuming $\dot{y} \in [-1.5, 1.5]$m/s and using the same approach as that in Tanaka et al. (1996), a two-rule continuous-time T–S fuzzy system to approximate (7.107) is given as

$$\begin{aligned} &\text{IF } \dot{y} \text{ is } F_1^1, \text{ THEN } \ddot{y} = -15\dot{y} - 20y + 1.4387u, \\ &\text{IF } \dot{y} \text{ is } F_1^2, \text{ THEN } \ddot{y} = -15\dot{y} - 20y + 0.5613u, \end{aligned}$$

with the membership functions describing "F_1^1" and "F_1^2" chosen as

$$F_1^1(\dot{y}) = 0.5 + \dot{y}^3/6.75, \quad F_1^2(\dot{y}) = 0.5 - \dot{y}^3/6.75. \tag{7.108}$$

If the sampling time T is chosen small enough, we can approximate $\dot{y}$ and $\ddot{y}$ with $\dot{y} = [y(t+1) - y(t)]/T$ and $\ddot{y} = [y(t+2) - 2y(t+1) + y(t)]/T^2$. Then, a discrete-time model can be obtained as

$$\begin{aligned} R^i : \ &\text{IF } \xi_1(t) \text{ is } F_1^i, \text{THEN} \\ &y(t+2) + a_{i,1}y(t+1) + a_{i,2}y(t) = b_{i,0}u(t), \end{aligned} \tag{7.109}$$

where $\xi_1(t) = [y(t+1) - y(t)]/T$, $a_{i,1} = 15T - 2$, $a_{i,2} = 1 - 15T + 20T^2$, $i = 1, 2$ and $b_{1,0} = 1.4387T^2$, $b_{2,0} = 0.5613T^2$.

Based on a standard fuzzy modeling technique, we obtain the global fuzzy model of (7.109) as

$$y(t+2) = \sum_{i=1}^{2} \mu_i(t)[-a_{i,1}y(t+1) - a_{i,2}y(t)] + \sum_{i=1}^{2} \mu_i(t)b_{i,0}u(t), \tag{7.110}$$

where $\mu_i(t)$ is the normalized membership function:

$$\mu_i(t) = \frac{\lambda_i(\xi_1(t))}{\sum_{i=1}^{2} \lambda_i(\xi_1(t))}, \quad \lambda_i(t) = \prod_{j=1}^{1} F_j^i(\xi_1(t)). \tag{7.111}$$

Since $y(t+1)$ in (7.110) is not available at time t, we need to develop a prediction model in the form (7.57) for the control design by following the procedures in Sect. 7.3.1. With (7.110), we can express $y(t+1)$ as

$$\begin{aligned} y(t+1) = &\sum_{i=1}^{2} \mu_i(t-1)[-a_{i,1}y(t) - a_{i,2}y(t-1)] \\ &+ \sum_{i=1}^{2} \mu_i(t-1)b_{i,0}u(t-1), \end{aligned} \tag{7.112}$$

Replacing $y(t+1)$ in (7.110) by (7.112), we obtain the following global 2-step prediction fuzzy system model:

$$y(t+2) = \theta_y^T \phi_y(t) + \theta_{u0}^T \phi_{u0}(t) + \theta_{u1}^T \phi_{u1}(t), \tag{7.113}$$

where

$$\begin{aligned}
\theta_{u0} &= [b_{1,0}, b_{2,0}]^T, \quad \phi_{u0}(t) = [\mu_1(t)u(t), \mu_2(t)u(t)]^T \\
\theta_{u1} &= [-a_{1,1}b_{1,0}, -a_{1,1}b_{2,0}, -a_{2,1}b_{1,0}, -a_{2,1}b_{2,0}]^T \\
\theta_{y1} &= [a_{1,1}^2, a_{1,1}a_{1,2}, a_{1,1}a_{2,1}, a_{1,1}a_{2,2}, -a_{1,2}]^T \\
\theta_{y2} &= [a_{2,1}^2, a_{2,1}a_{2,2}, a_{2,1}a_{1,1}, a_{2,1}a_{1,2}, -a_{2,2}]^T
\end{aligned}$$

$$\begin{aligned}
\phi_{u1}(t) &= [\mu_1(t)\mu_1(t-1), \mu_1(t)\mu_2(t-1), \mu_2(t)\mu_1(t-1), \mu_2(t)\mu_2(t-1)]^T u(t-1) \\
\phi_{y1}(t) &= \mu_1(t)[\mu_1(t-1)y(t), \mu_1(t-1)y(t-1), \mu_2(t-1)y(t), \mu_2(t-1)y(t-1), y(t)]^T \\
\phi_{y2}(t) &= \mu_2(t)[\mu_2(t-1)y(t), \mu_2(t-1)y(t-1), \mu_1(t-1)y(t), \mu_1(t-1)y(t-1), y(t)]^T \\
\theta_y &= [\theta_{y1}^T, \theta_{y2}^T]^T, \quad \phi_y(t) = [\phi_{y1}^T(t), \phi_{y2}^T(t)]^T.
\end{aligned}$$

Based on (7.113), we choose the adaptive global fuzzy control law as

$$u(t) = \frac{1}{\sum_{i=1}^{2} \mu_i(t)\hat{b}_{i0}(t)} \left(y_m(t+2) - \hat{\theta}_y^T(t)\phi_y(t) - \hat{\theta}_{u1}^T(t)\phi_{u1}(t) \right), \tag{7.114}$$

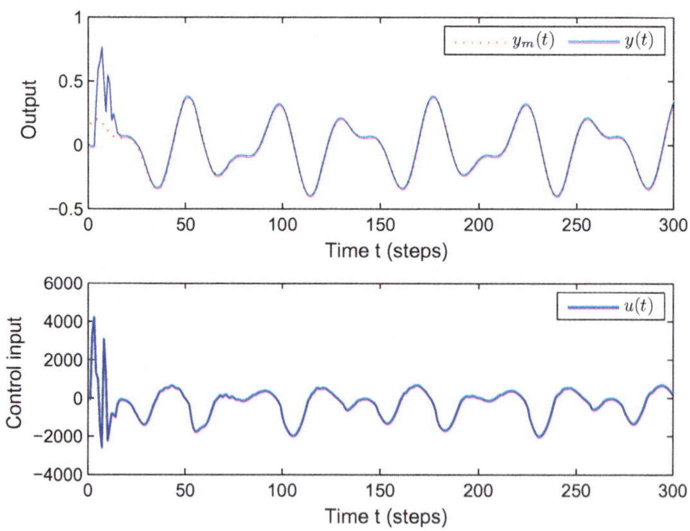

Fig. 7.5 Adaptive T–S fuzzy system response and control input

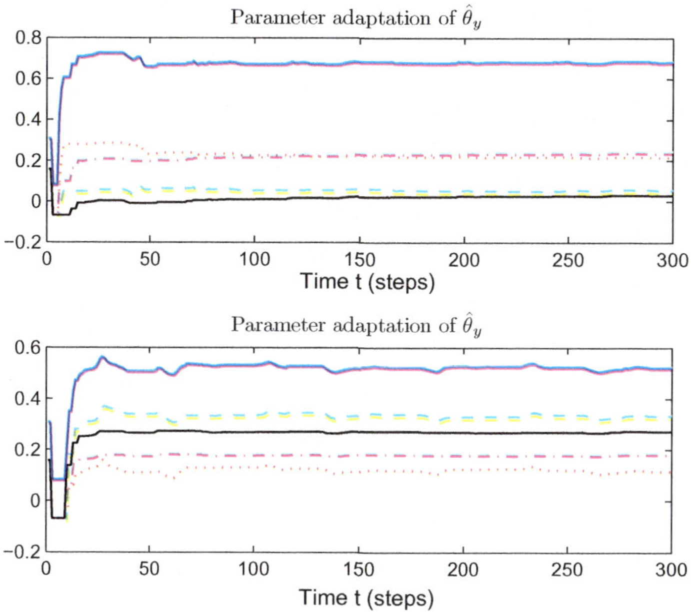

Fig. 7.6 Parameter adaptation of $\hat{\theta}_y$

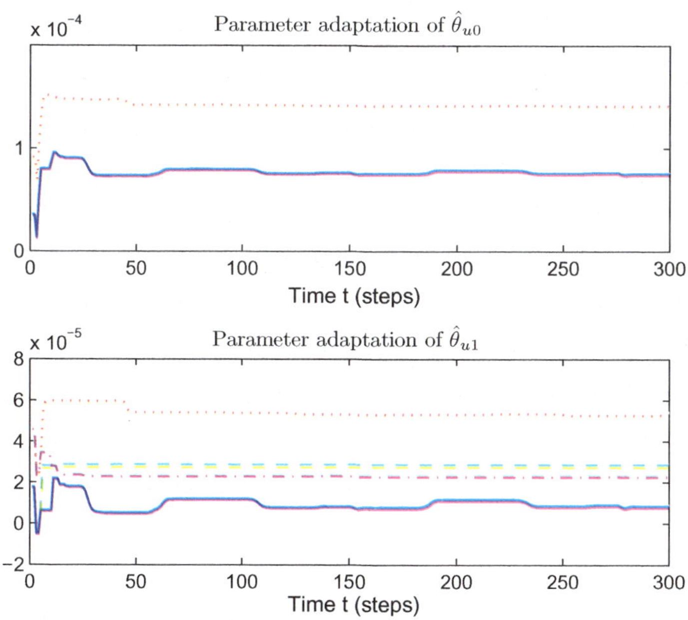

Fig. 7.7 Parameter adaptation of $\hat{\theta}_{u0}$ and $\hat{\theta}_{u1}$

and the parameter estimates $\hat{\theta}(t) = [\hat{\theta}_y^T(t), \hat{\theta}_{u0}^T(t), \hat{\theta}_{u1}^T(t)]^T$ are updated by (7.76).

Simulation results. The initial parameter values are set as 65% of their true values. Other parameters are chosen as $T = 0.1$ s, $\gamma(t) = 0.8$, and $c = 0.01$. The T–S fuzzy system response for tracking a sinusoidal signal $y_m(t) = 0.25\sin(0.15t) + 0.15\cos(0.25t)$ with the adaptive controller (7.114) is shown in Fig. 7.5, which verifies the desired system performance with an adaptive controller based on the general prediction model (7.113). The parameter adaptation results are given in Fig. 7.6 and 7.7. All the parameters remain bounded under the designed adaptive control scheme.

7.4 Summary

In this chapter, we have developed two approaches for adaptive fuzzy control for T–S fuzzy systems with multiple input–output delays. In particular, we have derived two nonlinear prediction models for T–S fuzzy systems with multiple input–output delays and developed stable adaptive control schemes with complete stability and convergence analysis. Those results are along the direction of adaptive fuzzy control using an input–output approach for output tracking control of systems (with relative degree one and with general relative degree).

For adaptive control of a dynamic system with d-step delays (relative degree d), it is crucial to derive a global prediction model. For a T–S fuzzy system with d-step delays, naturally there are two ways to obtain its prediction model: (i) deriving a local prediction model from the local linear model and then constructing the global prediction model by fuzzily blending all the local prediction models as we did in Sect. 7.2 and (ii) deriving the global prediction model directly from the global T–S fuzzy system as we did in Sect. 7.3. This second way of deriving a global prediction model considers the effects of the time-varying firing strength $\mu_i(t)$ and reduces the modeling error between the prediction model and the T–S fuzzy system, leading to more effective feedback control of the T–S fuzzy system. We have developed desired adaptive feedback control schemes for both approaches, providing a solution framework for adaptive control of dynamic systems with multiple input–output delays through T–S fuzzy input–output modeling.

References

Angelov PP, Filev DP (2004) An approach to online identification of Takagi–Sugeno fuzzy models. IEEE Trans Syst Man Cybern Part B Cybern 34(1):484–498

Barada S, Singh H (1998) Generating optimal adaptive fuzzy-neural models of dynamical systems with applications to control. IEEE Trans Syst Man Cybern Part C Appl Rev 28(3):297–313

Feng G (2010) Analysis and synthesis of fuzzy control systems: a model-based approach. CRC Press, Boca Raton

Franklin GF, Powell JD, Workman M (1998) Digit Control Dyn Syst. Addison-Wesley, Reading

Ghorbel F, Hung JH, Spong MW (1989) Adaptive control of flexible joint manipulators. IEEE Control Syst Mag 9(7):9–13

Goodwin GC, Sin KS (1984) Adaptive filtering prediction and control. Prentice-Hall, Englewood Cliffs

Ioannou PA, Sun J (1996) Robust adaptive control. Prentice-Hall, Englewood Cliffs

Qi R, Brdys M (2008) Stable indirect adaptive control based on discrete-time T–S fuzzy model. Fuzzy Sets Syst 159(8):900–925

Qi R, Tao G, Jiang B, Tan C (2012a) Adaptive control schemes for discrete-time T–S fuzzy systems with unknown parameters and actuator failures. IEEE Trans Fuzzy Syst 20:471–486

Qi R, Tao G, Tan C, Yao X (2012b) Adaptive prediction and control of discrete-time T–S fuzzy systems. Int J Adapt Control Signal Process 26(7):560–575

Shi W (2008) Indirect adaptive fuzzy control for a class of nonlinear discrete-time systems. J Syst Eng Electron 19(6):1203–1207

Tanaka K, Wang HO (2001) Fuzzy control system design and analysis: a LMI approach. Wiley, New York

Tanaka K, Ikeda T, Wang HO (1996) Robust stabilization of a class of uncertain nonlinear systems via fuzzy control: quadratic stabilizability, H^{∞} control theory, and linear matrix inequalities. IEEE Trans Fuzzy Syst 4:1–13

Tao G (2003) Adaptive control design and analysis. Wiley, New York

Tseng C-S (2006) Model reference output feedback fuzzy tracking control design for nonlinear discrete-time systems with time-delay. IEEE Trans Fuzzy Syst 14(1):58–70

Ying H (1999) Analytical analysis and feedback linearization tracking control of the general Takagi–Sugeno fuzzy dynamic systems. IEEE Trans Syst Man Cybern 29(1):290–298

Chapter 8
Adaptive T–S Fuzzy Control Using Output Feedback: MIMO Case

In Chap. 7, we developed adaptive output feedback fuzzy control schemes of single-input single-output (SISO) discrete-time nonlinear systems with multiple input–output delays based on their T–S fuzzy approximation models. The goal of this chapter is to extend the results in Chap. 7 to multiple-input multiple-output (MIMO) nonlinear systems. We will develop a solution framework for adaptive fuzzy control of MIMO discrete-time nonlinear systems, by modeling them using discrete-time T–S fuzzy systems, parameterizing T–S fuzzy systems with uncertain parameters, designing and analyzing an adaptive control scheme for such systems, and establishing and evaluating desired adaptive control system properties.

The results of this chapter include the derivation of a global prediction fuzzy system model for MIMO systems, parametrization and parameter estimation of MIMO fuzzy systems, development of an adaptive control scheme for MIMO fuzzy systems, stability and tracking analysis of such an adaptive control system, and illustration of new features and concepts of adaptive control for MIMO fuzzy systems. The design and analysis of this chapter are not only for MIMO fuzzy systems but also applicable to a class of time-varying MIMO dynamic systems with characterizable time variations extended from the fuzzy membership functions (Qi et al. 2014).

We will describe the main problems in Sect. 8.1, give the solutions including the derivation of a MIMO T–S fuzzy system prediction model in Sect. 8.2 and the design and analysis of the adaptive control scheme in Sect. 8.3, and present a simulation study in Sect. 8.4.

R. Qi et al., *Fuzzy System Identification and Adaptive Control*, Communications and Control Engineering,
https://doi.org/10.1007/978-3-030-19882-4_8

8.1 Problem Statement

Consider a MIMO nonlinear system in its discrete-time input–output form

$$y(t) = f(y(t-1), \ldots, y(t-n), u(t-d_0), \ldots, u(t-n)), \tag{8.1}$$

where $f(\cdot, \ldots, \cdot) \in R^r$ is a vector of nonlinear functions, $y(\cdot) \in R^r$ is the system output signal, $u(\cdot) \in R^r$ is the system input signal, $t = 0, 1, 2, \ldots$, is the discrete-time time variable, and d_0 is the number of nominal system input–output delays. In this chapter, we will address the general case with $d_0 \geq 1$, for the MIMO case characterized by a general system delay structure.

A MIMO prediction model of (8.1), in the form

$$\begin{aligned} y(t+d_0) = \; & f_d(y(t), y(t-1), \ldots, y(t-n+1), \\ & u(t), u(t-1), \ldots, u(t-n+1)), \end{aligned} \tag{8.2}$$

for some function $f_d(\cdot, \ldots, \cdot)$, is useful, as shown in the literature, for feedback control. If the function f is known, such a prediction model may be directly derived by iterations for the SISO case. For the MIMO case, the situation is more complicated, because of the interactions between system inputs and outputs. In Sect. 8.2, using a fuzzy system modeling method, we derive one such model and demonstrate that such a model has a sophisticated form for MIMO systems.

To handle the nonlinearity and uncertainty of f in deriving a parameterized model of (8.1), an effective method is to approximate the nonlinear plant (8.1) by some well-defined approximation functions. In this chapter, we employ the fuzzy system approximation theory, to develop such an approximate system model, using the following MIMO local T–S fuzzy model with the ith rule:

$$\begin{aligned} R^i: \; & \text{IF } \xi_1 \text{ is } F_1^i \text{ and } \ldots \text{ and } \xi_L \text{ is } F_L^i, \\ & \text{THEN } A^i(z^{-1})[y](t) = B^i(z^{-1})[u](t), \end{aligned} \tag{8.3}$$

where $i = 1, \ldots, N$, N is the number of fuzzy rules, F_j^i being typically an interval of real numbers, called a fuzzy set associated with which there is a membership function $F_j^i(\xi_j(t))$ to indicate the membership degree of $\xi_j(t)$ in F_j^i, and the system dynamics matrices are

$$A^i(z^{-1}) = I + A_1^i z^{-1} + \cdots + A_n^i z^{-n} \tag{8.4}$$

$$\begin{aligned} B^i(z^{-1}) &= z^{-d_0^i}(B_0^i + B_1^i z^{-1} + \cdots + B_{n_b - d_0^i}^i z^{-n_b + d_0^i}) \\ &\triangleq z^{-d_0^i} \bar{B}^i(z^{-1}), \; d_0^i \leq n_b \leq n \end{aligned} \tag{8.5}$$

with $A_j^i \in R^{r \times r}$ and $B_j^i \in R^{r \times r}$ being constant matrices, and $d_0^i \geq 1$ being the nominal delay of the system (8.3).

Remark 8.1 It should be noted that in the MIMO case, the matrix B_0^i is nonzero but may be singular (that is, $\det[B_0^i] = 0$). Hence, the delay d_0^i is only a nominal delay of the system and the essential delay of a MIMO system is characterized by a delay structure called an interactor matrix defined in Lemma 5.2.3 in Goodwin and Sin (1984). For a SISO system, there is no difference between the nominal delay and the essential delay. One can simply consider the delay d_0^i so that the leading coefficient of $\bar{B}^i(z^{-1})$ is nonzero, as we considered in Chap. 7.

In this chapter, we will solve the following two problems:

Problem I: Derive a MIMO T–S fuzzy system prediction model with a general input–output delay structure, based on the fuzzy rules (8.3), for a MIMO nonlinear system.

Problem II: Design an adaptive control scheme for the MIMO T–S fuzzy prediction model, to achieve the control objective: closed-loop stability and asymptotic tracking of a desired bounded output $y_m(t)$ by the plant output $y(t)$.

8.2 MIMO T–S Fuzzy System Prediction Model

In this section,[1] we shall first derive a MIMO fuzzy system prediction model by using MIMO system interactor matrix.

8.2.1 Interactor Matrix

Consider the transfer matrix $T^i(z) = (A^i(z^{-1}))^{-1}B^i(z^{-1})$ of the local system model in (8.3). Since $d_0^i \geq 1$, the transfer matrices $T^i(z)$, $i = 1, \ldots, N$, are all strictly proper. For output tracking control, it is assumed that

Assumption 8.1 $T^i(z), i = 1, \ldots, N$, all have full rank.

The above condition means that there is a delay of at least one unit between each input and each output, and, by Assumption 8.1, that the output function controllability of each local model is ensured.

We first present the following MIMO system characterization which plays an important role in MIMO system parametrization and control and will be used for MIMO fuzzy systems.

Proposition 8.1 (Goodwin and Sin 1984) *For any $r \times r$ strictly proper full rank rational transfer matrix $T(z)$ there exists a unique lower triangular polynomial matrix $\xi_d(z)$, defined as the interactor matrix of $T(z)$, of the form*

[1] Parts of Sect. 8.2 are reprinted from Qi et al. (2014), Copyright 2014, with permission from Elsevier.

$$\xi_d(z) = \begin{bmatrix} z^{d_1} & 0 & \cdots & \cdots & 0 \\ z^{d_1}h_{21}(z) & z^{d_2} & 0 & \cdots & 0 \\ \vdots & \vdots & \vdots & \vdots & \vdots \\ z^{d_1}h_{r1}(z) & z^{d_2}h_{r2}(z) & \cdots & z^{d_{r-1}}h_{r\,r-1}(z) & z^{d_r} \end{bmatrix}, \tag{8.6}$$

where $h_{ij}(z)$ are polynomials divisible by z, and $d_i \geq 1$ are integers, $j = 1, \ldots, r - 1$, $i = 2, \ldots, r$, such that

$$\lim_{z\to\infty} \xi_d(z)T(z) = K_p \tag{8.7}$$

is finite and nonsingular.

An important property of $\xi_d(z)$ is that $\xi_d^{-1}(z)$ is a *strictly proper and stable* transfer matrix. In the above MIMO characterization, K_p is called the high frequency gain matrix of $T(z)$, and the inverse of the interactor matrix $\xi_d(z)$ can be seen as the delay structure of $T(z)$. For example, when $\xi_d(z)$ is diagonal:

$$\xi_d(z) = \text{diag}\{z^{d_1}, z^{d_2}, \ldots, z^{d_r}\}, \tag{8.8}$$

we have

$$\xi_d^{-1}(z) = \text{diag}\{z^{-d_1}, z^{-d_2}, \ldots, z^{-d_r}\}, \tag{8.9}$$

which represents the r delays in the r input–output channels.

This chapter considers the local prediction models (8.3) with a general (non-diagonal) interactor matrix (input–output delay structure), and assumes the following.

Assumption 8.2 The local model transfer matrices $T^i(z)$, $i = 1, \ldots, N$, have a common known interactor matrix $\xi_d(z)$.

The high frequency gain matrices associated with each $T^i(z)$ are

$$\lim_{z\to\infty} \xi_d(z)T^i(z) = K_p^i, \tag{8.10}$$

which are finite and nonsingular, $i = 1, 2, \ldots, N$. Note that it does not require all $T^i(z)$, $i = 1, \ldots, N$, to have a common high frequency gain matrix.

8.2.2 *Prediction Model*

We now derive a MIMO T–S fuzzy system prediction model based on a general plant interactor matrix (delay structure). Following the procedure in deriving Theorem 5.2.4 of Goodwin and Sin (1984), with the interactor matrix $\xi_d(z)$, we define a new variable

$$\bar{y}(t) = \xi_d(z)[y](t), \tag{8.11}$$

and express $\bar{y}(t)$ in the predictor form:

$$\bar{y}(t) = \alpha^i(z^{-1})[y](t) + \beta^i(z^{-1})[u](t), \tag{8.12}$$

where

$$\alpha^i(z^{-1}) = \alpha_0^i + \alpha_1^i z^{-1} + \cdots + \alpha_{n-1}^i z^{-(n-1)}, \tag{8.13}$$

$$\beta^i(z^{-1}) = \beta_0^i + \beta_1^i z^{-1} + \cdots + \beta_{n_b-1}^i z^{-(n_b-1)}, \tag{8.14}$$

with $\beta_0^i = K_p^i$ being nonsingular. In this expression,

$$\alpha^i(z^{-1}) = G^i(z^{-1}),\ \beta^i(z^{-1}) = F^i(z)z^{-d_0^i}\bar{B}^i(z^{-1}), \tag{8.15}$$

where $F^i(z)$ and $G^i(z^{-1})$ are the unique polynomial matrices satisfying

$$\xi_d(z) = F^i(z)A^i(z^{-1}) + G^i(z^{-1}) \tag{8.16}$$

with, for $\xi_d(z) = K_d z^d + \cdots + K_1 z$,

$$F^i(z) = F_0^i z^d + F_1^i z^{d-1} + \cdots + F_{d-1}^i z, \tag{8.17}$$

$$G^i(z^{-1}) = G_0^i + G_1^i z^{-1} + \cdots + G_{n-1}^i z^{-n+1}, \tag{8.18}$$

whose coefficient matrices, with $F_0^i = K_d$, can be computed as follows:

$$F_l^i = K_{d-l} - \sum_{j=0}^{l-1} F_j^i A_{l-j}^i,\ l = 1, \cdots, d-1, \tag{8.19}$$

$$G_l^i = -\sum_{j=0}^{d-1} F_j^i A_{l+d-j}^i,\ l = 0, \cdots, n-1, \tag{8.20}$$

with $A_k^i = 0$ for $k > n$. We first express $\beta^i(z^{-1}) = F^i(z)z^{-d_0^i}\bar{B}^i(z^{-1}) = \beta_{d_0^i-d}^i z^{d-d_0^i} + \cdots + \beta_{-1}^i z + \beta_0^i + \beta_1^i z^{-1} + \cdots + \beta_{n_b-1}^i z^{-(n_b-1)}$. Since $T^i(z) = (\xi_d(z) - \alpha^i(z))^{-1} \beta^i(z)$ and $\lim_{z\to\infty} \xi_d(z)T^i(z) = K_p^i = \beta_0^i$ is nonsingular, it follows that $\beta_{d_0^i-d}^i = \cdots = \beta_{-1}^i = 0$ so that we have $\beta^i(z^{-1})$ as shown in (8.14).

In our study, we will use the local models (8.12) to form a global fuzzy system prediction model. To proceed, based on (8.12), using the standard technique of *singleton fuzzification*, *product inference*, and *weighted average*, we obtain the following global MIMO T–S fuzzy system prediction model (Qi et al. 2011).

Proposition 8.2 *Following a standard fuzzy modeling procedure, a nonlinear dynamic system (8.1), via the local fuzzy system model (8.3), can be approximated by a global fuzzy system prediction model:*

$$\bar{y}(t) = \sum_{i=1}^{N} \mu_i(\xi(t))(\alpha^i(z^{-1})[y](t) + \beta^i(z^{-1})[u](t)), \tag{8.21}$$

where $\mu_i(\xi(t))$ is the normalized firing strength:

$$\mu_i(\xi(t)) = \frac{\lambda^i(\xi(t))}{\sum_{i=1}^{N} \lambda^i(\xi(t))}, \quad \lambda^i(\xi(t)) = \prod_{j=1}^{L} F_j^i(\xi_j(t)) \tag{8.22}$$

with $\xi(t) = [\xi_1(t), \ldots, \xi_L(t)]^T$, such that

$$0 \leq \mu_i(\xi(t)) \leq 1, \quad \sum_{i=1}^{N} \mu_i(\xi(t)) = 1. \tag{8.23}$$

In this following study, we treat the fuzzy system prediction model (8.21) as an approximation model of the original nonlinear system (8.1), based on which the feedback control law is designed. The control objective is to find an adaptive control law to generate the input signal $u(t)$ for the system (8.21) with unknown parameters in $\alpha_0^i, \alpha_1^i, \ldots, \alpha_{n-1}^i, \beta_0^i, \beta_1^i, \ldots, \beta_{m-1}^i$, $i = 1, \ldots, N$, to ensure closed-loop signal boundedness and asymptotic tracking of a given bounded reference signal $y_m(t)$ by the system output signal $y(t)$, under the following assumptions:

Assumption 8.3 The system (8.21) is minimum phase.

Assumption 8.4 $\sum_{i=1}^{N} \mu_i \beta_0^i$ is nonsingular, for all possible μ_i.

Assumption 8.5 The system order n is known.

Assumption 8.3 assumes the minimum phase property of the system (8.21), which will be further clarified in Sect. 8.2.3.

Assumption 8.4 is necessary to ensure the controllability of the system (8.21).

Assumption 8.5 assumes the knowledge of n. However, the knowledge of an upper bound of n would be sufficient. Here, for simplicity, we assume n is known.

Remark 8.2 It should be noted that there are naturally approximation and modeling errors $\Delta(y(\cdot), u(\cdot), t)$ in representing the original nonlinear system (8.1) by the fuzzy system prediction model (8.21):

$$\bar{y}(t) = \sum_{i=1}^{N} \mu_i \alpha^i(z^{-1})[y](t) + \sum_{i=1}^{N} \mu_i \beta^i(z^{-1})[u](t) + \Delta(y(t), y(t-1), \ldots, u(t-1), u(t-2), \ldots, t).$$

Robust adaptive control and nonlinear damping/bounding design tools can be used to deal with approximation and modeling errors. In this study, the focus is on the design and analysis of some baseline adaptive control scheme for discrete-time MIMO fuzzy systems with general delay matrices in the form (8.21).

8.2.3 Minimum Phase Property

The control design is based on the assumption that the fuzzy system prediction model (8.21) is minimum phase. To clarify the minimum phase property of a fuzzy system, we first recall the minimum phase property for a regular SISO linear time-invariant (LTI) system.

Minimum phase SISO LTI system. Consider the following SISO LTI system :

$$A(z^{-1})[y](t) = z^{-d}\bar{B}(z^{-1})[u](t), \ \ \bar{B}(0) = b_0 \neq 0, \tag{8.24}$$

where $A(z^{-1})$ and $\bar{B}(z^{-1})$ being polynomials in z^{-1}. The system (8.24) is minimum phase if all zeros of $\bar{B}(z^{-1})$ are within a unit circle $|z| < 1$, which implies the minimum phase property:

$$|u(t-d)| \leq c_1|y(t)| + c_2 \sum_{\tau=0}^{t-1} \lambda^{t-\tau-1}|y(\tau)|, \ t \geq d, \tag{8.25}$$

for some constants $c_1 > 0$, $c_2 > 0$, and $\lambda \in (0, 1)$.

Minimum phase MIMO LTI system. For a regular MIMO LTI system

$$A(z^{-1})[y](t) = z^{-d_0}\bar{B}(z^{-1})[u](t), \ \ \bar{B}(0) \neq 0 \tag{8.26}$$

with $A(z^{-1})$ and $\bar{B}(z^{-1})$ similar to $A^i(z^{-1})$ and $\bar{B}^i(z^{-1})$ in (8.5) and with an interactor matrix $\xi_d(z)$ for $T(z) = A^{-1}(z^{-1})z^{-d_0}\bar{B}(z^{-1})$: $\lim_{z\to\infty} \xi_d(z)T(z) = K_p$ is finite and nonsingular. Under the condition that $T(z)$ has full rank (and so does $\bar{B}(z^{-1})$), the values of z such that $\det[\bar{B}(z^{-1})] = 0$ define the zeros of $T(z)$.

Under the assumption that all zeros of $T(z)$ are in $|z| < 1$ and with d being the maximum degree of $\xi_d(z)$, we can derive some similar relationship between $u(t)$ and $y(t)$ to that given in (8.25) but, due to the MIMO system nature, different from (8.25), that is, based on the expression

$$u(t) = G^{-1}(z)[y](t) = (\xi_d(z)T(z))^{-1}\xi_d(z)[y](t) \tag{8.27}$$

and for $\bar{y}(t) = \xi_d(z)[y](t)$ and with $(\xi_d(z)T(z))^{-1}$ being proper transfer matrix, we have

$$\|u(t-d)\| \leq c_1\|\bar{y}(t-d)\| + c_2 \sum_{\tau=0}^{t-1} \lambda^{t-\tau-1}\|\bar{y}(\tau-d)\|, \ \ t \geq d, \tag{8.28}$$

for some constants $c_1 > 0$, $c_2 > 0$, and $\lambda \in (0, 1)$, where $\|\cdot\|$ is the l^2 vector norm. Since $z^{-d}\xi_d(z)$ is a proper transfer matrix, the inequality (8.28) implies that

$$\|u(t-d)\| \leq c_3\|y(t)\| + c_4 \sum_{\tau=0}^{t-1} \lambda^{t-\tau-1}\|y(\tau)\|, \ \forall t \geq d, \tag{8.29}$$

for some constants $c_3 > 0$ and $c_4 > 0$. The property (8.29) has the same form as that in (8.25) for the SISO case but is different from the property (8.28) which characterizes the MIMO system in terms of its delay structure $\xi_d(z)$ and thus is more relevant as a system property than (8.28).

Minimum phase MIMO fuzzy system. Analogous to this nature of minimum phase regular MIMO LTI systems, we use the following minimum phase definition for the general MIMO fuzzy system (8.21) (Qi et al. 2011).

Definition 8.1 The fuzzy system (8.21) is minimum phase if the condition (8.29) is satisfied.

While the minimum phase condition of a regular LTI system (8.26) can be checked using the knowledge of the zeros of $\bar{B}(z^{-1})$, the fuzzy system (8.21) is nonlinear in nature and the zeros of $\bar{B}^i(z^{-1})$ cannot completely determine its minimum phase property which also depends on the normalized firing strength μ_i.

In this following study, we use a simple illustrative example to show how the zeros of the fuzzy system are affected by μ_i.

Example 8.1 Consider the system (8.21) with $n = 2$, $N = 2$ and $r = 2$:

$$A^i(z^{-1}) = \begin{bmatrix} 1 & 0 \\ 0 & 1 \end{bmatrix} + \begin{bmatrix} a_{11}^i & 0 \\ 0 & a_{14}^i \end{bmatrix} z^{-1} + \begin{bmatrix} a_{21}^i & 0 \\ 0 & a_{24}^i \end{bmatrix} z^{-2} \tag{8.30}$$

$$B^i(z^{-1}) = z^{-1}\left(\begin{bmatrix} b_{01}^i & b_{02}^i \\ 0 & 0 \end{bmatrix} + \begin{bmatrix} b_{11}^i & b_{12}^i \\ 0 & b_{14}^i \end{bmatrix} z^{-1}\right) \tag{8.31}$$

with $b_{01}^i \neq 0$ and $b_{14}^i \neq 0$. Its transfer matrix is

$$\begin{aligned} T^i(z) &= (A^i(z^{-1}))^{-1} B^i(z^{-1}) \\ &= \begin{bmatrix} \dfrac{b_{01}^i z^{-1} + b_{11}^i z^{-2}}{1 + a_{11}^i z^{-1} + a_{21}^i z^{-2}} & \dfrac{b_{02}^i z^{-1} + b_{12}^i z^{-2}}{1 + a_{11}^i z^{-1} + a_{21}^i z^{-2}} \\ 0 & \dfrac{b_{14}^i z^{-2}}{1 + a_{14}^i z^{-1} + a_{24}^i z^{-2}} \end{bmatrix} \end{aligned} \tag{8.32}$$

whose interactor matrix is selected as

$$\xi_d(z) = \begin{bmatrix} z & 0 \\ 0 & z^2 \end{bmatrix}, \tag{8.33}$$

which leads to the high frequency gain matrices

$$K_p^i = \lim_{z\to\infty} \xi_d(z) T^i(z) = \begin{bmatrix} b_{01}^i & b_{02}^i \\ 0 & b_{14}^i \end{bmatrix}, \quad i = 1, 2, \ldots, N$$

which is nonsingular. With this $\xi_d(z)$, the local prediction equation (8.12) can be derived as follows. For $\xi_d(z)$ in the form

$$\xi_d(z) = K_2 z^2 + K_1 z = \begin{bmatrix} 0 & 0 \\ 0 & 1 \end{bmatrix} z^2 + \begin{bmatrix} 1 & 0 \\ 0 & 0 \end{bmatrix} z, \tag{8.34}$$

we have $F_0^i = K_2$, and from (8.17)–(8.20), we have

$$F^i(z) = \begin{bmatrix} 0 & 0 \\ 0 & 1 \end{bmatrix} z^2 + \begin{bmatrix} 1 & 0 \\ 0 & -a_{14}^i \end{bmatrix} z \tag{8.35}$$

$$G^i(z^{-1}) = \begin{bmatrix} -a_{11}^i & 0 \\ 0 & -a_{24}^i + (a_{14}^i)^2 \end{bmatrix} + \begin{bmatrix} -a_{21}^i & 0 \\ 0 & a_{14}^i a_{24}^i \end{bmatrix} z^{-1}.$$

Finally, with (8.15), we obtain

$$\alpha^i(z^{-1}) = \begin{bmatrix} -a_{11}^i & 0 \\ 0 & -a_{24}^i + (a_{14}^i)^2 \end{bmatrix} + \begin{bmatrix} -a_{21}^i & 0 \\ 0 & a_{14}^i a_{24}^i \end{bmatrix} z^{-1}$$

$$\beta^i(z^{-1}) = \begin{bmatrix} b_{01}^i & b_{02}^i \\ 0 & b_{14}^i \end{bmatrix} + \begin{bmatrix} b_{11}^i & b_{12}^i \\ 0 & -a_{14}^i b_{14}^i \end{bmatrix} z^{-1}. \tag{8.36}$$

The zeros of $\beta^i(z^{-1})$ are those values of z making $\det[\beta^i(z^{-1})] = (b_{01}^i + b_{11}^i z^{-1})$ $(b_{14}^i - a_{14}^i b_{14}^i z^{-1}) = 0$, which can be calculated by

$$z_1^i = -\frac{b_{11}^i}{b_{01}^i}, \; z_2^i = a_{14}^i. \tag{8.37}$$

For each local system with $\beta^i(z^{-1})$ to be stable, we need $|z_1^i| < 1$ and $|z_2^i| < 1$. Due to the nonlinear nature (characterized by the presence of $\mu_i = \mu_i(\xi(t))$) of the global system model (8.21), the condition that all zeros of $\beta^i(z^{-1})$ are in $|z| < 1$ for all $i = 1, 2, \ldots, N$ does not necessarily mean that the global system model (8.21) is minimum phase. However, for this simple system example with first-order and triangular $\beta^i(z^{-1})$, $i = 1, 2, \ldots, N$, this condition is sufficient for (8.21) to be minimum phase, given that $\mu_i \in [0, 1]$, $i = 1, 2, \ldots, N$. □

8.3 Adaptive Control Design and Analysis

In this section,[2] we will present a nominal controller parameterized with the system parameters, develop an adaptive scheme to update the estimates of the nominal controller parameters, and analyze the stability and tracking performance of the adaptive control scheme for the fuzzy system with unknown parameters.

[2] Parts of Sect. 8.3 are reprinted from Qi et al. (2014), Copyright 2014, with permission from Elsevier.

8.3.1 Nominal Controller

The control objective is to make the output $y(t)$ of the fuzzy system (8.21) track a given bounded reference signal $y_m(t)$. When all the parameters in (8.21) are known, the feedback control law can be derived from the following equation:

$$\sum_{i=1}^{N} \mu_i \beta^i(z^{-1})[u](t) = \bar{y}_m(t) - \sum_{i=1}^{N} \mu_i \alpha^i(z^{-1})[y](t), \tag{8.38}$$

or equivalently,

$$u(t) = \left(\sum_{i=1}^{N} \mu_i \beta_0^i\right)^{-1} \left(\bar{y}_m(t) - \sum_{i=1}^{N} \mu_i \alpha^i(z^{-1})[y](t) - \sum_{i=1}^{N} \mu_i \bar{\beta}^i(z^{-1})[u](t)\right),$$

where $\bar{\beta}^i(z^{-1}) = \beta_1^i z^{-1} + \cdots + \beta_{n_b-1}^i z^{-(n_b-1)}$, $\bar{y}_m(t) = \xi_d(z)[y_m](t)$. Solvability for $u(t)$ is guaranteed by Assumption 8.4, in which the coefficient matrix $\sum_{i=1}^{N} \mu_i \beta_0^i$ is nonsingular.

The resulting closed-loop system with (8.38) is

$$\bar{y}(t) = \bar{y}_m(t). \tag{8.39}$$

With Assumption 8.3 and $\xi_d^{-1}(z)$ being a stable operator, the closed-loop system characterized by (8.38) and (8.39) is stable in the sense that all signals are bounded, and $\lim_{t\to\infty}(y(t) - y_m(t)) = 0$, with a transient response produced by the dynamics of $\xi_d(z)$ and the initial conditions.

For the case of unknown system parameters, we shall employ a parameter estimation algorithm to generate adaptive estimates of the parameters in (8.38), based on which an adaptive controller is implemented.

8.3.2 Parameter Estimation

For parameter estimation, the first step is to derive a parametrized model of (8.21). Then, certain adaptive parameter laws can be designed.

The fuzzy system model (8.21) can be parametrized as

$$\bar{y}(t) = \Theta^T \phi(t), \tag{8.40}$$

where the parameter Θ and regressor $\phi(t)$ are defined as

$$\phi(t) = [\phi_1^T(t), \phi_2^T(t), \ldots, \phi_N^T(t)]^T \tag{8.41}$$

$$\Theta = [\Theta_1^T, \Theta_2^T, \ldots, \Theta_N^T]^T \tag{8.42}$$

$$\phi_i(t) = [\mu_i y(t)^T, \mu_i y(t-1)^T, \ldots, \mu_i y(t-n+1)^T, \tag{8.43}$$

$$\mu_i u(t)^T, \mu_i u(t-1)^T, \ldots, \mu_i u(t-n_b+1)^T]^T \tag{8.44}$$

$$\Theta_i = [\alpha_0^i, \alpha_1^i, \ldots, \alpha_{n-1}^i, \beta_0^i, \beta_1^i, \ldots, \beta_{n_b-1}^i]^T. \tag{8.45}$$

Define the *estimation error*

$$\varepsilon(t) = \bar{y}(t-d) - \hat{\Theta}^T(t-1)\phi(t-d), \tag{8.46}$$

where $\hat{\Theta}(t-1)$ is the estimate of the unknown parameter Θ at $t-1$.

Recall that for $\xi_d(z) = K_d z^d + \cdots + K_1 z$,

$$\bar{y}(t) = \xi_d(z)[y](t) = K_d y(t+d) + \cdots + K_1 y(t+1), \tag{8.47}$$

which means $\bar{y}(t-d)$ and all the components in the regressor $\phi(t-d)$ are available at time t. Thus, the estimation error $\varepsilon(t)$ can be calculated by (8.46) at time t.

Parameter adaptive law. Based on (8.46), the parameter estimate $\hat{\Theta}$ can be updated by the following recursive algorithm:

$$\hat{\Theta}(t) = \hat{\Theta}(t-1) + \frac{\gamma(t)\phi(t-d)\varepsilon(t)}{c + \phi^T(t-d)\phi(t-d)}, \tag{8.48}$$

where $\gamma(t) : R^+ \to (\gamma_0, 2-\gamma_0)$ is an adaptation gain, for some constant $\gamma_0 \in (0, 1)$, $c > 0$ is a small design parameter, and an initial estimate $\hat{\Theta}(0)$ is chosen to make the matrix $\sum_{i=1}^N \mu_i(\xi(0))\hat{\beta}_0^i(0)$ nonsingular.

This parameter adaptive law (8.48) has the following desired properties:

Lemma 8.1 *The parameter adaptive law (8.48), when applied to the fuzzy system (8.40), has the properties:*

(i) $\|\hat{\Theta}(t) - \Theta\| \le \|\hat{\Theta}(t-1) - \Theta\| \le \|\hat{\Theta}(0) - \Theta\|$, *for the matrix norm* $\|\hat{\Theta}(t) - \Theta\| = \sqrt{\mathrm{tr}[(\hat{\Theta}(t) - \Theta)^T(\hat{\Theta}(t) - \Theta)]}$;

(ii) $\frac{\varepsilon(t)}{\sqrt{c+\phi^T(t-d)\phi(t-d)}} \in L^\infty \bigcap L^2$;

(iii) $\lim_{t\to\infty} \frac{\varepsilon(t)}{\sqrt{c+\phi^T(t-d)\phi(t-d)}} = 0$;

(iv) $\|\hat{\Theta}(t) - \hat{\Theta}(t-t_1)\| \in L^2$; *and*

(v) $\lim_{t\to\infty} \|\hat{\Theta}(t) - \hat{\Theta}(t-t_1)\| = 0$, *for any finite* $t_1 > 0$.

The proof of this lemma is standard, based on the positive definite function

$$V(\tilde{\Theta}) = \mathrm{tr}[\tilde{\Theta}^T\tilde{\Theta}], \ \tilde{\Theta} = \hat{\Theta} - \Theta \tag{8.49}$$

whose time increment satisfies

$$V(\tilde{\Theta}(t)) - V(\tilde{\Theta}(t-1)) = \operatorname{tr}[\tilde{\Theta}^T(t)\tilde{\Theta}(t)] - \operatorname{tr}[\tilde{\Theta}^T(t-1)\tilde{\Theta}(t-1)]$$
$$\leq -\frac{a_0\|\varepsilon(t)\|^2}{c+\phi^T(t-d)\phi(t-d)} \leq 0, \quad (8.50)$$

for some $a_0 > 0$. The details of the proof is given in Appendix A.

It should be noted that Assumption 8.5 ensures the nonsingularity of $\sum_{i=1}^N \mu_i \beta_0^i$. However, the estimated parameters are used to implement the controller (8.38), where the inverse of $\sum_{i=1}^N \mu_i \hat{\beta}_0^i(t)$ is used. Particular attention should be given on how to ensure the nonsingularity of $\sum_{i=1}^N \mu_i \hat{\beta}_0^i(t)$ while the parameter adaptive laws generate the online estimate $\hat{\Theta}(t)$.

In the following section, we shall discuss on how to ensure $\sum_{i=1}^N \mu_i \hat{\beta}_0^i(t)$ is nonsingular during parameter adaptation.

8.3.3 Nonsingularity of $\sum_{i=1}^N \mu_i \hat{\beta}_0^i(t)$

From the parameter estimate matrices

$$\hat{\Theta} = [\hat{\Theta}_1^T, \hat{\Theta}_2^T, \ldots, \hat{\Theta}_N^T]^T \quad (8.51)$$
$$\hat{\Theta}_i = [\hat{\alpha}_0^i, \hat{\alpha}_1^i, \ldots, \hat{\alpha}_{n-1}^i, \hat{\beta}_0^i, \hat{\beta}_1^i, \ldots, \hat{\beta}_{n_b-1}^i]^T, \quad (8.52)$$

we form the matrix $\sum_{i=1}^N \mu_i \hat{\beta}_0^i(t)$ which will be the coefficient matrix of the control signal $u(t)$ in the adaptive control law to be designed. Such a matrix needs to be ensured nonsingular for all t, so that an adaptive control equation can be solved online, to avoid control singularity.

A similar issue was encountered in Goodwin and Sin (1984) and Tao and Ioannou (1989) for the case of adaptive control of a regular MIMO linear system with $N = 1$ and $\mu_1 = 1$ so that $\sum_{i=1}^N \mu_i \hat{\beta}_0^i(t) = \hat{\beta}_0^1(t)$ as the coefficient matrix of $u(t)$. For this case, as shown in Goodwin and Sin (1984) and Tao and Ioannou (1989), some suitable choice of $\gamma(t)$ can be made for the parameter adaptive law to ensure that $\hat{\beta}_0^1(t)$ is nonsingular for any $t \geq 0$, if $\hat{\beta}_0^1(0)$ is nonsingular.

In our current new problem of dealing with MIMO fuzzy systems, the issue with the nonsingularity of $\sum_{i=1}^N \mu_i \hat{\beta}_0^i(t)$ is more complicated than that in Goodwin and Sin (1984) and Tao and Ioannou (1989), due to the presence of $\mu_i(\xi(t))$ and multiple parameter estimates $\hat{\beta}_0^i(t)$. The main difficulty is caused by the time variations of $\mu_i(\xi(t))$ in determining the parameter estimates $\hat{\beta}_0^i(t)$, $i = 1, 2, \ldots, N$, to make $\sum_{i=1}^N \mu_i(\xi(t))\hat{\beta}_0^i(t)$ nonsingular for all $t \geq 0$.

To make the notions in the following discussions more concise, we use $\mu_i(t)$ to represent the value of $\mu_i(\xi(t)$ at time t in the following discussion.

For example, under the condition (C1): the nonsingularity of $\sum_{i=1}^N \mu_i(t-1)\hat{\beta}_0^i(t)$ implies the nonsingularity of $\sum_{i=1}^N \mu_i(t)\ \hat{\beta}_0^i(t)$ for any t, the nonsingularity of

$\sum_{i=1}^{N} \mu_i(t)\hat{\beta}_0^i(t)$ can be ensured by the adaptive law (8.48) for all $t \geq 0$. To see this, from (8.45) and (8.48), we have

$$\hat{\beta}_0^i(t) = \hat{\beta}_0^i(t-1) + \frac{\gamma(t)\mu_i(t-d)u(t-d)\varepsilon^T(t)}{c + \phi^T(t-d)\phi(t-d)}. \tag{8.53}$$

Introducing the matrix

$$g(t-1) = \frac{u(t-d)\varepsilon^T(t)}{c + \phi^T(t-d)\phi(t-d)}, \tag{8.54}$$

multiplying (8.53) by $\mu_i(t-1)$, and summing up the resulting equation for $i = 1, 2, \ldots, N$, we obtain

$$\begin{aligned}\sum_{i=1}^{N} \mu_i(t-1)\hat{\beta}_0^i(t) &= \sum_{i=1}^{N} \mu_i(t-1)\hat{\beta}_0^i(t-1) \\ &\quad + \gamma(t)\sum_{i=1}^{N} \mu_i(t-1)\mu_i(t-d)g(t-1).\end{aligned} \tag{8.55}$$

Given that the initial coefficient matrix $\sum_{i=1}^{N} \mu_i(0)\hat{\beta}_0^i(0)$ is nonsingular, from (8.55), we have

$$\sum_{i=1}^{N} \mu_i(0)\hat{\beta}_0^i(1) = \sum_{i=1}^{N} \mu_i(0)\hat{\beta}_0^i(0) + \gamma(1)\sum_{i=1}^{N} \mu_i(0)\mu_i(1-d)g(0), \tag{8.56}$$

which can be written in the form:

$$\begin{aligned}X(1) &= Y(0) + \gamma(1)Z(0) \\ &= Y(0)\gamma(1)(I\gamma^{-1}(1) + Y^{-1}(0)Z(0)),\end{aligned} \tag{8.57}$$

with $X(1) = \sum_{i=1}^{N} \mu_i(0)\hat{\beta}_0^i(1)$, $Y(0) = \sum_{i=1}^{N} \mu_i(0)\hat{\beta}_0^i(0)$ and $Z(0) = \sum_{i=1}^{N} \mu_i(0)\mu_i(1-d)g(0)$. Hence, the matrix $X(1)$ is nonsingular as long as $\gamma^{-1}(1)$ is not an eigenvalue of $-Y^{-1}(0)Z(0)$, which can be easily satisfied by some $\gamma(1)$ with $\gamma_0 < \gamma(1) < 2 - \gamma_0$, $0 < \gamma_0 < 1$, that is, a suitable choice of $\gamma(1)$ can be made to ensure that the matrix $\sum_{i=1}^{N} \mu_i(0)\hat{\beta}_0^i(1)$ is nonsingular. With the above condition (C1), the matrix $\sum_{i=1}^{N} \mu_i(1)\hat{\beta}_0^i(1)$ is then also nonsingular, and the process can be continued to ensure that $\sum_{i=1}^{N} \mu_i(t)\hat{\beta}_0^i(t)$ is nonsingular, for all $t \geq 0$. Without the condition (C1), a search of $\gamma(1) \in (\gamma_0, 2 - \gamma_0)$ would be needed to ensure the nonsingularity of $\sum_{i=1}^{N} \mu_i(1)\hat{\beta}_0^i(1)$.

Such a result can be similarly derived, under the condition (C2): the nonsingularity of $\sum_{i=1}^{N} \mu_i(t-1)\hat{\beta}_0^i(t-1)$ implies that of $\sum_{i=1}^{N} \mu_i(t)\hat{\beta}_0^i(t-1)$. In this case, the related matrix equation is

$$\sum_{i=1}^{N}\mu_i(t)\hat{\beta}_0^i(t)=\sum_{i=1}^{N}\mu_i(t)\hat{\beta}_0^i(t-1)$$
$$+\gamma(t)\sum_{i=1}^{N}\mu_i(t)\mu_i(t-d)g(t-1), \tag{8.58}$$

obtained from multiplying (8.53) by $\mu_i(t)$, and summing up the resulting equation for $i=1,2,\ldots,N$.

The above conditions (C1) and (C2) would require either that the time variation of $\mu_i(t)$, that is, $\mu_i(t)-\mu_i(t-1)$, is small, or that the parameter matrices $\hat{\beta}_0^i(t)$ have some additional properties. Smallness of time variations has been an assumption commonly used in the literature when dealing with time-varying systems. It should be noted that $\mu_i(t)\in[0,1]$ for all time t based on its definition. Therefore, the time variation $|\mu_i(t)-\mu_i(t-1)|$ is actually bounded by 1. In addition, if the sampling time is set small enough, the different between $\xi(t)$ and $\xi(t-1)$ will be small, which results in small time variation $\mu_i(t)-\mu_i(t-1)$.

Remark 8.3 Under the framework developed in this chapter, the condition (C1) or (C2) is needed only when dealing with the uncertainty of the gain matrix $\sum_{i=1}^{N}\mu_i(t)\beta_0^i$, not the uncertainties of other parameters.

Next, we will focus our attention on some classes of systems with which the parameter matrices $\hat{\beta}_0^i(t)$ have some useful properties for ensuring the nonsingularity of $\sum_{i=1}^{N}\mu_i(t)\hat{\beta}_0^i(t)$, and also develop a general parameter projection algorithm to ensure such a key property.

A simple case. Consider the case when all parameter matrices β_0^i are the same, that is, $\beta_0^i=\beta_0, i=1,2,\ldots,N$, while other system matrices can all be different for different i. This corresponds to the system (8.21): $\bar{y}(t)=\sum_{i=1}^{N}\mu_i(\xi(t))(\alpha^i(z^{-1})[y](t)$ $+\beta^i(z^{-1})[u](t))$, with $\alpha^i(z^{-1})=\alpha_0^i+\alpha_1^i z^{-1}+\cdots+\alpha_{n-1}^i z^{-(n-1)}$ and $\beta^i(z^{-1})=$ $\beta_0^i+\beta_1^i z^{-1}+\cdots+\beta_{n_b-1}^i z^{-(n_b-1)}$ for $\beta_0^i=\beta_0, i=1,2,\ldots,N$.

In this case, there is only one parameter estimate $\hat{\beta}_0^i(t)=\hat{\beta}_0(t)$ whose nonsingularity can be ensured by some choice of $\gamma(t)$ from (8.53) with $\sum_{i=1}^{N}\mu_i(0)\hat{\beta}_0^i(0)=$ $\hat{\beta}_0(0)$ being nonsingular. Then, the matrix $\sum_{i=1}^{N}\mu_i(t)\hat{\beta}_0^i(t)=\hat{\beta}_0(t)$ is ensured to be nonsingular for all $t\geq 0$.

General case with parameter projection. In general, additional information would be needed for ensuring $\sum_{i=1}^{N}\mu_i(t)\hat{\beta}_0^i(t)$ being nonsingular, without requiring that the time variations $\mu_i(t)-\mu_i(t-1)$ of $\mu_i(t)$ are small. Such information can be given in a form suitable for performing parameter projection on the adaptive law (8.48) to make $\hat{\beta}_0^i(t)$ stay in some desired intervals.

To add the parameter projection function, the adaptive law (8.48) is modified as

$$\hat{\Theta}(t)=\hat{\Theta}(t-1)+\frac{\gamma(t)\phi(t-d)\varepsilon^T(t)}{c+\phi^T(t-d)\phi(t-d)}+F(t), \tag{8.59}$$

where $F(t)$ is the projection matrix whose other elements are all zeros except for those (denoted as $F^i(t) \in R^{r\times r}$) corresponding to $\hat{\beta}_0^i(t) \in R^{r\times r}$ in $\hat{\Theta}_i(t)$ of $\hat{\Theta}(t)$. More precisely, in terms of (8.53), the parameter projection adaptive law for $\hat{\beta}_0^i(t)$ is

$$\hat{\beta}_0^i(t) = \hat{\beta}_0^i(t-1) + \frac{\gamma(t)\mu_i(t-d)u(t-d)\varepsilon^T(t)}{c + \phi^T(t-d)\phi(t-d)} + F^i(t). \tag{8.60}$$

The specification of $F^i(t)$ depends on the needed property of the true parameters β_0^i, and in this case, the nonsingularity of $\sum_{i=1}^N \mu_i(t)\beta_0^i$, for all possible $\mu_i(t)$ and any $t \geq 0$ (see Assumption 8.4).

Let the jkth element of $\beta_0^i \in R^{r\times r}$ be β_{0jk}^i, $j, k = 1, 2, \ldots, r$, and, based on Assumption 8.4, assume the following.

Assumption 8.6 $\beta_{0jk}^i \in [\beta_{0jk}^{ia}, \beta_{0jk}^{ib}]$ for some known constants β_{0jk}^{ia} and β_{0jk}^{ib} such that for any $\hat{\beta}_{0jk}^i \in [\beta_{0jk}^{ia}, \beta_{0jk}^{ib}]$, with $\hat{\beta}_0^i \in R^{r\times r}$ whose jkth element is $\hat{\beta}_{0jk}^i$, $j, k = 1, 2, \ldots, r$, the matrix $\sum_{i=1}^N \mu_i(t)\hat{\beta}_0^i$ is nonsingular, for all possible $\mu_i(t)$ and any $t \geq 0$.

Under this assumption, we can construct a desired parameter projection algorithm. Denoting the jkth element of $\frac{\gamma(t)\mu_i(t-d)u(t-d)\varepsilon(t)}{c+\phi^T(t-d)\phi(t-d)}$ in (8.60) as $g_{jk}^i(t)$ and that of $F^i(t)$ as $f_{jk}^i(t)$, and choosing the initial parameter estimates as

$$\hat{\beta}_{0jk}^i(0) \in [\beta_{0jk}^{ia}, \beta_{0jk}^{ib}], \tag{8.61}$$

the projection function components can be set as

$$f_{jk}^i(t) = \begin{cases} 0 & \text{if } h_{jk}^i(t) \in [\beta_{0jk}^{ia}, \beta_{0jk}^{ib}], \\ \beta_{0jk}^{ib} - h_{jk}^i(t) & \text{if } h_{jk}^i(t) > \beta_{0jk}^{ib}, \\ \beta_{0jk}^{ia} - h_{jk}^i(t) & \text{if } h_{jk}^i(t) < \beta_{0jk}^{ia}, \end{cases} \tag{8.62}$$

where $h_{jk}^i(t) = \hat{\beta}_{0jk}^i(t-1) + g_{jk}^i(t)$, for $j, k = 1, 2, \ldots, r$, and $i = 1, 2, \ldots, N$.

This algorithm has the key properties: (i) $\hat{\beta}_{0jk}^i(t) \in [\beta_{0jk}^{ia}, \beta_{0jk}^{ib}]$ for any $t \geq 0$, $j, k = 1, 2, \ldots, r$, so that $\sum_{i=1}^N \mu_i(t)\hat{\beta}_0^i(t)$ is nonsingular, for all possible $\mu_i(t)$ and any $t \geq 0$, by Assumption 8.6, and (ii)[3]

$$\begin{aligned} q_{jk}^i(t) = f_{jk}^i(t)(\hat{\beta}_{0jk}^i(t-1) - \beta_{0jk}^i + g_{jk}^i(t) + f_{jk}^i(t)) \leq 0, \\ j, k = 1, 2, \ldots, r, i = 1, 2, \ldots, N \end{aligned} \tag{8.63}$$

so that the desired properties of Lemma 8.1 are still valid. To see this, we obtain the time increment of the positive definite function V defined in (8.49) as

---

[3]From (8.62), if $h_{jk}^i(t) > \beta_{0jk}^{ib}$, then $f_{jk}^i(t) = \beta_{0jk}^{ib} - h_{jk}^i(t) < 0$ so that $q_{jk}^i(t) = f_{jk}^i(t)(\beta_{0jk}^{ib} - \beta_{0jk}^i) \leq 0$ as $\beta_{0jk}^{ib} - \beta_{0jk}^i \geq 0$ by definition of β_{0jk}^{ib} (it is similar when $h_{jk}^i(t) < \beta_{0jk}^{ia}$).

$$
\begin{aligned}
&V(\tilde{\Theta}(t)) - V(\tilde{\Theta}(t-1)) \\
&\leq -\frac{a_0\|\varepsilon(t)\|^2}{c+\phi^T(t-d)\phi(t-d)} + 2\sum_{i=1}^{N}\sum_{j=1}^{r}\sum_{k=1}^{r} q^i_{jk}(t) \\
&\quad - \sum_{i=1}^{N}\sum_{j=1}^{r}\sum_{k=1}^{r}(f^i_{jk})^2,
\end{aligned} \tag{8.64}
$$

so that the desired adaptive law property (8.50) holds.

While in general the needed parameter bounds β^{ia}_{0jk} and β^{ib}_{0jk} in Assumption 8.6 may be difficult to specify (while their existence is based on Assumption 8.4), they may be figured out specifically for individual cases. For systems whose control gain parameter matrices β^i_0 have some special form, the specification of such bounds can be carried out without difficulty.

A special case with parameter projection. We consider the case when the control gain submatrices β^i_0 are all lower triangular, that is, their elements β^i_{0jk} are such that $\beta^i_{0jk} = 0$ for $k > j$. Since β^i_0 is nonsingular, $\beta^i_{0jj} \neq 0$ for $j = 1, 2, \ldots, r$. In this case, Assumption 8.4 holds if and only if $\sum_{i=1}^{N} \mu_i(t)\beta^i_{0jj} \neq 0$ for all $t \geq 0$, $j = 1, 2, \ldots, r$, and for parameter projection, one can simply set $\hat{\beta}^i_{0jk}(t) = 0$ for $k > j$, any $t \geq 0$, let $\hat{\beta}^i_{0jk}(t)$ free for $k < j$, that is, let the corresponding $f^i_{jk}(t) = 0$, any $t \geq 0$, and only specify the needed $f^i_{jj}(t)$, that is, the needed parameter bounds β^{ia}_{0jj} and β^{ib}_{0jj} for Assumption 8.6 based on Assumption 8.4, to make $\sum_{i=1}^{N} \mu_i(t)\hat{\beta}^i_{0jj}(t) \neq 0$ for all $t \geq 0$, $j = 1, 2, \ldots, r$. This can be fulfilled, as it is a scalar case individually. (The case when the control gain submatrices β^i_0 are all upper triangular can be dealt with through the similar way.)

In summary, the nonsingularity of $\sum_{i=1}^{N} \mu_i(t)\hat{\beta}^i_0(t)$ (the control coefficient matrix) is a new technical issue of parameter estimation for adaptive control of MIMO fuzzy systems. In this subsection, we have clarified this issue, conducted several case studies, and proposed some solution schemes.

8.3.4 Adaptive Control Scheme

With the parameter estimation algorithm (8.59), we use the following adaptive feedback control law:

$$
\sum_{i=1}^{N} \mu_i(t)\hat{\beta}^i(z^{-1})[u](t) = \bar{y}_m(t) - \sum_{i=1}^{N} \mu_i(t)\hat{\alpha}^i(z^{-1})[y](t), \tag{8.65}
$$

or equivalently,

$$u(t) = \left(\sum_{i=1}^{N} \mu_i \hat{\beta}_0^i\right)^{-1} \left(\bar{y}_m(t) - \sum_{i=1}^{N} \mu_i \hat{\alpha}^i(z^{-1})[y](t) - \sum_{i=1}^{N} \mu_i \hat{\bar{\beta}}^i(z^{-1})[u](t)\right),$$

where $\hat{\bar{\beta}}^i(z^{-1}) = \hat{\beta}_1^i z^{-1} + \cdots + \hat{\beta}_{n_b-1}^i z^{-(n_b-1)}$.

The control law (8.65) can be formulated in terms of $\hat{\Theta}(t)$:

$$\hat{\Theta}(t)^T \phi(t) = \bar{y}_m(t), \ \ \bar{y}_m(t) = \xi_d(z)[y_m](t). \tag{8.66}$$

The coefficient matrix of $u(t)$ in (8.65) is $\sum_{i=1}^{N} \mu_i(t)\hat{\beta}_0^i(t)$ whose nonsingularity, addressed in Sect. 8.3.3 and ensured by special designs of parameter adaptation, is crucial for solving the control Eq. (8.65).

Before establishing the stability and tracking properties of the adaptive control system, we first prove the following bounding property for the regressor vector $\phi(t)$.

Lemma 8.2 *Under Assumption 8.3 and with $\bar{e}(t) = \xi_d(z)[e](t)$ for $e(t) = y(t) - y_m(t)$ and $\xi_d(z)$ having degree d, the regressor $\phi(t)$ defined in (8.41) satisfies*

$$\|\phi(t-d)\| \leq \rho_1 + \rho_2 \max_{0 \leq \tau \leq t} \|\bar{e}(\tau - d)\|, \tag{8.67}$$

for some positive constants ρ_1 and ρ_2.

Proof For $\phi(t)$ in (8.41) and μ_i satisfying $\sum_{i=1}^{N} \mu_i = 1$, we have

$$\|\phi(t)\| \leq \kappa_1 \|\psi(t)\|, \tag{8.68}$$

for some constant $\kappa_1 > 0$, where $\psi(t) = [y^T(t), y^T(t-1), \ldots, y^T(t-n+1), u^T(t), u^T(t-1), \ldots, u^T(t-n_b+1)]^T$.

Since $y(t) = e(t) + y_m(t)$, $\psi(t)$ can be expressed as

$$\begin{aligned}\psi(t) = & [e^T(t), \ldots, e^T(t-n+1), u^T(t), \ldots, u^T(t-n_b+1)]^T \\ & + [y_m^T(t), \ldots, y_m^T(t-n+1), 0, 0, \ldots, 0]^T,\end{aligned} \tag{8.69}$$

we have

$$\|\psi(t)\| \leq \kappa_2 \max_{t-n+1 \leq \tau \leq t} \|e(\tau)\| + \kappa_3 \max_{t-n_b+1 \leq \tau \leq t} \|u(\tau)\| + \kappa_4, \tag{8.70}$$

where κ_i, $i = 2, 3, 4$, are some positive constants. Since $e(t) = \xi_d^{-1}(z)[\bar{e}](t)$ with $\xi_d^{-1}(z)$ being stable and strictly proper, we have from (8.70) that

$$\|\psi(t)\| \leq \bar{\kappa}_2 \max_{t-n+1 \leq \tau \leq t} \|\bar{e}(\tau)\| + \kappa_3 \max_{t-n_b+1 \leq \tau \leq t} \|u(\tau)\| + \kappa_4, \tag{8.71}$$

for some constant $\bar{\kappa}_2 > 0$.

The system (8.21) under Assumption 8.3 and (8.28) satisfies

$$\begin{aligned}\|u(t-d)\| &\le \kappa_5\|\bar{e}(t-d)\| + \kappa_6 \sum_{\tau=0,1,\ldots,t-1} \lambda^{t-\tau-1}\|\bar{e}(\tau-d)\| + \kappa_7 \\ &\le \kappa_8 \max_{0\le\tau\le t} \|\bar{e}(\tau-d)\| + \kappa_7, \end{aligned} \tag{8.72}$$

where κ_i, $i = 5, 6, 7, 8$, are some positive constants.

Finally, using (8.68), (8.70), and (8.72), we obtain

$$\|\phi(t-d)\| \le \rho_1 + \rho_2 \max_{0\le\tau\le t} \|\bar{e}(\tau-d)\|, \tag{8.73}$$

where ρ_1 and ρ_2 are some positive constants. ∇

We now establish the stability and asymptotic tracking properties of the closed-loop control system.

Theorem 8.1 *The adaptive feedback control law (8.65), updated by the parameter adaptive law (8.59) and applied to the system (8.21) subject to Assumptions 8.1–8.5, ensures that all the closed-loop signals are bounded and* $\lim_{t\to\infty}(y(t) - y_m(t)) = 0$.

Proof From (8.40) and (8.46), we have

$$\varepsilon(t) = -(\hat{\Theta}(t-1) - \Theta)^T\phi(t-d), \tag{8.74}$$

and from (8.40) and (8.66), for $\bar{e}(t) = \bar{y}(t) - \bar{y}_m(t)$, we have

$$\begin{aligned}\bar{e}(t-d) &= -(\hat{\Theta}(t-d) - \Theta)^T\phi(t-d) \\ &= \varepsilon(t) - (\hat{\Theta}(t-d) - \hat{\Theta}(t-1))^T\phi(t-d) \\ &= \bar{\varepsilon}(t)\sqrt{c + \phi^T(t-d)\phi(t-d)} \\ &\quad -(\hat{\Theta}(t-d) - \hat{\Theta}(t-1))^T\bar{\phi}(t-d)m_\phi(t), \end{aligned} \tag{8.75}$$

(under the solvability of (8.66)), where

$$\begin{aligned} m_\phi(t) &= \sqrt{c + \phi^T(t-d)\phi(t-d)}, \\ \bar{\varepsilon}(t) &= \frac{\varepsilon(t)}{\sqrt{c + \phi^T(t-d)\phi(t-d)}}, \\ \bar{\phi}(t-d) &= \frac{\phi(t-d)}{\sqrt{c + \phi^T(t-d)\phi(t-d)}}, \end{aligned} \tag{8.76}$$

with $\bar{\varepsilon}(t) \in L_2 \cap L_\infty$, $\hat{\Theta}(t-d) - \hat{\Theta}(t-1) \in L_2 \cap L_\infty$, and $\|\bar{\phi}(t-d)\| \le 1$. Using the inequality

$$m_\phi(t) \le \sqrt{c} + \|\phi^T(t-d)\|, \tag{8.77}$$

we express $\bar{e}(t-d)$ from (8.75) as

$$\|\bar{e}(t-d)\| \le c_1 + \|\bar{\varepsilon}(t)\| \|\phi^T(t-d)\| \\ +\|\hat{\Theta}(t-d) - \hat{\Theta}(t-1)\| \, \|\phi^T(t-d)\|, \tag{8.78}$$

for some constant $c_1 > 0$. Using Lemma 8.2, we obtain

$$\|\bar{e}(t-d)\| \le c_2 + c_3\|\bar{\varepsilon}(t)\| \max_{0\le\tau\le t} \|\bar{e}(\tau-d)\| \\ + c_4\|\hat{\Theta}(t-d) - \hat{\Theta}(t-1)\| \max_{0\le\tau\le t} \|\bar{e}(\tau-d)\|, \tag{8.79}$$

for some constants $c_i > 0$, $i = 2, 3, 4$.

From Lemma 8.1, we have that $\lim_{t\to\infty} \bar{\varepsilon}(t) = 0$ and $\lim_{t\to\infty} \|\hat{\Theta}(t) - \hat{\Theta}(t-d)\| = 0$, and with these properties, it follows from (8.79) that $\bar{e}(t)$ is bounded, which implies that $e(t)$ and in turn $y(t)$ are bounded, and from the system's minimum phase property that $u(t)$ is bounded. Hence, all signals in the closed-loop system are bounded, based on which, from (8.75) in which $\bar{\varepsilon}(t) \in L_2$ and $\hat{\Theta}(t-d) - \hat{\Theta}(t-1) \in L_2$, we have that $\bar{e}(t) \in L_2$ so that $\lim_{t\to\infty} \bar{e}(t) = 0$. With $e(t) = \xi_d^{-1}(z)[\bar{e}](t)$ and $\xi_d^{-1}(z)$ being a strictly proper and stable transfer function, we have $\lim_{t\to\infty} e(t) = 0$. ∇

Thus far, we have solved Problem II, the adaptive control problem formulated in Sect. 8.1, for the system (8.21) with uncertain parameters. Next, we present an illustrative example with simulation results to show the key steps of the adaptive control design and the desired adaptive control system performance.

8.4 Simulation Study

In this section,[4] we present a numerical example to show the proposed system modeling and parametrization and the feedback control designs, and to demonstrate the effectiveness of the proposed adaptive control scheme.

8.4.1 Simulation System

Considering the two-input-two-output T–S fuzzy model:

$$R^i: \text{ IF } \; y_1(t) \text{ is } F_1^i \text{ and } y_2(t) \text{ is } F_2^i, \\ \text{THEN } \; A^i(z^{-1})[y](t) = B^i(z^{-1})[u](t), \tag{8.80}$$

[4]Parts of Sect. 8.4 are reprinted from Qi et al. (2014), Copyright 2014, with permission from Elsevier.

where, in the same form as that of Example 8.1 introduced in Sect. 8.2.3, the system dynamics matrices are

$$
\begin{aligned}
A^1(z^{-1}) &= \begin{bmatrix} 1 & 0 \\ 0 & 1 \end{bmatrix} + \begin{bmatrix} 2 & 0 \\ 0 & 0.5 \end{bmatrix} z^{-1} + \begin{bmatrix} 1 & 0 \\ 0 & 3 \end{bmatrix} z^{-2} \\
B^1(z^{-1}) &= z^{-1} \left(\begin{bmatrix} 1 & 2 \\ 0 & 0 \end{bmatrix} + \begin{bmatrix} 0.5 & 2 \\ 0 & 1 \end{bmatrix} z^{-1} \right) \\
A^2(z^{-1}) &= \begin{bmatrix} 1 & 0 \\ 0 & 1 \end{bmatrix} + \begin{bmatrix} 4 & 0 \\ 0 & 0.2 \end{bmatrix} z^{-1} + \begin{bmatrix} 2 & 0 \\ 0 & 4 \end{bmatrix} z^{-2} \\
B^2(z^{-1}) &= z^{-1} \left(\begin{bmatrix} 2 & 3 \\ 0 & 0 \end{bmatrix} + \begin{bmatrix} 1 & 2 \\ 0 & 1 \end{bmatrix} z^{-1} \right)
\end{aligned}
\tag{8.81}
$$

and the membership functions for $\mathscr{F}_1^i$ and $\mathscr{F}_2^i$ are

$$
F_1^1(y_1(t)) = \begin{cases} 0 & y_1(t) \leq -6 \\ 0.5 + \frac{y_1(t)}{12} & -6 < y_1(t) < 6 \\ 1 & y_1(t) \geq 6 \end{cases} \tag{8.82}
$$

$$
F_1^2(y_1(t)) = \begin{cases} 1 & y_1(t) \leq -6 \\ 0.5 - \frac{y_1(t)}{12} & -6 < y_1(t) < 6 \\ 0 & y_1(t) \geq 6 \end{cases} \tag{8.83}
$$

$$
F_2^1(y_2(t)) = \begin{cases} 0 & y_2(t) \leq -4 \\ 0.5 + \frac{y_2(t)}{8} & -4 < y_2(t) < 4 \\ 1 & y_2(t) \geq 4 \end{cases} \tag{8.84}
$$

$$
F_2^2(y_2(t)) = \begin{cases} 1 & y_2(t) \leq -4 \\ 0.5 - \frac{y_2(t)}{8} & -4 < y_2(t) < 4 \\ 0 & y_2(t) \geq 4. \end{cases} \tag{8.85}
$$

As derived in Sect. 8.2.3, with the interactor matrix

$$
\xi_d(z) = \begin{bmatrix} z & 0 \\ 0 & z^2 \end{bmatrix}, \tag{8.86}
$$

from (8.36) and (8.81), we have

$$
\begin{aligned}
\alpha^1(z^{-1}) &= \begin{bmatrix} -2 & 0 \\ 0 & -2.75 \end{bmatrix} + \begin{bmatrix} -1 & 0 \\ 0 & 1.5 \end{bmatrix} z^{-1} \\
\beta^1(z^{-1}) &= \begin{bmatrix} 1 & 2 \\ 0 & 1 \end{bmatrix} + \begin{bmatrix} 0.5 & 2 \\ 0 & -0.5 \end{bmatrix} z^{-1} \\
\alpha^2(z^{-1}) &= \begin{bmatrix} -4 & 0 \\ 0 & 0 \end{bmatrix} + \begin{bmatrix} -2 & 0 \\ 0 & 8 \end{bmatrix} z^{-1} \\
\beta^2(z^{-1}) &= \begin{bmatrix} 2 & 3 \\ 0 & 1 \end{bmatrix} + \begin{bmatrix} 1 & 2 \\ 0 & -0.2 \end{bmatrix} z^{-1}.
\end{aligned}
$$

As in (8.21) with $\mu_i(t) = \mu_i(\xi(t))$, we obtain the global fuzzy system prediction system model:

$$\bar{y}(t) = \sum_{i=1}^{2} \mu_i(\xi(t))(\alpha^i(z^{-1})[y](t) + \mu_i \beta^i(z^{-1})[u](t)), \tag{8.87}$$

where $\mu_i(\xi(t))$ is the normalized firing strength:

$$\mu_i(\xi(t)) = \frac{\lambda^i(\xi(t))}{\sum_{i=1}^{2} \lambda^i(\xi(t))}, \quad \lambda^i(\xi(t)) = \prod_{j=1}^{2} F_j^i(\xi_j(t)) \tag{8.88}$$

with $\xi(t) = [\xi_1(t), \xi_2(t)]^T = [y_1(t), y_2(t)]^T$.

From the expressions of $\beta^1(z^{-1})$ and $\beta^2(z^{-1})$ below (8.86), we can calculate the zeros of $\beta^1(z^{-1})$ and $\beta^2(z^{-1})$ as

$$z_1^1 = 0.5, \quad z_2^1 = 0.5, \quad z_1^2 = 0.5, \quad z_2^2 = 0.2. \tag{8.89}$$

Since in this case, $\beta^i(z^{-1}), i = 1, 2$, are first order and triangular, it can be verified that the condition that all the zeros of $\beta^i(z^{-1}), i = 1, 2$, are in the unit circle $|z| < 1$ which ensures that the global fuzzy system prediction model (8.87) is minimum phase, given that $\mu_i(\xi(t)) \in [0, 1], i = 1, 2$.

If the parameter of $\alpha^i(z^{-1})$ and $\beta^i(z^{-1})$ are known, the nominal control law (8.38) can be applied and the control signal $u(t)$ can be derived from

$$\sum_{i=1}^{2} \mu_i \beta^i(z^{-1})[u](t) = \bar{y}_m(t) - \sum_{i=1}^{2} \mu_i \alpha^i(z^{-1})[y](t). \tag{8.90}$$

If the parameters of $\alpha^i(z^{-1})$ and $\beta^i(z^{-1})$ are unknown, the adaptive control scheme (8.65) is applied with the parameter estimation (8.59).

8.4.2 Simulation Results

In the simulation, the initial parameter values are set as 50% of the true values and the adaptive gain $\gamma(t)$ is chosen as 0.5 (in this simulation, there was no need to adjust $\gamma(t)$ for solving (8.65)). The control objective is for the system output $y(t)$ to track the reference signal $y_m(t) = [2\sin(0.2t), \sin(0.2t)]^T$. The adaptive system responses and control signals are shown in Figs. 8.1 and 8.2, respectively. Figures 8.3, 8.4, 8.5, and 8.6 present the parameter adaptation results.

The simulation results show that the developed MIMO adaptive fuzzy control scheme can achieve the desired system performance: closed-loop system stability and asymptotic tracking of $y_m(t)$ by $y(t)$, despite the system parameter uncertainties.

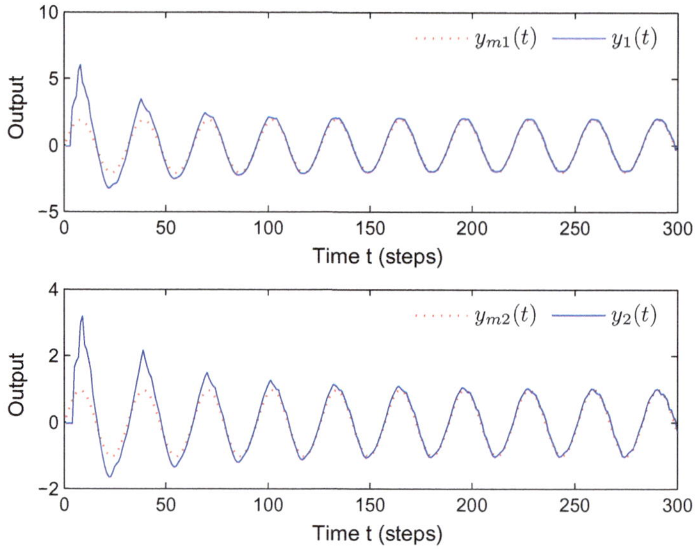

Fig. 8.1 Adaptive system responses. Reprinted from Qi et al. (2014), Copyright 2014, with permission from Elsevier

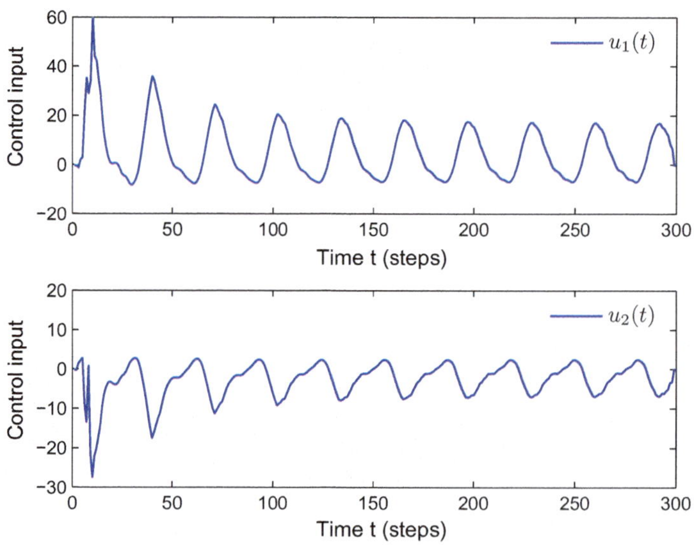

Fig. 8.2 Control signals. Reprinted from Qi et al. (2014), Copyright 2014, with permission from Elsevier

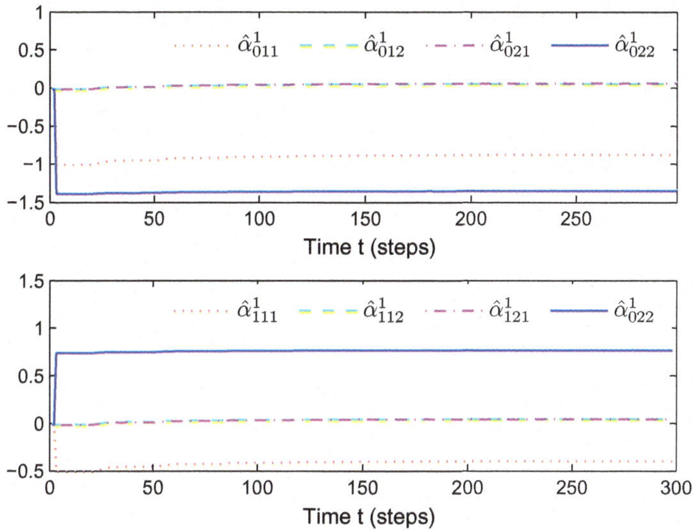

Fig. 8.3 Parameter estimation of $\alpha^1(z^{-1})$. Reprinted from Qi et al. (2014), Copyright 2014, with permission from Elsevier

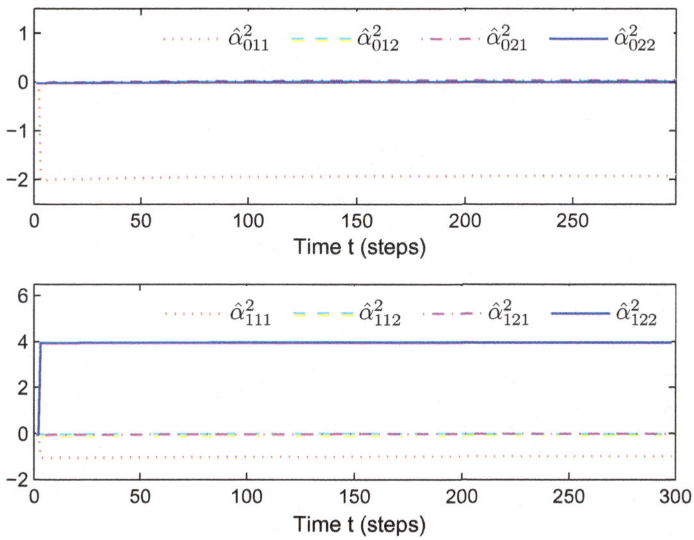

Fig. 8.4 Parameter estimation of $\alpha^2(z^{-1})$. Reprinted from Qi et al. (2014), Copyright 2014, with permission from Elsevier

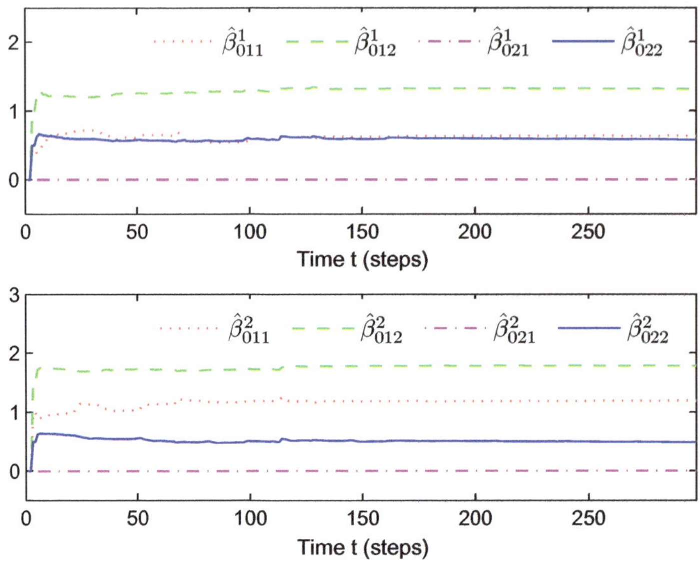

Fig. 8.5 Parameter estimation of β_0^1 and β_0^2. Reprinted from Qi et al. (2014), Copyright 2014, with permission from Elsevier

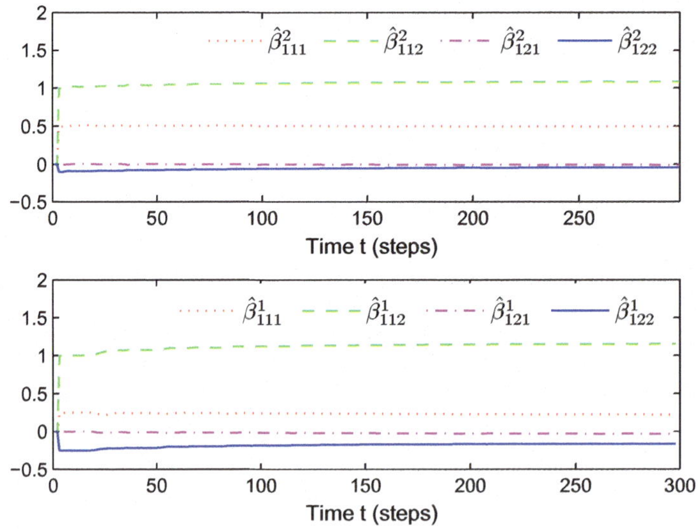

Fig. 8.6 Parameter estimation of β_1^1 and β_1^2. Reprinted from Qi et al. (2014), Copyright 2014, with permission from Elsevier

8.5 Summary

In this chapter, we have studied an adaptive fuzzy control scheme for nonlinear MIMO systems in the input–output form based on discrete-time MIMO T–S fuzzy models with general delay matrices. Similar to that for a regular adaptive control problem, a prediction model is crucial for solving a fuzzy adaptive control problem, which has been derived as a fuzzy prediction model with a general delay structure. A model-based adaptive controller is then designed and closed-loop stability and tracking performance analysis have been given. Simulation results have demonstrated the desired stability and tracking performance of the developed adaptive fuzzy control systems.

References

Goodwin GC, Sin KS (1984) Adaptive filtering prediction and control. Prentice-Hall, Englewood Cliffs, NJ

Qi R, Tao G, Jiang B, Gong H (2011) Adaptive control of MIMO T-S fuzzy systems with general delay matrice. In: Proceeding of the 18th IFAC world congress. Milano, Italy, pp 3446–3450

Qi R, Tao G, Jiang B (2014) Adaptive control of MIMO time-varying systems with indicator function based parametrization. Automatica 50(5):1369–1380

Tao G, Ioannou PA (1989) Robust stability and performance improvement of multivariable adaptive control systems. Int J Control 50(5):1825–1855

Chapter 9
Adaptive T–S Fuzzy Control with Unknown Membership Functions

In the previous chapter, the T–S fuzzy dynamic systems were formulated into linearly parametrized models, based on which parameter adaptive laws were designed and closed-loop system stability was analyzed. However, there are two group of parameters in a T–S fuzzy system: consequent parameters and membership functions parameters. The former are usually linearly dependent parameters while the latter are usually nonlinearly dependent, i.e., parameters of Gaussian or Sigmoidal membership functions. Most adaptive control approaches assume the parameters of membership functions are accurate enough so that only the uncertainties in consequent parameters are considered. However, in practice, it is difficult to set the membership function parameters accurately in advance. Thus, it is of major interest to develop effective schemes to deal with membership parameter uncertainties.

In this chapter, we address such an issue using an adaptive estimation method with a gradient algorithm derived based on a nonlinearly parameterized error model resulted from the parameter nonlinearity of membership functions. Our study is conducted based on a discrete-time input–output multiple-delay prediction T–S fuzzy dynamic system model. Such a model has been used in Sects. 7.2 and 7.3, for adaptive fuzzy control, and their designs make use of the knowledge of the membership parameters. In this section, we solve the adaptive control problem without using such membership parameter knowledge, and our adaptive control design use adaptation to estimate the membership parameters, in addition to local system parameters.

R. Qi et al., *Fuzzy System Identification and Adaptive Control*, Communications and Control Engineering,
https://doi.org/10.1007/978-3-030-19882-4_9

9.1 Problem Statement

Consider the fuzzy system model (7.13), that is

$$y(t+d) = \sum_{i=1}^{N} \mu_i \alpha_i(z^{-1})[y](t) + \sum_{i=1}^{N} \mu_i \beta_i(z^{-1})[u](t), \tag{9.1}$$

where μ_i is the normalized firing strength satisfying

$$\mu_i(\xi) = \frac{\lambda_i(\xi)}{\sum_{i=1}^{N} \lambda_i(\xi)}, \quad \lambda_i(\xi) = \prod_{j=1}^{L} F_j^i(\xi_j),$$
$$\mu_i(\xi) \geq 0, \quad \sum_{i=1}^{N} \mu_i(\xi) = 1. \tag{9.2}$$

The goal of this chapter is to address the issue of parameter uncertainties of the membership functions $F_j^i(\xi_j(t))$ for adaptive control of the fuzzy system (9.1).

Parameters in $\mu_i(\xi)$. The normalized firing strength $\mu_i(\xi)$ depends on the selection of the membership functions $F_j^i(\xi_j)$, $i = 1, 2, \ldots, N$, $j = 1, 2, \ldots, L$. For example, for the Gaussian membership function

$$F_j^i(\xi_j) = \exp\left\{ -\frac{(\xi_j - c_j^i)^2}{\sigma_j^{i2}} \right\}, \tag{9.3}$$

the parameters are σ_j^{i2} and c_j^i, and for the sigmoidal-shape membership function

$$F_j^i(\xi_j) = \frac{1}{1 + \exp\{-\kappa_j^i(\xi_j - c_j^i)\}}, \tag{9.4}$$

the parameters are c_j^i and κ_j^i. In the adaptive control design of Chap. 7, those membership function parameters are assumed known. In this chapter, we consider the adaptive control design with unknown membership function parameters under the following assumptions:

Assumption 9.1 The fuzzy system (9.1) is minimum phase.

The minimum phase property of a fuzzy system is defined in Definition 7.1.

We will first derive an adaptive parameter estimation algorithm for the system (9.1) with uncertain parameters in $\mu_i(\xi(t))$, $\alpha_i(z^{-1})$, and $\beta_i(z^{-1})$, and then design an adaptive control law and analyze its baseline and robustness properties.

9.2 Parameter Estimation Algorithm

In this section, we develop a solution to the adaptive parameter estimation problem involving a nonlinearly parameterized prediction system model when the parameters in the membership functions μ_i are unknown. We first derive an error model based on an estimation error signal introduced from the nonlinear parametrization, and then design a parameter adaptive law based on a gradient algorithm to estimate the prediction system parameters.

9.2.1 Nonlinearly Parameterized Model

To estimate the system parameters in both $\mu_i(\xi(t)) = \mu_i(\theta_\mu; t)$ (for some parameter vector θ_μ to be defined) and $(\alpha_i(z^{-1}), \beta_i(z^{-1}))$, we need to develop a parameterized model. With the knowledge of n and d, the fuzzy system (9.1) can be expressed as

$$y(t+d) = F(\theta_s, \theta_\mu; t) \triangleq \sum_{i=1}^{N} f_i^T(\theta_i, \theta_\mu; t)\phi_0(t), \tag{9.5}$$

where

$$f_i(\theta_i, \theta_\mu; t) = \mu_i(\xi(t))\theta_i = \mu_i(\theta_\mu; t)\theta_i, \tag{9.6}$$

$$\begin{aligned} \phi_0(t) &= [y(t), y(t-1), \ldots, y(t-n+1), \\ &\qquad u(t), u(t-1), \ldots, u(t-n+1)]^T, & (9.7) \\ \theta_s &= [\theta_1^T, \theta_2^T, \ldots, \theta_N^T]^T, & (9.8) \\ \theta_i &= [\alpha_{i,0}, \alpha_{i,1}, \ldots, \alpha_{i,n-1}, \beta_{i,0}, \beta_{i,1}, \ldots, \beta_{i,n-1}]^T, & (9.9) \end{aligned}$$

and the parameter vector θ_μ depends on the selection of the membership functions $F_j^i(\xi_j)$, $i = 1, 2, \ldots, N$, $j = 1, 2, \ldots, L$. For the Gaussian membership functions in (9.3), with $\nu_j^i = \sigma_j^{i2}$, for μ_i in (9.2), we have

$$\theta_\mu = [c_1^1, \nu_1^1, \ldots, c_L^1, \nu_L^1, \ldots, c_1^N, \nu_1^N, \ldots, c_L^N, \nu_L^N]^T. \tag{9.10}$$

For the sigmoidal membership functions in (9.4), we have

$$\theta_\mu = [c_1^1, \kappa_1^1, \ldots, c_L^1, \kappa_L^1, \ldots, c_1^N, \kappa_1^N, \ldots, c_L^N, \kappa_L^N]^T. \tag{9.11}$$

In both cases, θ_μ is an $n_\mu = 2NL$-dimensional vector.

Remark 9.1 The model (9.5) is a nonlinearly parameterized model when θ_μ is unknown. When the parameters of θ_μ are known, this model reduces to the following linearly parameterized model (as given in Sect. 7.2.3)

$$y(t+d) = \theta_s^T \phi(t), \tag{9.12}$$

where

$$\phi(t) = [\phi_1^T(t), \phi_2^T(t), \ldots, \phi_N^T(t)]^T, \tag{9.13}$$
$$\phi_i(t) = [\mu_i y(t), \mu_i y(t-1), \ldots, \mu_i y(t-n+1), \tag{9.14}$$
$$\mu_i u(t), \mu_i u(t-1), \ldots, \mu_i u(t-n+1)]^T. \tag{9.15}$$

In this linearly parameterized model, θ_s is unknown and $\phi(t)$ is known, for which linear parameter estimation algorithms (such as a gradient algorithm or least squares algorithm), can be used to estimate θ. A key step is to introduce an estimation error

$$\varepsilon(t) = y(t) - \hat{\theta}_s^T(t-1)\phi(t-d), \tag{9.16}$$

where $\hat{\theta}_s(t-1)$ is the estimate of θ_s at time $t-1$. For the linear parametrization case with (9.12), this estimation error $\varepsilon(t)$ leads to a linear error model

$$\varepsilon(t) = -(\hat{\theta}_s(t-1) - \theta_s)^T \phi(t-d). \tag{9.17}$$

For the nonlinearly parameterized model (9.5), a gradient algorithm can still be used but it needs to be designed in a nonlinear parametrization form. □

9.2.2 Estimation Error Model

Let $\hat{\theta}(t)$ and $\hat{\theta}_\mu(t)$ be the estimates of the unknown parameter vectors θ and θ_μ, respectively. Then we introduce the estimation error signal

$$\begin{aligned}\varepsilon(t) &= y(t) - \sum_{i=1}^{N} f_i^T(\hat{\theta}_i(t-1), \hat{\theta}_\mu(t-1); t-d)\phi_0(t-d)\\ &= y(t) - F(\hat{\theta}_s(t-1), \hat{\theta}_\mu(t-1); t-d).\end{aligned} \tag{9.18}$$

From (9.5) and (9.18), we obtain the estimation error model

$$\varepsilon(t) = \sum_{i=1}^{N} \Big(f_i(\theta_i, \theta_\mu; t-d) - f_i(\hat{\theta}_i(t-1), \hat{\theta}_\mu(t-1); t-d) \Big)^T \phi_0(t-d). \tag{9.19}$$

Note that, (9.18) gives a "physical" expression (in the sense that all of its signals are available) of $\varepsilon(t)$ useful for implementation, while (9.19) gives a mathematical expression (with θ and θ_μ unknown) of $\varepsilon(t)$ useful for analysis.

For the nonlinear parametrization case with (9.5), an approximation is needed to reach an approximated linear error model suitable for adaptive law development. To proceed, using Taylor series expansion, we express

$$\begin{aligned} f_i(\theta_i, \theta_\mu; t-d) &= f_i(\hat{\theta}_i(t-1), \hat{\theta}_\mu(t-1); t-d) \\ &+ \frac{\partial f_i}{\partial \theta_i}|_{(\hat{\theta}_i(t-1), \hat{\theta}_\mu(t-1); t-d)} (\theta_i - \hat{\theta}_i(t-1)) \\ &+ \frac{\partial f_i}{\partial \theta_\mu}|_{(\hat{\theta}_i(t-1), \hat{\theta}_\mu(t-1); t-d)} (\theta_\mu - \hat{\theta}_\mu(t-1)) + \delta_i(t), \end{aligned} \tag{9.20}$$

where $\delta_i(t)$ includes all "higher order partial derivative or error terms" in a standard Taylor series.

For $f_i(\theta_i, \theta_\mu; t) = \mu_i(\theta_\mu; t)\theta_i$ in (9.6) with $\mu_i(\theta_\mu; t) \in R$ and $\theta_i \in R^{2n}$, we have

$$\frac{\partial f_i}{\partial \theta_i}|_{(\hat{\theta}_i(t-1), \hat{\theta}_\mu(t-1); t-d)} = \mu_i(\hat{\theta}_\mu(t-1); t-d) I_{2n}, \tag{9.21}$$

$$\begin{aligned} \frac{\partial f_i}{\partial \theta_\mu}|_{(\hat{\theta}_i(t-1), \hat{\theta}_\mu(t-1); t-d)} &= \theta_i \otimes \frac{\partial \mu_i}{\partial \theta_\mu}|_{(\hat{\theta}_i(t-1), \hat{\theta}_\mu(t-1); t-d)} \\ &\triangleq \hat{\theta}_i \otimes \frac{\partial \hat{\mu}_i}{\partial \hat{\theta}_\mu}, \end{aligned} \tag{9.22}$$

where $\otimes$ is the Kronecker product, and

$$\frac{\partial \mu_i}{\partial \theta_\mu} = \left[\frac{\partial \mu_i}{\partial \theta_{\mu 1}}, \ldots, \frac{\partial \mu_i}{\partial \theta_{\mu n_\mu}} \right], \tag{9.23}$$

for $\theta_\mu = [\theta_{\mu 1}, \ldots, \theta_{\mu n_\mu}]^T$, that is, the jth row of $\theta_i \otimes \frac{\partial \mu_i}{\partial \theta_\mu}$ is $\theta_{ij} \left[\frac{\partial \mu_i}{\partial \theta_{\mu 1}}, \ldots, \frac{\partial \mu_i}{\partial \theta_{\mu n_\mu}} \right]$, $j = 1, 2, \ldots, 2n$, for

$$\theta_i = [\theta_{i1}, \theta_{i2}, \ldots, \theta_{i\,2n}]^T. \tag{9.24}$$

Moreover, from the linearity of $f_i(\theta_i, \theta_\mu; t)$ in θ_i, we have

$$\frac{\partial^k f_i}{\partial \theta_i^k} = 0, \; k > 1, \tag{9.25}$$

$$\frac{\partial^{k+j} f_i}{\partial \theta_i^k \partial \theta_\mu^j} = 0, \ k > 1, \ j > 0. \tag{9.26}$$

Hence, the higher order terms in $\delta_i(t)$ only include $(\theta_{ij} - \hat{\theta}_{ij})(\theta_{\mu l} - \hat{\theta}_{\mu l})^k, k \geq 1$, and $(\theta_{\mu l} - \hat{\theta}_{\mu l})^k$, $k > 1$, etc., that is, no $(\theta_{ij} - \hat{\theta}_{ij})^k$ $(k > 1)$ related terms are included, where θ_{ij} and $\theta_{\mu l}$ are the components of θ_i and θ_μ, with their estimates $\hat{\theta}_{ij}$ and $\hat{\theta}_{\mu l}$ in $\hat{\theta}_i$ and $\hat{\theta}_\mu$.

To simplify the notation, we denote

$$\frac{\partial \hat{f}_i}{\partial \hat{\theta}_i} = \frac{\partial f_i}{\partial \theta_i}|_{(\hat{\theta}_i(t-1), \hat{\theta}_\mu(t-1); t-d)}, \tag{9.27}$$

$$\frac{\partial \hat{f}_i}{\partial \hat{\theta}_\mu} = \frac{\partial f_i}{\partial \theta_\mu}|_{(\hat{\theta}_i(t-1), \hat{\theta}_\mu(t-1); t-d)}. \tag{9.28}$$

Assuming $\delta_i(t) = 0$ in (9.20), we have

$$f_i(\theta_i, \theta_\mu; t-d) = f_i(\hat{\theta}_i(t-1), \hat{\theta}_\mu(t-1); t-d) + \frac{\partial \hat{f}_i}{\partial \hat{\theta}_i}(\theta_i - \hat{\theta}_i(t-1)) + \frac{\partial \hat{f}_i}{\partial \hat{\theta}_\mu}(\theta_\mu - \hat{\theta}_\mu(t-1)). \tag{9.29}$$

Then, $\varepsilon(t)$ in (9.19) can be expressed as

$$\varepsilon(t) = \sum_{i=1}^{N} \left((\theta_i - \hat{\theta}_i(t-1))^T \left(\frac{\partial \hat{f}_i}{\partial \hat{\theta}_i} \right)^T \phi_0(t-d) + (\theta_\mu - \hat{\theta}_\mu(t-1))^T \left(\frac{\partial \hat{f}_i}{\partial \hat{\theta}_\mu} \right)^T \phi_0(t-d) \right). \tag{9.30}$$

This is a linearly parameterized model as the basic error equation for the adaptive parameter estimation problem, with which a baseline adaptive algorithm can be designed.

9.2.3 Adaptive Law

Based on the estimation error model (9.30), we design the adaptive law for updating the parameter estimates $\hat{\theta}_i$ and $\hat{\theta}_\mu$ as

$$\hat{\theta}_i(t) = \hat{\theta}_i(t-1) + \frac{\gamma(t)\left(\frac{\partial \hat{f}_i}{\partial \hat{\theta}_i}\right)^T \phi_0(t-d)\varepsilon(t)}{m^2(t)}, \tag{9.31}$$

$$\hat{\theta}_\mu(t) = \hat{\theta}_\mu(t-1) + \frac{\gamma(t)\sum_{i=1}^{N}\left(\frac{\partial \hat{f}_i}{\partial \hat{\theta}_\mu}\right)^T \phi_0(t-d)\varepsilon(t)}{m^2(t)}, \tag{9.32}$$

where $\gamma(t)$ is the adaptation gain which, based on the linear-like error model (9.30), should satisfy: $\gamma(t) \in (\gamma_0, 2-\gamma_0)$ for some constant $\gamma_0 \in (0, 1)$, and

$$m^2(t) = c + \sum_{i=1}^{N} \| \left(\frac{\partial \hat{f}_i}{\partial \hat{\theta}_i}\right)^T \phi_0(t-d)\|^2 + \| \sum_{i=1}^{N} \left(\frac{\partial \hat{f}_i}{\partial \hat{\theta}_\mu}\right)^T \phi_0(t-d)\|^2 \tag{9.33}$$

with $c > 0$ and $\|\cdot\|$ being the l^2 vector norm.

This adaptive law is a gradient algorithm of the form

$$\hat{\theta}(t) = \hat{\theta}(t-1) - \gamma \frac{\partial J}{\partial \hat{\theta}(t-1)}, \tag{9.34}$$

for J being a quadratic function of ε (Ioannou and Sun 1996), applied to our T–S fuzzy control case with a nonlinear parametrization (9.5). For the adaptive laws (9.31) and (9.32), we have used $J = \frac{1}{2}\frac{\varepsilon^2}{m^2(t)}$ and such a normalized cost function J enables us to choose an adaptation gain $\gamma(t)$ to ensure the stability of parameter adaptation, as shown next.

9.2.4 Stability Analysis

Stability analysis for error model (9.30): $\delta_i(t) = 0$. We first analyze the stability of the adaptive laws (9.31) and (9.32) for the approximated estimation error model (9.30) without $\delta_i(t)$. Define the parameter errors

$$\begin{aligned}\tilde{\theta}_i(t) &= \hat{\theta}_i(t) - \theta_i, \quad i = 1, 2, \ldots, N \\ \tilde{\theta}_\mu(t) &= \hat{\theta}_\mu(t) - \theta_\mu.\end{aligned} \tag{9.35}$$

Then, the error model (9.30) can be written into

$$\varepsilon(t) = -\sum_{i=1}^{N}\left(\tilde{\theta}_i^T(t-1)\left(\frac{\partial \hat{f}_i}{\partial \hat{\theta}_i}\right)^T \phi_0(t-d) - \tilde{\theta}_\mu^T(t-1)\left(\frac{\partial \hat{f}_i}{\partial \hat{\theta}_\mu}\right)^T \phi_0(t-d)\right) \tag{9.36}$$

which can be further formulated as

$$\varepsilon(t) = -\tilde{\theta}^T(t-1)\psi(t-d), \tag{9.37}$$

where

$$\begin{aligned}
\tilde{\theta}(t-1) &= [\tilde{\theta}_1^T(t-1), \ldots, \tilde{\theta}_N^T(t-1), \tilde{\theta}_\mu^T(t-1)]^T, \\
\psi(t-d) &= \left[\phi_0^T(t-d)\left(\frac{\partial \hat{f}_1}{\partial \hat{\theta}_1}\right), \ldots, \phi_0^T(t-d)\left(\frac{\partial \hat{f}_N}{\partial \hat{\theta}_N}\right),\right. \\
&\quad \left.\sum_{i=1}^N \phi_0^T(t-d)\left(\frac{\partial \hat{f}_i}{\partial \hat{\theta}_\mu}\right)\right]^T.
\end{aligned} \tag{9.38}$$

Choose the Lyapunov function $V(t) = \tilde{\theta}^T(t)\tilde{\theta}(t)$ and define $\triangle V(t) = V(t) - V(t-1)$. Then, we have

$$\begin{aligned}
\triangle V(t) &= \tilde{\theta}^T(t)\tilde{\theta}(t) - \tilde{\theta}^T(t-1)\tilde{\theta}(t-1) \\
&= \sum_{i=1}^N \left(\tilde{\theta}_i^T(t)\tilde{\theta}_i(t) - \tilde{\theta}_i^T(t-1)\tilde{\theta}_i(t-1)\right) \\
&\quad + \tilde{\theta}_\mu^T(t)\tilde{\theta}_\mu(t) - \tilde{\theta}_\mu^T(t-1)\tilde{\theta}_\mu(t-1).
\end{aligned} \tag{9.39}$$

Subtracting θ_i from both sides of (9.31) and θ_μ from both sides of (9.32), we obtain

$$\tilde{\theta}_i(t) = \tilde{\theta}_i(t-1) + \frac{\gamma(t)\left(\frac{\partial \hat{f}_i}{\partial \hat{\theta}_i}\right)^T \phi_0(t-d)\varepsilon(t)}{m^2(t)}, \tag{9.40}$$

$$\tilde{\theta}_\mu(t) = \tilde{\theta}_\mu(t-1) + \frac{\gamma(t)\sum_{i=1}^N \left(\frac{\partial \hat{f}_i}{\partial \hat{\theta}_\mu}\right)^T \phi_0(t-d)\varepsilon(t)}{m^2(t)}. \tag{9.41}$$

Substituting (9.40) and (9.41) into (9.39) yields

$$\begin{aligned}
\triangle V(t) &= \sum_{i=1}^N \left(\frac{2\gamma(t)\tilde{\theta}_i^T(t-1)\left(\frac{\partial \hat{f}_i}{\partial \hat{\theta}_i}\right)^T \phi_0(t-d)\varepsilon(t)}{m^2(t)}\right. \\
&\quad \left. + \frac{\gamma^2(t)\varepsilon^T(t)\phi_0^T(t-d)\left(\frac{\partial \hat{f}_i}{\partial \hat{\theta}_i}\right)\left(\frac{\partial \hat{f}_i}{\partial \hat{\theta}_i}\right)^T \phi_0(t-d)\varepsilon(t)}{m^4(t)}\right) \\
&\quad + \frac{2\gamma(t)\tilde{\theta}_\mu^T(t-1)\sum_{i=1}^N \left(\frac{\partial \hat{f}_i}{\partial \hat{\theta}_\mu}\right)^T \phi_0(t-d)\varepsilon(t)}{m^2(t)}
\end{aligned}$$

$$+\frac{\gamma^2(t)\varepsilon^T(t)\sum_{i=1}^N \phi_0^T\left(\frac{\partial \hat{f}_i}{\partial \hat{\theta}_\mu}\right)\sum_{i=1}^N\left(\frac{\partial \hat{f}_i}{\partial \hat{\theta}_\mu}\right)^T \phi_0\varepsilon(t)}{m^4(t)}. \tag{9.42}$$

With (9.38), the above equation can be written as

$$\Delta V(t) = \frac{2\gamma(t)\tilde{\theta}^T(t-1))\psi(t-d)\varepsilon(t)}{m^2(t)} + \frac{\gamma^2(t)\varepsilon^T(t)\psi^T(t-d)\psi(t-d)\varepsilon(t)}{m^4(t)}. \tag{9.43}$$

Making both sides of (9.37) multiplied by $\varepsilon(t)$ produces

$$\varepsilon^2(t) = -\tilde{\theta}^T(t-1)\psi(t-d)\varepsilon(t). \tag{9.44}$$

Substituting (9.44) into (9.43) yields

$$\Delta V(t) \le \frac{-2\gamma(t)\varepsilon^2(t)}{m^2(t)} + \frac{\gamma^2(t)\|\psi(t-d)\|^2\varepsilon^2(t)}{m^4(t)}. \tag{9.45}$$

For $m^2(t)$ and $\Psi(t-d)$ in (9.33) and (9.38), we have

$$\|\Psi(t-d)\|^2 < m^2(t). \tag{9.46}$$

Hence, it follows that

$$\Delta V(t) < \frac{-2\gamma(t)\varepsilon^2(t)}{m^2(t)} + \frac{\gamma^2(t)\varepsilon^2(t)}{m^2(t)} = -\gamma(t)(2-\gamma(t))\frac{\varepsilon^2(t)}{m^2(t)}. \tag{9.47}$$

With $\gamma(t) \in (\gamma_0, 2-\gamma_0)$ for a constant $\gamma_0 \in (0, 1)$, we have

$$\Delta V(t) \le -\gamma_0(2-\gamma_0)\frac{\varepsilon^2(t)}{m^2(t)} \le 0. \tag{9.48}$$

From this inequality, we can derive the following properties (Qi et al. 2011):

Lemma 9.1 *The parameter adaptive laws (9.31) and (9.32), when applied to the error model (9.30), have the properties:*

(i) $\|\hat{\theta}(t)-\theta\| \le \|\hat{\theta}(t-1)-\theta\| \le \|\hat{\theta}(0)-\theta\|$, *for the* l^2*-vector norm* $\|\cdot\|$;

(ii) $\frac{\varepsilon(t)}{m(t)} \in L_2$;

(iii) $\lim_{t\to\infty}\frac{\varepsilon(t)}{m(t)} = 0$;

(iv) $\|\hat{\theta}(t) - \hat{\theta}(t - t_1)\| \in L_2$, $t_1 > 0$; *and*

(vi) $\lim_{t\to\infty} \|\hat{\theta}(t) - \hat{\theta}(t - t_1)\| = 0$, $t_1 > 0$.

These are the baseline properties of the designed adaptive parameter estimation scheme with the estimation error (9.18), under the condition that the approximation error $\delta_i = 0$.

Robustness with respect to nonlinearity error $\delta_i(t)$. To ensure the adaptive law's robustness with respect to the nonlinearity approximation errors $\delta_i(t)$ in (9.20), we need to modify the adaptive laws with certain design signals (Tao 2003); for example, a projection design signal to ensure that the parameter estimates (the components of $\hat{\theta}_i$ and $\hat{\theta}_\mu$) are in certain prespecified ranges. To ensure smallness of the estimation error signal $\varepsilon(t)$ and parameter variations $\hat{\theta}_i(t) - \hat{\theta}_i(t-1)$ and $\hat{\theta}_\mu(t) - \hat{\theta}_\mu(t-1)$, relative to the modeling errors $\delta_i(t)$, important for closed-loop system stability, the normalized errors $\frac{\delta_i(t)}{m(t)}$ needs to be small. This can be ensured for $\hat{\theta}(t)$ initialized in a certain neighborhood of θ. Furthermore, certain characterizations of $\delta_i(t)$ can also be used to redesign the adaptive laws (9.31) and (9.32) to enhance its robustness. In Sect. 9.4, a simulation study will be presented to show the desired system performance despite the presence of $\delta_i(t)$.

9.3 Adaptive Control Design

In this section, we design and analyze an adaptive control scheme for the T–S fuzzy dynamic system (9.1), to solve the adaptive control problem with unknown membership function parameters. A critical stability condition is that the partial derivatives $\frac{\partial \hat{\mu}_i}{\partial \hat{\theta}_\mu}$ are bounded. We will establish such a condition for the Gaussian membership functions $F_j^i(\xi_j)$ in (9.3), and use it to prove some desired system stability and tracking properties.

9.3.1 Nominal Controller

If the system parameters are known, the control problem can be solved by the following nominal control scheme. For designing the nominal control law, we have the following assumption:

Assumption 9.2 $\sum_{i=1}^{N} \mu_i(\theta_\mu)\beta_{i0} \neq 0$.

Nominal control law. For a reference signal $y_m(t)$ to be tracked by the system output $y(t)$, the nominal control law for the system (9.1) is

$$u(t) = \frac{1}{\sum_{i=1}^{N} \mu_i(\theta_\mu)\beta_{i0}} \left[-\sum_{i=1}^{N} \mu_i(\theta_\mu)\alpha_i(z^{-1})[y](t) - \sum_{i=1}^{N} \mu_i(\theta_\mu)\bar{\beta}_i(z^{-1})[u](t) + y_m(t+d) \right], \quad (9.49)$$

where $\bar{\beta}_i(z^{-1}) = \beta_{i1}z^{-1} + \cdots + \beta_{i,n-1}z^{-n+1}$. Similar to that in Goodwin and Sin (1984) for a regular system, this control law, when applied to the fuzzy system (9.1) brings $y(t+d)$ to $y_m(t+d)$ in one step, and ensures closed-loop signal boundedness under Assumption 9.1 (the minimum phase condition).

The nominal control law (9.49) is a deadbeat control law, which generally requires large control signals. To reduce the possible deadbeat transient, we propose the following alternative control law:

An alternative control law. In stead of (9.49), we can also choose a more general nominal control law:

$$\begin{aligned} u(t) = \frac{1}{\sum_{i=1}^{N} \mu_i(\theta_\mu)\beta_{i0}} \Bigg[& -\sum_{i=1}^{N} \mu_i(\theta_\mu)\alpha_i(z^{-1})[y](t) \\ & -\sum_{i=1}^{N} \mu_i(\theta_\mu)\bar{\beta}_i(z^{-1})[u](t) + y_m(t+d)) \\ & +k_1(y_m(t+d-1) - y(t+d-1)) + \cdots \\ & +k_{d-1}(y_m(t+1) - y(t+1)) \\ & +k_d(y_m(t) - y(t))) \Bigg], \end{aligned} \quad (9.50)$$

where k_i, $i = 1, 2, \ldots, d$, are design parameters such that

$$k(z) = z^d + k_1 z^{d-1} + \cdots + k_{d-1}z + k_d \quad (9.51)$$

is a stable polynomial. To implement this control law, the signals $y(t+1), \ldots, y(t+d-1)$ when $d > 1$, not directly measured, can be reconstructed from (9.1).

Remark 9.2 For non-minimum phase systems, an adaptive pole placement control design may be a choice but it needs some extension new study for the T–S fuzzy dynamic systems. The control law (9.50) can also be applied to non-minimum phase systems, which may be considered as a kind of pole placement.

This nominal control law (9.49) or (9.50) provides the basic controller structure which can be adopted for adaptive control using adaptive estimates of the unknown system parameters, updated from the adaptive laws (9.31) and (9.32).

9.3.2 Adaptive Control Scheme

Let $\hat{\alpha}_i(z^{-1})$ and $\hat{\beta}_i(z^{-1})$ be the estimates of $\alpha_i(z^{-1})$ and $\beta_i(z^{-1})$ (whose expressions are given in (7.11) and (7.12)), and $\hat{\mu}_i = \mu_i(\hat{\theta}_\mu)$ be the estimate of $\mu_i = \mu_i(\theta_\mu)$, with parameters estimates $\hat{\alpha}_{ij}$, $\hat{\beta}_{ij}$ and $\hat{\theta}_\mu$. We choose the global fuzzy control law as

$$
\begin{aligned}
u(t) = \frac{1}{\sum_{i=1}^{N} \mu_i(\hat{\theta}_\mu)\hat{\beta}_{i0}} \Bigg[& -\sum_{i=1}^{N} \mu_i(\hat{\theta}_\mu)\hat{\alpha}_i(z^{-1})[y](t) \\
& -\sum_{i=1}^{N} \mu_i(\hat{\theta}_\mu)\hat{\bar{\beta}}_i(z^{-1})[u](t) + y_m(t+d) \Bigg],
\end{aligned} \tag{9.52}
$$

for the system (9.1) with unknown parameters α_{ij}, β_{ij}, and θ_μ.

The adaptive version of the alternative nominal control law (9.50) can also be developed with the estimates of $y(t+d-1), \ldots, y(t+1)$, parameterized in terms of some unknown parameters and known signals (current and past inputs and outputs).

A signal bounding property. We now show that the adaptive control system has desired stability and tracking properties. Substituting (9.52) into (9.1), we obtain the closed-loop system as

$$
y(t+d) = \sum_{i=1}^{N} [f_i(\theta_i, \theta_\mu; t) - f_i(\hat{\theta}_i(t), \hat{\theta}_\mu(t); t)]^T \phi_0(t) + y_m(t+d), \tag{9.53}
$$

which can be equivalently written into

$$
\begin{aligned}
y(t) = y_m(t) & + \sum_{i=1}^{N} f_i^T(\theta_i, \theta_\mu; t-d)\phi_0(t-d) \\
& - \sum_{i=1}^{N} f_i^T(\hat{\theta}_i(t-d), \hat{\theta}_\mu(t-d); t-d)\phi_0(t-d).
\end{aligned} \tag{9.54}
$$

With $e(t) = y(t) - y_m(t)$ and (9.20), we have

$$
\begin{aligned}
e(t) = & \sum_{i=1}^{N} \Big(f_i^T(\hat{\theta}_i(t-1), \hat{\theta}_\mu(t-1); t-d) - f_i^T(\hat{\theta}_i(t-d), \hat{\theta}_\mu(t-d); t-d) \Big) \phi_0(t-d) \\
& - \sum_{i=1}^{N} \left(\tilde{\theta}_i^T(t-1) \left(\frac{\partial \hat{f}_i}{\partial \hat{\theta}_i} \right)^T \phi_0(t-d) + \tilde{\theta}_\mu^T(t-1) \left(\frac{\partial \hat{f}_i}{\partial \hat{\theta}_\mu} \right)^T \phi_0(t-d) \right) \\
& + \sum_{i=1}^{N} \delta_i^T(t)\phi_0(t-d)
\end{aligned}
$$

$$= \sum_{i=1}^{N} \left(\mu_i(\hat{\theta}_\mu(t-1))\hat{\theta}_i^T(t-1) - \mu_i(\hat{\theta}_\mu(t-d))\hat{\theta}_i^T(t-d) \right) \phi_0(t-d)$$

$$+\varepsilon(t) + \sum_{i=1}^{N} \delta_i^T(t)\phi_0(t-d) \tag{9.55}$$

with $\varepsilon(t)$) defined in (9.37).

We first present a desired property for $\phi_0(t)$ (Qi et al. 2011).

Lemma 9.2 *Under Assumption 9.1, the regressor $\phi_0(t)$ defined in (9.7) satisfies*

$$\|\phi_0(t-d)\| \leq \rho_1 + \rho_2 \max_{0 \leq \tau \leq t} |e(\tau)|, \tag{9.56}$$

for some positive constants ρ_1 and ρ_2.

The proof of Lemma 9.2 is based on Assumption 9.1 and Definition 7.1 which is similar to the proof of Lemma 7.2.

A membership function property. For the adaptive control of (9.1) with uncertain membership function parameters, the property of membership function plays a key role in the stability analysis of the closed-loop adaptive control system. In the following proposition, we present the membership function property for Gaussian membership function.

Proposition 9.1 *For the Gaussian membership function (9.3), the partial derivative vector $\frac{\partial \hat{\mu}_i}{\partial \hat{\theta}_\mu}$ in (9.22), as a function of time t, has the property:*

$$\frac{\partial \hat{\mu}_i}{\partial \hat{\theta}_\mu} \in L_\infty. \tag{9.57}$$

Proof The membership function describing "F_j^i" is chosen as Gaussian function:

$$F_j^i(\xi_j) = \exp\left\{ -\frac{(\xi_j - c_j^i)^2}{\sigma_j^{i2}} \right\}. \tag{9.58}$$

From the normalized firing strength of Rule i:

$$\mu_i(\xi) = \frac{\lambda_i(\xi)}{\sum_{i=1}^{N} \lambda_i(\xi)}, \quad \lambda_i(\xi) = \prod_{j=1}^{L} F_j^i, \tag{9.59}$$

and with $v_j^i = \sigma_j^{i2}$, we can calculate $\frac{\partial \hat{\mu}_i}{\partial \hat{\theta}_\mu}$ as follows:

$$\frac{\partial \hat{\mu}_i}{\partial \hat{c}_j^i} = \frac{\lambda_i \sum_{l=1, l\neq i}^{N} \lambda_l}{\left(\sum_{i=1}^{N} \lambda_i\right)^2} \cdot \frac{2(\xi_j - \hat{c}_j^i)}{\hat{v}_j^i}, \tag{9.60}$$

$$\frac{\partial \hat{\mu}_i}{\partial \hat{c}_j^k} = \frac{-\lambda_i \lambda_k}{\left(\sum_{i=1}^{N} \lambda_i\right)^2} \cdot \frac{2\left(\xi_j - \hat{c}_j^k\right)}{\hat{v}_j^k}, \quad i \neq k \tag{9.61}$$

$$\frac{\partial \hat{\mu}_i}{\partial \hat{v}_j^i} = \frac{\lambda_i \sum_{l=1, l\neq i}^{N} \lambda_l}{\left(\sum_{i=1}^{N} \lambda_i\right)^2} \cdot \frac{\left(\xi_j - \hat{c}_j^i\right)^2}{(\hat{v}_j^i)^2}, \tag{9.62}$$

$$\frac{\partial \hat{\mu}_i}{\partial \hat{v}_j^k} = \frac{-\lambda_i \lambda_k}{\left(\sum_{i=1}^{N} \lambda_i\right)^2} \cdot \frac{\left(\xi_j - \hat{c}_j^k\right)^2}{\left(\hat{v}_j^k\right)^2}, \quad i \neq k. \tag{9.63}$$

From (9.58) and (9.59), we have

$$\lambda_i \frac{2\left(\xi_j - \hat{c}_j^i\right)}{\hat{v}_j^i} = \prod_{l=1, l\neq j}^{L} F_l^i(\xi) \cdot \rho_j^i(\xi), \tag{9.64}$$

where

$$\rho_j^i(\xi) = \frac{\frac{2(\xi_j - \hat{c}_j^i)}{\hat{v}_j^i}}{\exp\left\{\frac{(\xi_j - \hat{c}_j^i)^2}{\hat{v}_j^i}\right\}}. \tag{9.65}$$

From the expression of $\rho_j^i(\xi)$, we obtain that $\rho(\xi_j)$ is bounded. Furthermore, considering $\sum_{i=1}^{N} \lambda^i(\xi) > 0$ which ensures there is at least one rule activated for a valid fuzzy model, we have $\frac{\partial \hat{\mu}_i}{\partial \hat{c}_j^i} \in L_\infty$ in (9.60). Following the similar analysis, we can obtain $\frac{\partial \hat{\mu}_i}{\partial \hat{c}_j^k}$, $\frac{\partial \hat{\mu}_i}{\partial \hat{v}_j^i}$ and $\frac{\partial \hat{\mu}_i}{\partial \hat{v}_j^k}$ are all bounded. Therefore, we conclude that

$$\frac{\partial \hat{\mu}_i}{\partial \hat{\theta}_\mu} \in L_\infty. \tag{9.66}$$

∇

This proposition shows a key property of the Gaussian membership functions, which is important for establishing the stability of adaptive control of T–S fuzzy systems with membership function parameter uncertainties leading to a nonlinear parametrization. Such a property can be similarly shown for the sigmoidal-shape membership functions. This proved property provides a basic development for the application of the linear parameter adaptation technique to nonlinearly parameterized T–S fuzzy systems. The use of this property will be shown next for stability analysis.

Stability properties for error model (9.30) **when** $\delta_i(t) = 0$. We first study the ideal properties of the adaptive control system under the condition that the approximation error $\delta_i(t) = 0$ as in the error model (9.30).

The following results can be obtained (Qi et al. 2011).

Theorem 9.1 *The closed-loop system, with the plant (9.1) satisfying Assumptions 9.1 and 9.2 and the membership function satisfying (9.57), the controller (9.52) and the adaptation laws (9.31) and (9.32), under the condition that* $\delta_i(t) = 0$, *has the baseline properties: all system signals are bounded and the tracking error* $e(t) = y(t) - y_m(t)$ *converges to zero asymptotically with time.*

Proof From (9.55) with $\delta_i(t) = 0$, we have

$$\begin{aligned} e(t) = \sum_{i=1}^{N} \Big(& \mu_i(\hat{\theta}_\mu(t-1))\hat{\theta}_i^T(t-1) \\ & - \mu_i(\hat{\theta}_\mu(t-d))\hat{\theta}_i^T(t-d)\Big)\phi_0(t-d) + \bar{\varepsilon}(t)m(t) \end{aligned} \tag{9.67}$$

with $\bar{\varepsilon}(t) = \frac{\varepsilon(t)}{m(t)}$ and $m(t)$ defined in (9.33).

From (9.33), we have

$$\begin{aligned} m(t) &\leq \sqrt{c} + \sqrt{\sum_{i=1}^{N} \left\| \left(\frac{\partial \hat{f}_i}{\partial \hat{\theta}_i}\right)^T \phi_0(t-d) \right\|^2} + \sqrt{\left\| \sum_{i=1}^{N} \left(\frac{\partial \hat{f}_i}{\partial \hat{\theta}_\mu}\right)^T \phi_0(t-d) \right\|^2} \\ &\leq \sqrt{c} + \sum_{i=1}^{N} \left\| \left(\frac{\partial \hat{f}_i}{\partial \hat{\theta}_i}\right)^T \phi_0(t-d) \right\| + \left\| \sum_{i=1}^{N} \left(\frac{\partial \hat{f}_i}{\partial \hat{\theta}_\mu}\right)^T \phi_0(t-d) \right\| \\ &\leq \sqrt{c} + \sum_{i=1}^{N} \left\| \left(\frac{\partial \hat{f}_i}{\partial \hat{\theta}_i}\right) \right\| \left\| \phi_0(t-d) \right\| + \sum_{i=1}^{N} \left\| \frac{\partial \hat{f}_i}{\partial \hat{\theta}_\mu} \right\| \left\| \phi_0(t-d) \right\|. \end{aligned} \tag{9.68}$$

From (9.21) and $\mu_i(\hat{\theta}_\mu) \in (0, 1)$, we have $\frac{\partial \hat{f}_i}{\partial \hat{\theta}_i} \in L_\infty$. With Lemma 9.1 and (9.57), we obtain $\frac{\partial \hat{f}_i}{\partial \hat{\theta}_\mu} \in L_\infty$ from (9.22). Hence,

$$m(t) \leq \sqrt{c} + c_1 \left\| \phi_0(t-d) \right\|, \tag{9.69}$$

where c_1 is a positive constant.

With Lemma 9.1, we have $\bar{\varepsilon}(t) \in L_2 \bigcap L_\infty$ and $\hat{\theta}(t-d) - \hat{\theta}(t-1) \in L_2 \bigcap L_\infty$. Then using the inequality (9.69), we can express $e(t)$ in (9.67) as

$$\begin{aligned} |e(t)| \leq c_2 + \| \sum_{i=1}^{N} \Big(& \mu_i(\hat{\theta}_\mu(t-1))\hat{\theta}_i^T(t-1) \\ & -\mu_i(\hat{\theta}_\mu(t-d))\hat{\theta}_i^T(t-d)) \Big) \| \, \|\phi_0(t-d)\| \\ & +c_1|\bar{\varepsilon}(t)| \, \|\phi_0(t-d)\|, \end{aligned} \tag{9.70}$$

for some constant $c_2 > 0$. Using Lemma 9.2: $\|\phi_0(t-d)\| \leq \rho_1 + \rho_2 \max_{0 \leq \tau \leq t} |e(\tau)|$, we obtain

$$\begin{aligned} |e(t)| \leq c_3 + c_4 \| \sum_{i=1}^{N} \Big(& \mu_i(\hat{\theta}_\mu(t-1))\hat{\theta}_i^T(t-1) \\ & -\mu_i(\hat{\theta}_\mu(t-d))\hat{\theta}_i^T(t-d)) \Big) \| \max_{0 \leq \tau \leq t} |e(\tau)| \\ & +c_5|\bar{\varepsilon}(t)| \max_{0 \leq \tau \leq t} |e(\tau)|, \end{aligned} \tag{9.71}$$

for some constants $c_i > 0$, $i = 3, 4, 5$. From Lemma 9.1, we have that $\lim_{t\to\infty} \bar{\varepsilon}(t) = 0$ and $\lim_{t\to\infty} \|\hat{\theta}(t-1) - \hat{\theta}(t-d)\| = 0$, which, from continuity of μ_i, means

$$\lim_{t\to\infty} \| \sum_{i=1}^{N} \Big(\mu_i(\hat{\theta}_\mu(t-1))\hat{\theta}_i^T(t-1) - \mu_i(\hat{\theta}_\mu(t-d))\hat{\theta}_i^T(t-d)) \Big) \| = 0. \tag{9.72}$$

With these properties, it follows from (9.71) that $e(t)$ is bounded, which implies $y(t)$ is bounded, and in turn from the system's minimum phase property that $u(t)$ is bounded. Hence all signals in the closed-loop system are bounded. The asymptotic convergence property of the tracking error $e(t)$ follows from the L_2 and convergence properties of $\bar{\varepsilon}(t)$ and $\|\hat{\theta}(t-1) - \hat{\theta}(t-d)\|$. ∇

Robustness with respect to $\delta_i(t) \neq 0$. Although the above system stability properties with $\delta_i(t) = 0$ are crucial as the baseline properties of an adaptive T–S fuzzy control system, the robustness of such properties with respect to the nonlinearity error $\delta_i(t)$ in (9.55) is an important issue.

We can add robustifying signals in the adaptive laws (9.31) and (9.32) to enhance their robustness with respect to the nonlinearity error $\delta_i(t)$. Under the condition that the normalized errors $\frac{\delta_i(t)}{m(t)}$ is small (which can be ensured in a neighborhood of θ), the closed-loop system signals can be ensured to be bounded and the tracking error can be of the order of $\frac{\delta_i(t)}{m(t)}$ in a mean sense (Tao 2003).

Remark 9.3 To ensure the nonsingularity of the control law (9.52), it should be guaranteed that $\sum_{i=1}^{N} \mu_i(\hat{\theta}_\mu)\hat{\beta}_{i0} \neq 0$. This condition can be ensured by using the parameter projection algorithm (Ioannou and Sun 1996) on the parameter estimates $\hat{\theta}_\mu$ and β_{i0}, $i = 1, 2, \ldots, N$.

9.4 Simulation Study

In this section, we present an illustrative example with simulation results to show the control design and evaluation, based on a mass–spring–damper mechanical system (Tanaka et al. 1996).

9.4.1 Simulation System

The dynamic model of this mass–spring–damper system is given in (7.107) in Sect. 7.3.4. Following the same modeling method in Sect. 7.3.4, a discrete-time model can be obtained as

$$\begin{aligned} R^i : \text{IF} \quad & \xi_1(t) \quad \text{is} \quad F_1^i, \quad \text{THEN} \\ & y(t+2) + a_{i1} y(t+1) + a_{i2} y(t) = b_{i0} u(t. \end{aligned} \tag{9.73}$$

The membership functions describing "F_1^1" and "F_1^2" chosen as

$$F_j^i(\xi_j) = \exp\left\{ -\frac{(\xi_j - c_j^i)^2}{\sigma_j^{i2}} \right\}, j = 1; i = 1, 2 \tag{9.74}$$

with $[c_1^1, \sigma_1^1, c_1^2, \sigma_1^2] = [1.5, 1.5, -1.5, 1.5]$.

The discrete-time T–S fuzzy system model (9.73) can be equivalently written into the form of (7.4) with $n = 2$ and the system delay $d = 2$. With Proposition 1, we obtain the following global 2-step prediction fuzzy system model:

$$y(t+2) = \sum_{i=1}^{2} \mu_i \alpha_i (z^{-1})[y](t) + \sum_{i=1}^{2} \mu_i \beta_i (z^{-1})[u](t), \tag{9.75}$$

where $\alpha_i(z^{-1}) = \alpha_{i0} + \alpha_{i1} z^{-1} = a_{i1}^2 - a_{i2} + a_{i1} a_{i2} z^{-1}$, $\beta_i(z^{-1}) = \beta_{i0} + \beta_{i1} z^{-1} = b_{i0} - a_{i1} b_{i0} z^{-1}$.

From (9.75), we have

$$\sum_{i=1}^{2} \mu_i \beta_i (z^{-1})[u](t) = \sum_{i=1}^{2} \mu_i (b_{i0} - a_{i1} b_{i0} z^{-1})[u](t). \tag{9.76}$$

To make the global fuzzy system (9.75) be minimum phase, it is required that

$$\left| \frac{\sum_{i=1}^{2} \mu_i a_{i1} b_{i0}}{\sum_{i=1}^{2} \mu_i b_{i0}} \right| = \left| \frac{\mu_1 a_{11} b_{10} + \mu_2 a_{21} b_{20}}{\mu_1 b_{10} + \mu_2 b_{20}} \right| < 1. \tag{9.77}$$

For $T = 0.01s$, we obtain $a_{11} = a_{21} = -0.5$, which makes the condition (9.77) satisfied so that the system (9.75) is minimum phase. We can then use the adaptive control algorithm in Sect. 9.3 to design an adaptive controller that guarantees the closed-loop boundedness and asymptotic tracking of a bounded reference signal $y_m = 1.5\sin(0.2t)$.

The model (9.75) can be further written into the nonlinear parametrization (9.19):

$$y(t+2) = \sum_{i=1}^{2} f_i^T(\theta_i, \theta_\mu; t)\phi_0(t), \tag{9.78}$$

where, for $i = 1, 2$ and $j = 1$,

$$f_i(\theta_i, \theta_\mu; t) = \mu_i(\xi(t))\theta_i = \mu_i(\theta_\mu; t)\theta_i, \tag{9.79}$$

$$\phi_0(t) = [y(t), y(t-1), u(t), u(t-1)]^T, \tag{9.80}$$

$$\theta_i = [\alpha_{i0}, \alpha_{i1}, \beta_{i0}, \beta_{i1}]^T, \tag{9.81}$$

$$\theta_\mu = [c_1^1, v_1^1, c_1^2, v_1^2]^T, \ v_j^i = (\sigma_j^i)^2, \tag{9.82}$$

with

$$\mu_i(\xi) = \frac{\lambda_i(\xi)}{\sum_{i=1}^2 \lambda_i(\xi)}, \quad \lambda_i(\xi) = \prod_{j=1}^{1} F_j^i(\xi_j). \tag{9.83}$$

To apply the adaptive laws (9.31) and (9.32) to estimate the unknown parameter vectors θ and θ_μ, we need to calculate $\frac{\partial \hat{f}_i}{\partial \hat{\theta}_i}$ and $\frac{\partial \hat{f}_i}{\partial \hat{\theta}_\mu}$ in the form

$$\frac{\partial \hat{f}_i}{\partial \hat{\theta}_i} = \mu_i(\hat{\theta}_\mu(t-1); t-2)I_4, \tag{9.84}$$

$$\frac{\partial \hat{f}_i}{\partial \hat{\theta}_\mu} = \hat{\theta}_i \otimes \frac{\partial \hat{\mu}_i}{\partial \hat{\theta}_\mu}, \tag{9.85}$$

where

$$\begin{aligned}\frac{\partial \hat{\mu}_i}{\partial \hat{\theta}_\mu} &= \left[\frac{\partial \mu_i}{\partial c_1^1}, \frac{\partial \mu_i}{\partial v_1^1}, \frac{\partial \mu_i}{\partial c_1^2}, \frac{\partial \mu_i}{\partial v_1^2}\right]|_{(\hat{\theta}_\mu(t-1);t-2)} \\ &= \left[\frac{\partial \hat{\mu}_i}{\partial \hat{c}_1^1}, \frac{\partial \hat{\mu}_i}{\partial \hat{v}_1^1}, \frac{\partial \hat{\mu}_i}{\partial \hat{c}_1^2}, \frac{\partial \hat{\mu}_i}{\partial \hat{v}_1^2}\right].\end{aligned} \tag{9.86}$$

With (9.74) and (9.83), the components of $\frac{\partial \hat{\mu}_i}{\partial \hat{\theta}_\mu}$ can be calculated as

$$\frac{\partial \hat{\mu}_1}{\partial \hat{c}_1^1} = \bar{\lambda}\frac{2(\xi_1 - \hat{c}_1^1)}{\hat{v}_1^1}, \quad \frac{\partial \hat{\mu}_1}{\partial \hat{v}_1^1} = \bar{\lambda}\frac{(\xi_1 - \hat{c}_1^1)^2}{(\hat{v}_1^1)^2}$$
$$\frac{\partial \hat{\mu}_1}{\partial \hat{c}_1^2} = -\bar{\lambda}\frac{2(\xi_1 - \hat{c}_1^2)}{\hat{v}_1^2}, \quad \frac{\partial \hat{\mu}_1}{\partial \hat{v}_1^2} = -\bar{\lambda}\frac{(\xi_1 - \hat{c}_1^2)^2}{(\hat{v}_1^2)^2}$$
$$\frac{\partial \hat{\mu}_2}{\partial \hat{c}_1^1} = -\bar{\lambda}\frac{2(\xi_1 - \hat{c}_1^1)}{\hat{v}_1^1}, \quad \frac{\partial \hat{\mu}_2}{\partial \hat{v}_1^1} = -\bar{\lambda}\frac{(\xi_1 - \hat{c}_1^1)^2}{(\hat{v}_1^1)^2}$$
$$\frac{\partial \hat{\mu}_2}{\partial \hat{c}_1^2} = \bar{\lambda}\frac{2(\xi_1 - \hat{c}_1^2)}{\hat{v}_1^2}, \quad \frac{\partial \hat{\mu}_2}{\partial \hat{v}_1^2} = \bar{\lambda}\frac{(\xi_1 - \hat{c}_1^2)^2}{(\hat{v}_1^2)^2}, \tag{9.87}$$

where $\bar{\lambda} = \frac{\lambda_1\lambda_2}{(\lambda_1+\lambda_2)^2}|_{(\hat{\theta}_\mu(t-1);t-2)}$.

9.4.2 Simulation Results

For the simulation, the initial parameters are set as $\hat{\theta}_s(0) = 60\% \times \theta_s$ and $\hat{\theta}_\mu(0) = [0.7, 2.75, -2.5, 2.75]^T$. Other parameters are set as $\gamma(t) = 0.5$ and $c = 0.01$. The system outputs $y(t)$ and $y_m(t)$, and the adaptive control signal are shown in Fig. 9.1. The tracking error is shown in Fig. 9.2. The adaptations of consequent parameters and membership parameters are given in Figs. 9.3 and 9.4, respectively. It can be observed that the tracking error decreases as the parameters adapt online and converge to some small residual values, despite the initial parameter errors $\hat{\theta}_s(0) - \theta_s$ and $\hat{\theta}_\mu(0) - \theta_\mu$.

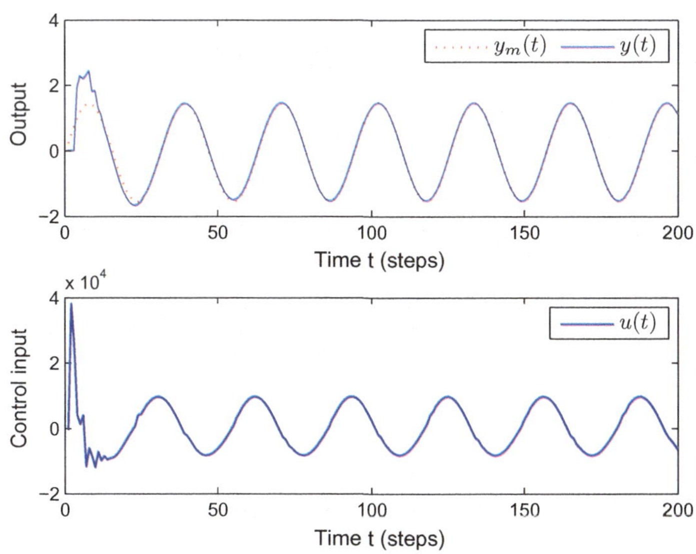

Fig. 9.1 Adaptive system response and control signal

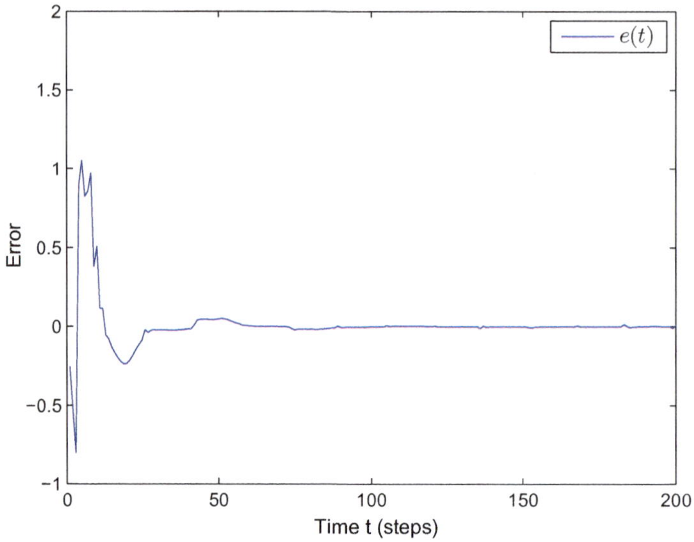

Fig. 9.2 Tracking error

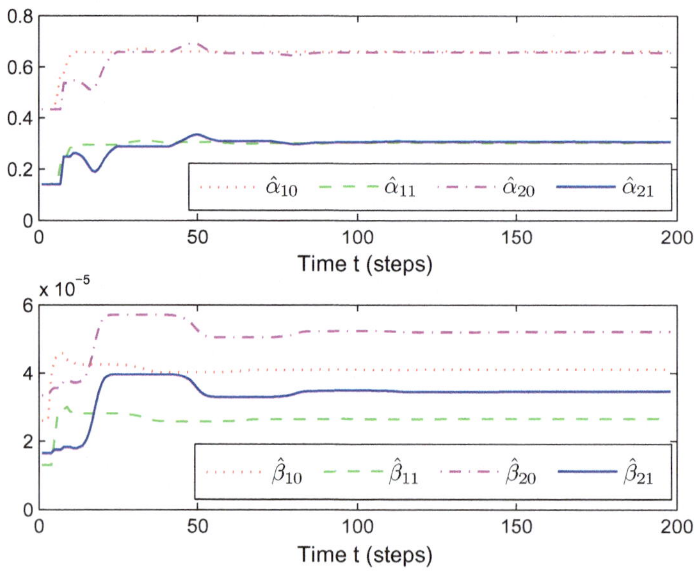

Fig. 9.3 Adaptation of consequent parameters

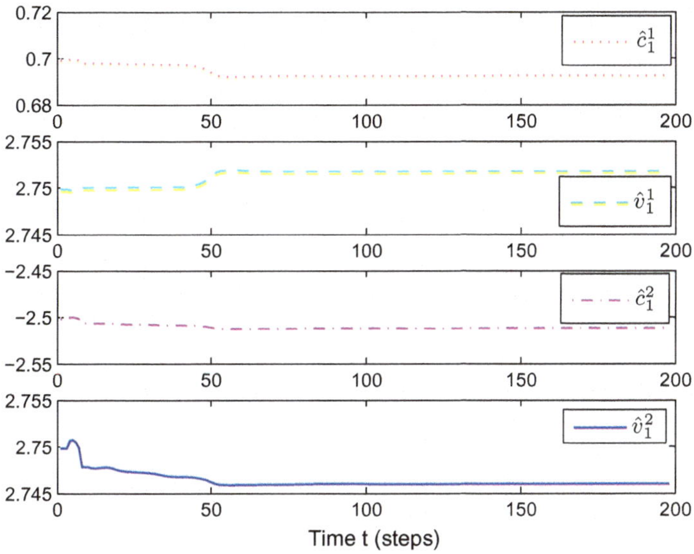

Fig. 9.4 Adaptation of membership parameters

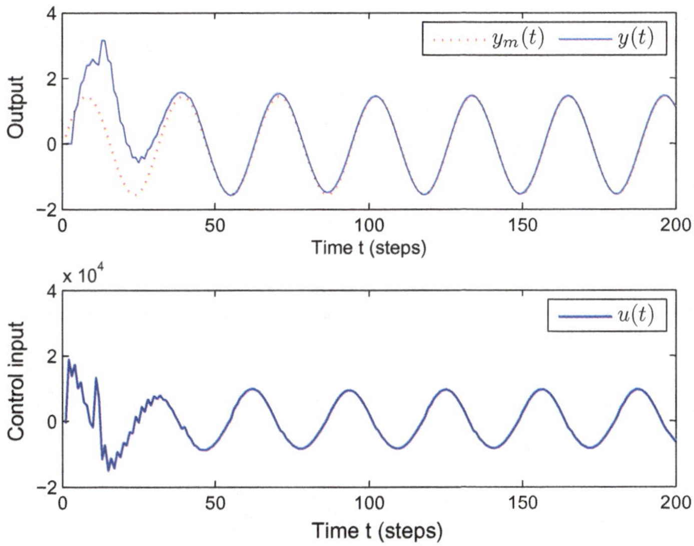

Fig. 9.5 Adaptive system response and alternative control signal

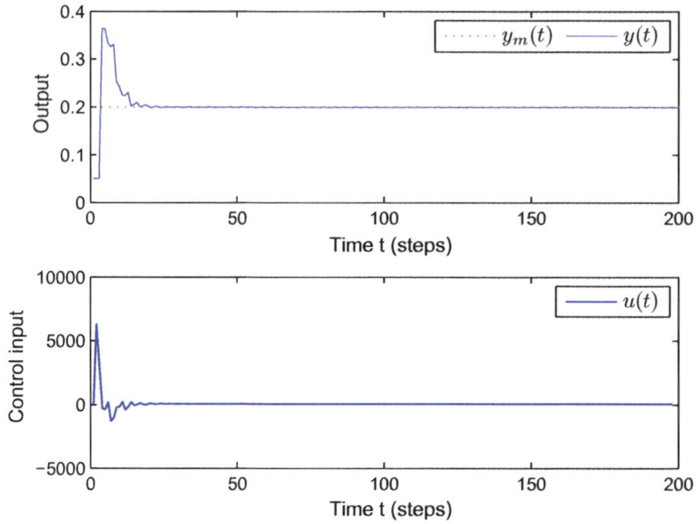

Fig. 9.6 Adaptive system response and control signal

The control law in Fig. 9.1 is a deadbeat control law which requires large control action. The deadbeat transient may be eased by using the adaptive version of the alternative control law (9.50) with $k_1 = -0.17$ and $k_2 = -0.72$, producing the corresponding tracking results as shown in Fig. 9.5.

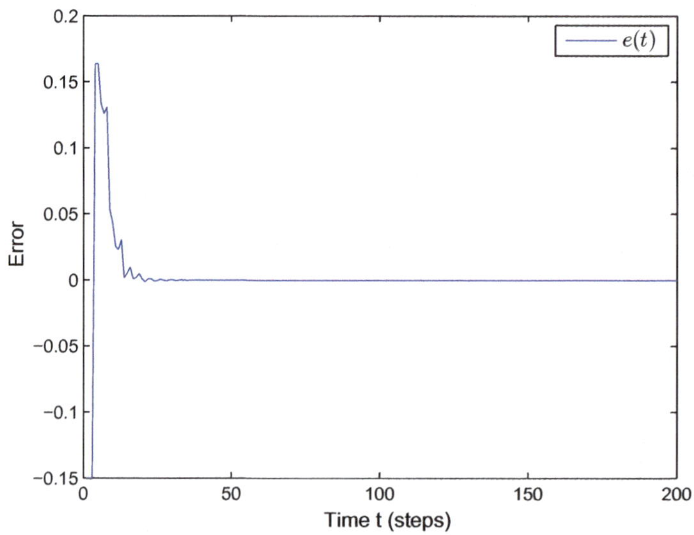

Fig. 9.7 Tracking error

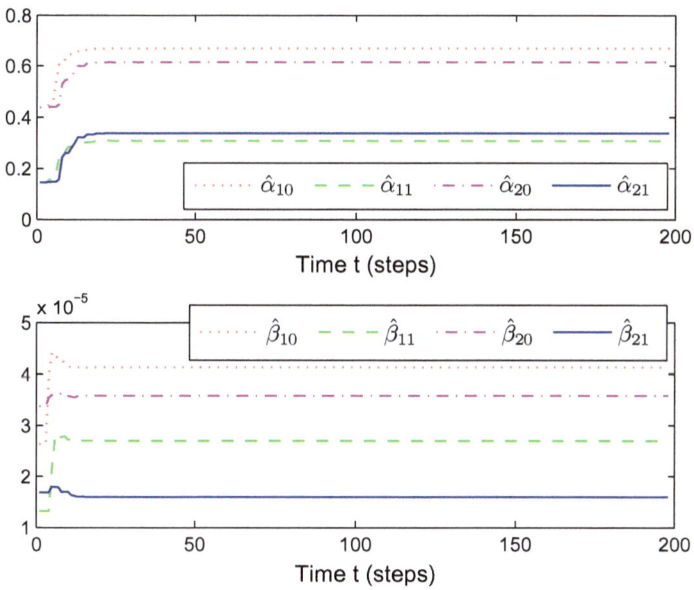

Fig. 9.8 Adaptation of consequent parameters

Figures 9.6, 9.7, 9.8 and 9.9 show the adaptive tracking control results for a constant reference signal $y_m(t) = 0.2$. Similar simulation results were also obtained for some other larger or smaller initial parameter errors $\hat{\theta}_s(0) - \theta_s$ and $\hat{\theta}_\mu(0) - \theta_\mu$.

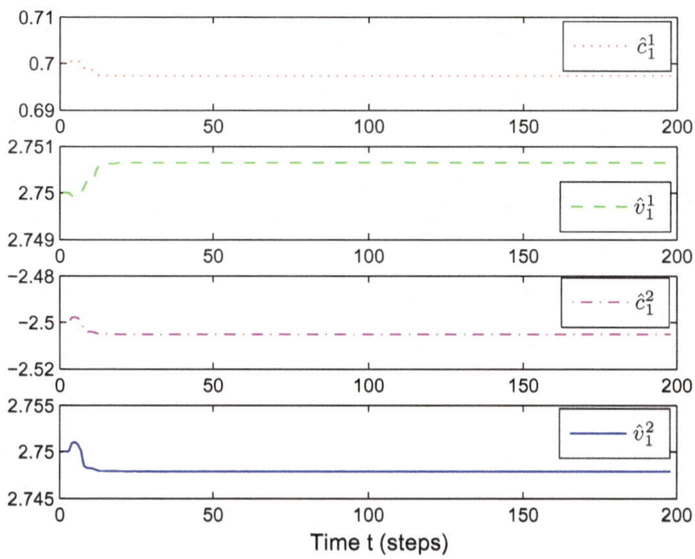

Fig. 9.9 Adaptation of membership parameters

9.5 Summary

In this chapter, we have formulated an adaptive control problem for discrete-time input–output multiple-delay T–S fuzzy systems with unknown membership function parameters and consequent parameters, and derived a solution for it. A fuzzy system with such uncertain parameters leads to a nonlinearly parametrized estimation error model. For such a nonlinear model, the gradient algorithm can be applied to derive an adaptive law to adaptively update the estimates of both the local system dynamics parameters and membership function parameters. Such parameter estimates can be used to implement the feedback control law for the uncertain T–S fuzzy system. The properties of the parameter adaptive laws and closed-loop stability have been studied with some key design conditions specified and established for the Gaussian membership functions. Simulation results have verified the desired adaptive fuzzy control system performance.

References

Goodwin GC, Sin KS (1984) Adaptive filtering prediction and control. Prentice-Hall, Englewood Cliffs
Ioannou PA, Sun J (1996) Robust adaptive control. Prentice-Hall, Englewood Cliffs
Qi R, Tao G, Tan C (2011) An adaptive control scheme for discrete-time T–S fuzzy systems with unknown membership parameters. In: Proceedings of the 8th Asia control conference. Taiwan, pp 1164–1169
Tanaka K, Ikeda T, Wang HO (1996) Robust stabilization of a class of uncertain nonlinear systems via fuzzy control: quadratic stabilizability, H^{∞} control theory, and linear matrix inequalities. IEEE Trans Fuzzy Syst 4:1–13
Tao G (2003) Adaptive control design and analysis. Wiley, New York

Chapter 10
Adaptive Control of T–S Fuzzy Systems with Actuator Faults

Recently, fault diagnosis (FD) and fault-tolerant control (FTC) have attracted more and more attentions (Chen and Patton 1999; Tao et al. 2004; Zhang and Jiang 2008). FD and FTC play an important role in ensuring the closed-loop system stability when faults occur in the system. When a nonlinear system is represented by a T–S fuzzy system, its fault-tolerant control design will be conducted based on the T–S fuzzy system. A number of papers have been published on T–S fuzzy model based FD (Ballé 1999; Ichtev et al. 2002; Zheng et al. 2006; Nguang Shi et al. 2007) and FTC (Jiang et al. 2010; Zhang et al. 2010; Jiang et al. 2011; Shen et al. 2012; Qi et al. 2012, 2013; Jiang and Qi 2016).

In the previous chapters, we have presented adaptive control schemes for three types of T–S fuzzy systems: continuous-time state-space T–S fuzzy systems, discrete-time state-space T–S fuzzy systems and discrete-time input/output form T–S fuzzy systems. In this chapter, we will develop two adaptive fault compensation schemes for two types of T–S fuzzy systems. One approach is designed for continuous-time T–S fuzzy systems subject to actuator faults and the other is for discrete-time input–output form T–S fuzzy systems. Detailed design and analysis for T–S fuzzy system based adaptive compensation schemes will be presented to deal with uncertain actuator faults (with unknown fault patterns, values and time instants) in closed-loop control. Fuzzy system based control, like most other control methodologies, designs control laws based on an approximate (nominal) plant subject to approximation errors. It is crucial to establish the desired stability and tracking properties for an adaptive control system with parameterized nominal plant model, as the foundations of a chosen fuzzy control technique, to be applied to a real plant with approximation errors which can be handled with robustification techniques (using robust adaptive laws for the adaptive control case).

R. Qi et al., *Fuzzy System Identification and Adaptive Control*, Communications and Control Engineering,
https://doi.org/10.1007/978-3-030-19882-4_10

10.1 Systems with Actuator Faults

Consider the following nonlinear system with redundant actuators:

$$\begin{aligned} \dot{x}_i &= x_{i+1}, \quad \text{for } i = 1, 2, \ldots, n-1 \\ \dot{x}_n &= f(x) + g(x)u, \end{aligned} \tag{10.1}$$

where $x = [x_1, x_2, \ldots, x_n]^T$ is the state vector and $u = [u_1, u_2, \ldots, u_m]^T$ represents the input vector with m actuators, $f(x)$ is a nonlinear function and $g(x) = [g_1(x), g_2(x), \ldots, g_m(x)]^T$ with $g_j(x),\ j = 1, 2, \ldots, m$ being nonlinear functions.

Actuator fault model. In presence of actuator faults, the input vector $u(t)$ to the plant (10.7) can be described as

$$u(t) = v(t) + \sigma(\bar{u}(t) - v(t)), \tag{10.2}$$

where $v = [v_1, v_2, \ldots, v_m]^T$ is the vector of applied control signals, $\bar{u} = [\bar{u}_1, \bar{u}_2, \ldots, \bar{u}_m]^T$ is the vector of possible actuator faults, and

$$\begin{aligned} \sigma &= \text{diag}\{\sigma_1, \sigma_2, \ldots, \sigma_m\} \\ \sigma_j &= \begin{cases} 1, \text{ if } \ u_j \ \text{ fails at time } t, \text{ i.e., } u_j(\tau) = \bar{u}_j, \ \tau \geq \text{t} \\ 0, \qquad \text{otherwise} \end{cases} \end{aligned} \tag{10.3}$$

represents the pattern of actuator faults.

The actuator faults can be described by

$$\bar{u}_j(t) = \bar{u}_{j0} + \sum_{l=1}^{n_j} \bar{u}_{jl} f_{jl}(t), \ t \geq t_j, \tag{10.4}$$

for some unknown index $j \in \{1, 2, \ldots, m\}$, unknown time instant t_j and unknown constants $\bar{u}_{j0}$ and $\bar{u}_{jl}$, $l = 1, \ldots, n_j$, and known bounded signals $f_{jl}(t)$ and $n_j \geq 1$ (Tao et al. 2004). This parameterized actuator fault model can be used to closely approximate a large class of practical faults, by a proper selection of these "basis" functions $f_{jl}(t)$. The selection of the basis functions of the actuator faults depends on individual applications; for example, a common fault is that the actuator is stuck at an unknown constant value $\bar{u}_{j0}$, for which those $f_{jl}(t)$ can be set to zero.

A basic existence assumption for a nominal solution to the actuator fault compensation problem is

Assumption 10.1 The system (10.1) is designed that for any up to $m-1$ actuator faults, the remaining actuators can still achieve a desired control performance.

The key task of adaptive control is to adjust the remaining controls to achieve the desired system performance when there are up to $m-1$ actuator faults, whose parameters are unknown.

Remark 10.1 At a given time t, there may be $p \in \{0, 1, 2, \ldots, m-1\}$ malfunctioned actuators with fault values $\bar{u}_j(t)$, $j = j_1, j_2, \ldots, j_p$, for $\{j_1, j_2, \ldots, j_p\} \subset \{1, 2, \ldots, m\}$. Different σ represents different fault patterns and such a σ is thus called a fault pattern. Let all fault patterns of interest be $\sigma_{(j)}$, $j = 1, 2, \ldots, N_p$, and denote the *fault pattern set* as $\Sigma = \{\sigma_{(j)}, j = 1, \ldots, N_p\}$ of interest for fault compensation, where $\sigma_{(1)} = \mathrm{diag}\{0, 0, \ldots, 0\}$ represents the no-fault case.

10.2 Adaptive Fault Compensation Control Using State Feedback

When a T–S fuzzy system is employed to represent a nonlinear system, we assume it has the same actuators as the nonlinear system. Therefore, when an actuator fault occurs, T–S fuzzy system based adaptive fault compensation control designs provide solutions to fault compensation control designs for uncertain nonlinear systems.

In this section, we consider a T–S fuzzy system approximation for the canonical nonlinear system (10.1) with actuator faults and study the adaptive fault compensation design of the T–S fuzzy system.

10.2.1 Problem Statement

Consider the following T–S fuzzy in canonical form with redundant actuators:

$$\begin{aligned} &\text{IF} \quad \xi_1(t) \text{ is } F_1^i \text{ and } \ldots \text{ and } \xi_L(t) \text{ is } F_L^i, \\ &\text{THEN} \quad \dot{x}(t) = A_i x(t) + B_i u(t), \end{aligned} \tag{10.5}$$

where $x = [x_1, x_2, \ldots, x_n]^T$, $u = [u_1, u_2, \ldots, u_m]^T$,

$$A_i = \begin{bmatrix} 0 & 1 & \cdots & 0 & 0 \\ \vdots & \vdots & & \vdots & \vdots \\ 0 & 0 & \cdots & 0 & 1 \\ a_1^i & a_2^i & \cdots & a_{n-1}^i & a_n^i \end{bmatrix}, \quad B_i = \begin{bmatrix} 0 & 0 & \cdots & 0 \\ \vdots & \vdots & \cdots & \vdots \\ 0 & 0 & \cdots & 0 \\ b_1^i & b_2^i & \cdots & b_m^i \end{bmatrix}. \tag{10.6}$$

Using the standard technique of *singleton fuzzification*, *product inference*, and *weighted average*, we obtain the following global T–S fuzzy system model from (10.5):

$$\dot{x}(t) = \sum_{i=1}^{N} \mu_i (A_i x(t) + B_i u(t)), \tag{10.7}$$

where μ_i is the normalized firing strength satisfying

$$\mu_i = \frac{\lambda_i}{\sum_{i=1}^{N} \lambda_i}, \ \lambda_i = \prod_{j=1}^{L} F_j^i(\xi_j(t)), \ \mu_i \geq 0, \ \sum_{i=1}^{N} \mu_i = 1. \tag{10.8}$$

Control objective. The control objective is to design a feedback control vector $v(t)$ for the fuzzy system (10.7) with unknown parameter matrices A_i, B_i, and subject to unknown actuator faults (10.4) belonging to a fault pattern set $\sum$ (which are characterized by some $s = j_1, j_2, \ldots, j_p$, with $\{j_1, j_2, \ldots, j_p\} \subset \{1, 2, \ldots, m\}$ and $p = 0, 1, \ldots, q < m$ for some $q > 0$, such that the sth actuator fails: $u_s(t) = \bar{u}_s(t)$, $t \geq t_j$), to ensure closed-loop signal boundedness and the state vector $x(t)$ asymptotically tracks of a given reference state vector $x_m(t)$ generated from the reference model

$$\dot{x}_m(t) = A_m x_m(t) + B_m r(t), \tag{10.9}$$

where $x_m(t) \in R^n$, $r(t) \in R$, $A_m \in R^{n\times n}$ is a Hurwitz matrix and $B_m \in R^n$:

$$A_m = \begin{bmatrix} 0 & 1 & \cdots & 0 & 0 \\ \vdots & \vdots & & \vdots & \vdots \\ 0 & 0 & \cdots & 0 & 1 \\ a_1^m & a_2^m & \cdots & a_{n-1}^m & a_n^m \end{bmatrix}, \quad B_m = \begin{bmatrix} 0 \\ \vdots \\ 0 \\ b^m \end{bmatrix}. \tag{10.10}$$

To meet the control objective, the following assumption is also needed:

Assumption 10.2 The signs of b_1^i, b_2^i, ..., b_m^i in B_i are known and the same for all $i = 1, 2, \ldots, N$.

Remark 10.2 Assumption 10.2 requires b_1^1, b_1^2, ..., b_1^N have the same sign, but does not require b_1^1 and b_2^1 have the same sign. In practice, the control directions from an actuator to each local model are usually the same, which makes Assumption 10.2 a practical assumption.

10.2.2 Nominal Controller

In this section, we propose a control design to compensate actuator faults for the T–S fuzzy system (10.7). For simplicity of presentation and without loss of generality, we focus our study on the T–S fuzzy systems (10.7) with two actuators (that is, with $m = 2$) and then give the main design ideas for the general case of $m \geq 2$. Before developing an adaptive control scheme for (10.7) with unknown parameters and unknown faults, it is important to make sure there exists a solution to meet the desired control objective when all the parameters are known.

Matching conditions. Assuming the sth actuator fails, $s = j_1, j_2, \ldots, j_p$, we design the nominal control law for the healthy actuator j, $j \neq j_1, j_2, \ldots, j_p$, as

$$v_j(t) = \frac{\sum_{i=1}^{N} \alpha_{ij}(-K_{ij}x(t) + l_{ij}r(t) + k_{ij} + \sum_{s=j_1,\ldots,j_p} \sum_{l=1}^{n_j} g_{ijsl} f_{sl}(t))}{\sum_{i=1}^{N} \alpha_{ij}}, \tag{10.11}$$

where α_{ij} satisfies $\alpha_{ij} = \left|b_j^i\right| \lambda_i$, $K_{ij} \in R^{1\times n}$ and $l_{ij} \in R$ are controller parameters to achieve state tracking, and $k_{ij} \in R$ and $g_{ijsl} \in R$ are parameters to compensate actuator faults.

Lemma 10.1 *Consider the fuzzy system (10.7) with m actuators subject to p actuator faults characterized by* $s = j_1, j_2, \ldots, j_p$ *with the actuator fault model described by (10.4), if the control law (10.11) is applied and the following matching conditions are satisfied:*

$$\begin{aligned}
&\sum_{j\neq j_1,\ldots,j_p} b_j^i K_{ij} = \left[a_1^i - a_1^m \cdots a_{n-1}^i - a_{n-1}^m \; a_n^i - a_n^m \right] \\
&\sum_{j\neq j_1,\ldots,j_p} b_j^i l_{ij} = b_m, \quad \sum_{j\neq j_1,\ldots,j_p} b_j^i k_{ij} + \sum_{s=j_1,\ldots,j_p} b_s^i \bar{u}_{s,0} = 0 \\
&\sum_{j\neq j_1,\ldots,j_p} b_j^i \sum_{s=j_1,\ldots,j_p} \sum_{l=1}^{n_j} g_{ijsl} + \sum_{s=j_1,\ldots,j_p} b_s^i \sum_{l=1}^{n_j} \bar{d}_{sl} = 0, \quad i = 1, 2, \ldots, N,
\end{aligned} \tag{10.12}$$

the closed-loop fuzzy system becomes $\dot{x}(t) = A_m x(t) + B_m r(t)$.

Proof With (10.6) and (10.2), we have

$$\sum_{i=1}^{N} \lambda_i B_i u(t) = \sum_{i=1}^{N} \lambda_i \begin{bmatrix} 0 \\ \vdots \\ 0 \\ \sum_{j\neq j_1,\ldots,j_p} b_j^i v_j(t) + \sum_{s=j_1,\ldots,j_p} b_s^i \bar{u}_s(t) \end{bmatrix}. \tag{10.13}$$

Applying (10.11) and (10.4) into (10.13), we obtain

$$\begin{aligned}
&\sum_{i=1}^{N} \lambda_i \left(\sum_{j\neq j_1,\ldots,j_p} b_j^i v_j(t) + \sum_{s=j_1,\ldots,j_p} b_s^i \bar{u}_s(t) \right) \\
&= \sum_{i=1}^{N} \lambda_i \left[\sum_{j\neq j_1,\ldots,j_p} b_j^i \left(-K_{ij}x(t) + l_{ij}r(t) + k_{ij} + \sum_{s=j_1,\ldots,j_p} \sum_{l=1}^{n_j} g_{ijsl} f_{sl}(t) \right) \right. \\
&\quad \left. + \sum_{s=j_1,\ldots,j_p} b_s^i \left(\bar{u}_{s0} + \sum_{l=1}^{n_j} \bar{d}_{sl} f_{sl}(t) \right) \right].
\end{aligned} \tag{10.14}$$

Using (10.12), Eq. (10.14) can be calculated as

$$\sum_{i=1}^{N}\lambda_i\left(\sum_{j\neq j_1,\ldots,j_p} b_j^i v_j(t)+\sum_{s=j_1,\ldots,j_p} b_s^i\bar{u}_s(t)\right)$$
$$=-\sum_{i=1}^{N}\lambda_i\left[a_1^i-a_1^m\ \cdots\ a_{n-1}^i-a_{n-1}^m\ a_n^i-a_n^m\right]x(t)+\sum_{i=1}^{N}\lambda_i b^m r(t). \tag{10.15}$$

Substituting (10.15) and (10.13) into (10.7) yields

$$\begin{aligned}\dot{x}(t)&=\frac{1}{\sum_{i=1}^{N}\lambda_i}\left(\sum_{i=1}^{N}\lambda_i A_i x(t)-\sum_{i=1}^{N}\lambda_i\left(A_i-A_m\right)x(t)+\sum_{i=1}^{N}\lambda_i B_m r(t)\right)\\&=A_m x+B_m r(t).\end{aligned} \tag{10.16}$$

The proof has been completed. ∇

If we know which actuators fail and the values of faults, the controller parameters can be calculated from the matching conditions (10.12). However, when we do not know the pattern and values of the faults, the controller (10.11) cannot be implemented. In the following study, we will propose a controller structure that is capable of dealing with different fault situations without knowing the exact fault pattern.

We first present the design procedure for the T–S fuzzy systems with $m=2$ actuators by proposing a nominal control law for actuator fault compensation and developing an adaptive control scheme. Then, the proposed fault compensation scheme is generalized to $m\geq 2$ case.

Nominal control law. For the system (10.7) with $m=2$ actuators, there are three possible situations which we need to deal with: (i) both actuators u_1 and u_2 are healthy, (ii) the actuator u_1 is healthy while u_2 fails (that is, $u_2=\bar{u}_2$), and (iii) the actuator u_2 is healthy while u_1 fails (that is, $u_1=\bar{u}_1$). Our goal is to develop one unified controller structure which is suitable for all three situations.

We first consider the second case with $u_2(t)=\bar{u}_2(t)$. From (10.11), the nominal control law for $u_1(t)$ is

$$u_1(t)=v_1(t)=\frac{\sum_{i=1}^{N}\alpha_{i1}(-K_{i1}x(t)+l_{i1}r(t)+k_{i1}+\sum_{l=1}^{n_j}g_{i12l}f_{2l}(t))}{\sum_{i=1}^{N}\alpha_{i1}}. \tag{10.17}$$

If both the system parameters and fault parameters are known in this case, with Lemma 10.1, the controller parameters can be calculated as

$$K_{i1} = \frac{1}{b_1^i} \left[a_1^i - a_1^m \cdots a_{n-1}^i - a_{n-1}^m \; a_n^i - a_n^m \right]$$

$$l_{i1} = \frac{b_m}{b_1^i}, \quad k_{i1} = -\frac{b_2^i \bar{u}_{20}}{b_1^i}, \quad g_{i12l} = -\frac{b_2^i \bar{d}_{2l}}{b_1^i}. \tag{10.18}$$

Applying (10.17) to the system (10.7) with $u_2(t) = \bar{u}_2(t)$ results in the closed-loop dynamics: $\dot{x}(t) = A_m x(t) + B_m r(t)$, achieving state tracking objective. The third case can be similarly handled.

For the first case with no fault, that is, when both actuators u_1 and u_2 are healthy, there is a need of actuator coordination to meet a desired system output performance (otherwise, for example, $u_1(t)$ and $u_2(t)$ are against to each other, there would be a practical problem). Such an actuator coordination is characterized by the chosen actuation scheme

$$u_i(t) = \delta_i v_0(t), \; \delta_i > 0, \; i = 1, 2 \tag{10.19}$$

for an applied input signal $v_0(t)$ to be designed. Then the system (10.7) becomes

$$\dot{x}(t) = \frac{\sum_{i=1}^N \lambda_i [A_i x(t) + (\delta_1 b_1^i + \delta_2 b_2^i) v_0(t)]}{\sum_{i=1}^N \lambda_i}. \tag{10.20}$$

The control signal $v_0(t)$ can be designed as

$$v_0(t) = \frac{\sum_{i=1}^N \alpha_i (-K_i x(t) + l_i r(t))}{\sum_{i=1}^N \alpha_i}, \tag{10.21}$$

where

$$K_i = \frac{1}{\delta_1 b_1^i + \delta_2 b_2^i} \left[a_1^i - a_1^m \cdots a_{n-1}^i - a_{n-1}^m \; a_n^i - a_n^m \right]$$

$$l_i = -\frac{b_m}{\delta_1 b_1^i + \delta_2 b_2^i}, \quad \alpha_i = \left| \delta_1 b_1^i + \delta_2 b_2^i \right| \lambda_i. \tag{10.22}$$

Applying (10.22) into (10.20) also results in the closed-loop dynamics: $\dot{x}(t) = A_m x(t) + B_m u(t)$, achieving the state tracking.

When we do not know which actuator fails, we need to design a controller that is capable of handling all the above three cases. Under a chosen actuation scheme (10.19), we propose the following nominal controller $v_0^*(t)$ to accommodate all the above three cases:

$$v_0^*(t) = \frac{\sum_{i=1}^N |b_i^*| \lambda_i (-K_i^* x(t) + l_i^* r(t) + k_i^* + g_{i1}^*(t) + g_{i2}^*(t))}{\sum_{i=1}^N |b_i^*| \lambda_i}, \tag{10.23}$$

where

$$K_i^* = \frac{1}{b_i^*}\left[a_1^i - a_1^m \cdots a_{n-1}^i - a_{n-1}^m \; a_n^i - a_n^m\right], \quad l_i^* = \frac{b_m}{b_i^*}$$
$$k_i^* = -\frac{b_{i2}^*}{b_i^*}\bar{u}_{20} - \frac{b_{i1}^*}{b_i^*}\bar{u}_{10}, \quad g_{i12l}^* = -\frac{b_{i2}^*}{b_i^*}\bar{d}_{2l}, \quad g_{i21l}^* = -\frac{b_{i1}^*}{b_i^*}\bar{d}_{1l}$$
$$g_{i1}^*(t) = \sum_{l=1}^{n_j} g_{i12l}^* f_{2l}(t), \quad g_{i2}^*(t) = \sum_{l=1}^{n_j} g_{i21l}^* f_{1l}(t), \tag{10.24}$$

and

$$(b_i^*, \; b_{i1}^*, \; b_{i2}^*) = \begin{cases} (\delta_1 b_1^i + \delta_2 b_2^i, \; 0, \; 0) & \text{for no failure} \\ (\delta_1 b_1^i, \; 0, \; b_2^i) & \text{for } u_2(t) = \bar{u}_2(t) \\ (\delta_2 b_2^i, \; b_1^i, \; 0) & \text{for } u_1(t) = \bar{u}_1(t), \end{cases} \tag{10.25}$$

which are all piecewise constants that change their values when an actuator fault occurs.

With (10.19) and (10.25), the fuzzy system (10.7) can be written as

$$\dot{x}(t) = \frac{1}{\sum_{i=1}^N \lambda_i}\left(\sum_{i=1}^N \lambda_i A_i x + \sum_{i=1}^N \lambda_i \begin{bmatrix} 0 & 0 & 0 & 0 & 0 \\ \vdots & \vdots & \vdots & \vdots & \vdots \\ 0 & 0 & 0 & 0 & 0 \\ b_i^* & b_{i1}^*\bar{u}_{10} & b_{i2}^*\bar{u}_{20} & b_{i1}^*\bar{d}_1^T & b_{i2}^*\bar{d}_2^T \end{bmatrix} \begin{bmatrix} v_0^*(t) \\ 1 \\ 1 \\ f_1(t) \\ f_2(t) \end{bmatrix}\right), \tag{10.26}$$

where

$$\bar{d}_1 = \left[\bar{d}_{11} \; \bar{d}_{12} \cdots \bar{d}_{1n_1}\right]^T, \quad \bar{d}_2 = \left[\bar{d}_{21} \; \bar{d}_{22} \cdots \bar{d}_{2n_2}\right]^T \tag{10.27}$$
$$f_1(t) = \left[f_{11}(t) \; f_{12}(t) \cdots f_{1n_1}(t)\right]^T, \quad f_2(t) = \left[f_{21}(t) \; f_{22}(t) \cdots f_{2n_2}(t)\right]^T.$$

Let $b_{i1}^*\bar{u}_{10} + b_{i2}^*\bar{u}_{20} = b_{i,u_0}^* \in R$, $b_{i1}^*\bar{d}_1^T = b_{i1,d}^* \in R^{1\times n_1}$, $b_{i2}^*\bar{d}_2^T = b_{i2,d}^* \in R^{1\times n_2}$. and define

$$B_i^* = \begin{bmatrix} 0 & 0 & 0 & 0 & 0 \\ \vdots & \vdots & \vdots & \vdots & \vdots \\ 0 & 0 & 0 & 0 & 0 \\ b_i^* & b_{i1}^*\bar{u}_0 & b_{i2}^*\bar{u}_0 & b_{i1}^*\bar{d} & b_{i2}^*\bar{d} \end{bmatrix}, \quad \omega(t) = \left[v_0^*(t) \; 1 \; 1 \; f_1^T(t) \; f_2^T(t)\right]^T. \tag{10.28}$$

Then equation (10.26) can be formulated as

$$\dot{x}(t) = A_m x(t) + \frac{\sum_{i=1}^N \lambda_i[(A_i - A_m)x(t) + B_i^*\omega(t)]}{\sum_{i=1}^N \lambda_i}. \tag{10.29}$$

Applying (10.23) to (10.29) results in the closed-loop fuzzy system:

$$\dot{x}(t) = A_m x(t) + B_m r(t), \tag{10.30}$$

which is the same as the reference model.

The nominal control law (10.23) provides the basic controller structure, which can be parameterized for parameter adaptation when the system and fault parameters are unknown. In the following section, we will develop an adaptive version of (10.23) for unknown parameters.

10.2.3 Adaptive Control Scheme

In this section, we design an adaptive parameter estimation algorithm to estimate the unknown system parameters, and develop an adaptive control law and analyze the closed-loop system performance.

Adaptive control law. Since both the system and fault parameters are unknown, the nominal control law (10.23) cannot be applied directly. Using parameter estimation, an adaptive control law can be implemented as

$$\hat{v}_0(t) = \frac{\sum_{i=1}^{N} \left|\hat{b}_i\right| \lambda_i(-\hat{K}_i x(t) + \hat{l}_i r(t) + \hat{k}_i + \hat{g}_{i1}(t) + \hat{g}_{i2}(t))}{\sum_{i=1}^{N} \left|\hat{b}_i\right| \lambda_i}, \tag{10.31}$$

where $\hat{b}_i$, $\hat{K}_i$, $\hat{l}_i$, $\hat{k}_i$, $\hat{g}_{i1}(t)$, $\hat{g}_{i2}(t)$ are the estimates of b_i^*, K_i^*, l_i^*, k_i^*, $g_{i1}^*(t)$, $g_{i2}^*(t)$, respectively.

Defining the state tracking error

$$e(t) = x(t) - x_m(t) \tag{10.32}$$

and applying (10.31) into (10.29) and using (10.10), (10.24) and (10.30), we have the tracking error dynamics

$$\dot{e}(t) = A_m e(t) + \frac{\sum_{i=1}^{N} \lambda_i[-\tilde{A}(t)_i x(t) - \tilde{B}_i(t)\omega(t)]}{\sum_{i=1}^{N} \lambda_i}, \tag{10.33}$$

where $\tilde{A}_i = \hat{A}_i - A_i$, $\tilde{B}_i = \hat{B}_i - B_i^*$, with $\hat{A}_i$ and $\hat{B}_i$ being the estimates of A_i and B_i^*, respectively.

Parameter adaptive law. To estimate the parameters in (10.31), we design the following adaptive law:

$$\begin{aligned}
\dot{\hat{\mathrm{a}}}_{\mathrm{i}}(t) &= \gamma_{1i}\frac{\lambda_i}{\sum_{i=1}^{N}\lambda_i}x(t)e^T(t)P_1\\
\dot{\hat{b}}_i(t) &= \begin{cases}\gamma_{2i}\frac{\lambda_i}{\sum_{i=1}^{N}\lambda_i}P_1^Te(t)\hat{v}_0(t) & \text{if } \left|\hat{b}_i(t)\right| > b_{i0}^*\\ & \text{or } \left|\hat{b}_i(t)\right| = b_{i0}^* \text{ and } P_1^Te(t)\hat{v}_0(t)\mathrm{sgn}(b_i) \geq 0\\ 0 & \text{otherwise}\end{cases}\\
\dot{\hat{b}}_{i,u_0}(t) &= \gamma_{3i}\frac{\lambda_i}{\sum_{i=1}^{N}\lambda_i}P_1^Te(t),\\
\dot{\hat{b}}_{i1,d}(t) &= \gamma_{4i}\frac{\lambda_i}{\sum_{i=1}^{N}\lambda_i}P_1^Te(t)f_1^T(t),\quad \dot{\hat{b}}_{i2,d}(t) = \gamma_{4i}\frac{\lambda_i}{\sum_{i=1}^{N}\lambda_i}P_1^Te(t)f_2^T(t),
\end{aligned} \tag{10.34}$$

where b_{i0}^* is the lower bound of b_i^*, which is assumed to be known, and $\hat{b}_i(0)$ satisfies $\hat{b}_i(0)\mathrm{sgn}(b_i^*) \geq 0$. P_1 is the first column of a positive definite symmetrical matrix $P \in R^{n\times n}$, which is the solution of the following equation:

$$A_m^T P + PA_m = -I, \tag{10.35}$$

where $I \in R^{n\times n}$ is an identity matrix.

Based on the adaptive control law (10.31) and the parameter adaptation law (10.34), the closed-loop stability and asymptotic tracking results are proved and summarized in the following theorem.

Theorem 10.1 *For the T–S fuzzy system (10.7) with two redundant actuators and the actuator fault model (10.4), under Assumptions 10.1 and 10.2, the adaptive controller (10.31) with the adaptive law (10.34) guarantees:*

(i) all closed-loop system signals are bounded; and
(ii) the state tracking error $\lim_{t\to\infty} e(t) = 0$.

Proof Consider the following Lyapunov function candidate

$$V(e,\tilde{\mathrm{a}}_{\mathrm{i}},\tilde{b}_i) = e^TPe + \sum_{i=1}^{N}\frac{\tilde{\mathrm{a}}_{\mathrm{i}}^{\mathrm{T}}\tilde{\mathrm{a}}_i}{\gamma_{1i}} + \sum_{i=1}^{N}\left(\frac{\tilde{b}_i^{*2}}{\gamma_{2i}} + \frac{\tilde{b}_{i,u_0}^{*2}}{\gamma_{3i}} + \frac{\tilde{b}_{i1,d}^{*}\tilde{b}_{i1,d}^{*T} + \tilde{b}_{i2,d}^{*}\tilde{b}_{i2,d}^{*T}}{\gamma_{4i}}\right),$$

where $\tilde{\mathrm{a}}_i = \hat{\mathrm{a}}_i - \mathrm{a}_i, \mathrm{a}_i = [a_1^i, a_2^i, \cdots, a_n^i]^T, \tilde{b}_i = \hat{b}_i - b_i, \tilde{b}_{i,u_0} = \hat{b}_{i,u_0} - b_{i,u_0}^*, \tilde{b}_{i1,d} = \hat{b}_{i1,d} - b_{i1,d}^*, \tilde{b}_{i2,d} = \hat{b}_{i2,d} - b_{i2,d}^*$.

The differentiation of V along the error dynamics (10.33) is

$$\begin{aligned}
\dot{V} &= \dot{e}^T Pe + e^T P\dot{e} + 2\sum_{i=1}^{N}\frac{\dot{\tilde{\mathrm{a}}}_i^T\tilde{\mathrm{a}}_i}{\gamma_{1i}} + 2\sum_{i=1}^{N}\left(\frac{\dot{\tilde{b}}_i\tilde{b}_i}{\gamma_{2i}} + \frac{\dot{\tilde{b}}_{i,u_0}\tilde{b}_{i,u_0}}{\gamma_{3i}} + \frac{\tilde{b}_{i1,d}\dot{\tilde{b}}_{i1,d}{}^T}{\gamma_{4i}} + \frac{\tilde{b}_{i2,d}\dot{\tilde{b}}_{i2,d}^T}{\gamma_{4i}}\right)\\
&= -\dot{e}^T e - 2\frac{\sum_{i=1}^{N}\lambda_i P_1^T ex^T\tilde{\mathrm{a}}_\mathrm{i}}{\sum_{i=1}^{L}\lambda_i} + 2\sum_{i=1}^{N}\frac{\dot{\tilde{\mathrm{a}}}_i^T\tilde{\mathrm{a}}_i}{\gamma_{1i}}\\
&\quad -2\frac{\sum_{i=1}^{N}\lambda_i[P_1^T e\tilde{b}_i^*\hat{v}_0 + P_1^T e\tilde{b}_{i,u_0}^* + P_1^T e\tilde{b}_{i1,d}^* f_1(t) + P_1^T e\tilde{b}_{i2,d}^* f_2(t)]}{\sum_{i=1}^{N}\lambda_i}\\
&\quad +2\sum_{i=1}^{N}\frac{\dot{\tilde{b}}_i\tilde{b}_i}{\gamma_{2i}} + 2\sum_{i=1}^{N}\frac{\dot{\tilde{b}}_{i,u_0}\tilde{b}_{i,u_0}}{\gamma_{3i}} + 2\sum_{i=1}^{N}\frac{\tilde{b}_{i1,d}\dot{\tilde{b}}_{i1,d}^T}{\gamma_{4i}} + 2\sum_{i=1}^{N}\frac{\tilde{b}_{i2,d}\dot{\tilde{b}}_{i2,d}^T}{\gamma_{4i}}.
\end{aligned} \tag{10.36}$$

With the adaptive law (10.34), we have

$$\dot{V}(t) = -e^T e \le 0, \tag{10.37}$$

which means $e,\ \tilde{\mathrm{a}}_\mathrm{i},\ \tilde{b}_i,\ \tilde{b}_{i,u_0},\ \tilde{b}_{i1,d},\ \tilde{b}_{i2,d} \in L_\infty$.

Under the constraint condition $\hat{b}_i\mathrm{sgn}(b_i^*) \ge b_{i0}^*$, the adaptive law for $\hat{b}_i$ in (10.34) has a projection term. If the initial value of $\hat{b}_i(t)$ satisfies $\hat{b}_i(0)\mathrm{sgn}(b_i^*) \ge b_{i0}^*$, we have $\hat{b}_i(t)\mathrm{sgn}(b_i^*) = \left|\hat{b}_i(t)\right|$. It follows that $\dot{\hat{b}}_i(t)\hat{b}_i(t) \ge 0$. Then for all $t \ge 0$, we have $\left|\hat{b}_i(t)\right| \ge b_{i0}^*$. Thus, from (10.36), we obtain

$$\dot{V} = \begin{cases} -e^T(t)e(t) & \text{if } \left|\hat{b}_i(t)\right| > b_{i0}^* \\ & \text{or } \left|\hat{b}_i(t)\right| = b_{i0}^* \text{ and } P_1^T e(t)\hat{v}_0(t)\mathrm{sgn}(b_i^*) \ge 0 \\ -e^T(t)e(t) - 2\frac{\alpha_i\tilde{b}_i(t)}{\sum_{i=1}^{N}\lambda_i}P_1^T e(t)\hat{v}_0(t) & \text{otherwise.} \end{cases} \tag{10.38}$$

If $\left|\hat{b}_i(t)\right| = b_{i0}^*$, we have $(\hat{b}_i(t) - b_i^*)\mathrm{sgn}(b_i^*) \le 0$. Furthermore, if $\left|\hat{b}_i(t)\right| = b_{i0}^*$ and $P_1^T e(t)\hat{v}_0(t)\mathrm{sgn}(b_i^*) \le 0$, we have

$$\tilde{b}_i(t)P_1^T e(t)\hat{v}_0(t) = (\hat{b}_i(t) - b_i^*)\mathrm{sgn}(b_i^*)(P_1^T e(t)\hat{v}_0(t)\mathrm{sgn}(b_i^*)) \ge 0. \tag{10.39}$$

With (10.38) and (10.39), we finally obtain

$$\dot{V} \le -e^T(t)e(t), \tag{10.40}$$

which means the tracking error $e(t) \in L_\infty$ and parameter errors $\tilde{\mathrm{a}}_i,\ \tilde{b},\ \tilde{b}_{i,u_0},\ \tilde{b}_{i1,d},\ \tilde{b}_{i2,d} \in L_\infty$, hence the boundedness of $\hat{K}_i,\ \hat{l}_i,\ \hat{k}_i,\ \hat{g}_{i1},\ \hat{g}_{i2}$. With (10.32) and (10.10), we have $x(t) \in L_\infty$. From (10.22), we obtain $\hat{v}_0(t) \in L_\infty$. Hence, all the closed-loop signals are bounded.

From (10.33), we have $\dot{e}(t) \in L_\infty$. It follows from (10.40) and the boundedness of V that

$$\int_0^\infty e^T(\tau)e(\tau)d\tau = -\int_0^\infty \dot{V} d\tau = V(0) - V(\infty), \quad \forall t \geq 0, \tag{10.41}$$

that is, $e(t) \in L_2$. With Barbălat's lemma, we have the desired convergence property: $\lim_{t\to\infty} e(t) = 0$. ∇

Now we have completed our adaptive actuator fault compensation designs for T–S fuzzy systems with two actuators ($m = 2$). In the following section, we will give the main idea and design outline for the general $m \geq 2$ case.

Generalization to $m \geq 2$. When the T–S fuzzy system (10.7) has more than two actuators, the nominal control law for the jth healthy actuator can still be designed as (10.11) and the matching conditions (10.12) in Lemma 10.1 are still needed to be satisfied to achieve the actuator fault compensation. However, the solutions for the control parameters K_{ij}, l_{ij}, k_{ij} and g_{ijsl}, $i = 1, \ldots, N$, $j \in \{1, \ldots, m\}$, $j \neq j_1, \ldots, j_p$ $s = j_1, \ldots, j_p, l = 1, \ldots, n_j$ from (10.12), may not be unique.

With the similar design idea on obtaining (10.22) for the $m = 2$ case and using the same actuation scheme (10.19), the nominal controller can be chosen as

$$v_0^*(t) = \frac{\sum_{i=1}^N |b_i^*| \lambda_i (-K_i^* x(t) + l_i^* r(t) + k_i^* + \sum_{j=j_1,\ldots,j_p} g_{ij}^*(t))}{\sum_{i=1}^N |b_i^*| \lambda_i}, \tag{10.42}$$

where

$$K_i^* = \frac{1}{b_i^*} \left[a_1^i - a_1^m \cdots a_{n-1}^i - a_{n-1}^m \; a_n^i - a_n^m \right], \quad l_i^* = \frac{b_m}{b_i^*}$$

$$k_i^* = -\sum_{j=j_1,\ldots,j_p} \frac{b_{ij}^*}{b_i^*} \bar{u}_{j0}, \quad g_{ijsl}^* = -\frac{b_{is}^*}{b_i^*} \bar{d}_{sl}, \quad g_{ij}^*(t) = \sum_{s=j_1,\ldots,j_p} \sum_{l=1}^{n_j} g_{ijsl}^* f_{sl}(t),$$

with $b_i^* = \sum_{j=j_1,\ldots,j_p} (1 - \sigma_j) \delta_j b_j^i$ and $b_{is}^* = \sigma_s \delta_s b_s^i$, $s = j_1, \ldots, j_p$, which are all piecewise constants changing their values when an actuator fault occurs.

The nominal controller (10.42) provides the basic control structure for the actuator fault compensation, based on which an adaptive controller can be further developed by following the similar design procedure as presented for the case $m = 2$.

10.2.4 Simulation Study

In this section, we present a simulation study on the proposed adaptive control and adaptive actuator fault compensation schemes to illustrate the desired system performance and the unique advantages of our proposed control schemes.

Simulation system. Consider the following inverted-pendulum system (Wang et al. 1996):

$$\begin{aligned}\dot{x}_1 &= x_2\\ \dot{x}_2 &= \frac{g\sin x_1 - amlx_2^2\sin 2x_1/2 - a\cos x_1 u}{4l/3 - aml\cos^2 x_1},\end{aligned}\tag{10.43}$$

where x_1 and x_2 represent the position and velocity of the pendulum with respect to the vertical line, respectively, $u = [u_1, u_2]^T$ is the input force vector, which is assumed to be generated by two actuators, $a = 1/(m+M)$, M and m are masses of the cart and the pole, respectively, l is the half length of the pole, and g is the acceleration due to gravity.

In this example, a T–S fuzzy model is constructed by linearizing the nonlinear plant (10.43) at two operating points for $x_1 \in (-\pi/2, \pi/2)$. The ith rule of the T–S fuzzy model has the following form (Wang et al. 1996):

$$R^i: \text{ IF } \; x_1 \; \text{ is } \; F_1^i, \quad \text{THEN } \; \dot{x} = A_i x + B_i u,$$

where $i = 1, 2$, A_i and B_i are the local state-space matrices corresponding to the ith linearized model, and F_1^1 and F_1^2 are two fuzzy sets representing "about 0" and "about $\pm\pi/2$". The membership functions characterizing the fuzzy sets F_1^1 and F_1^2 are given in Fig. 10.1.

Simulation results. In the simulation, we use the following system parameters: $M = 8.0\,\text{kg}$, $m = 2.0\,\text{kg}$, $2l = 1\,\text{m}$, $g = 9.8\,\text{m/s}^2$. The reference model is chosen as $A_m = \begin{bmatrix} 0 & 1 \\ -4 & -4 \end{bmatrix}$ and $B_m = \begin{bmatrix} 0 & 1 \end{bmatrix}^T$ with the reference input $r = 1$. A stuck fault occurs on u_2, that is, $u_2(t) = 1$ for $t \geq 30\,\text{sec}$. The parameters and initial conditions used in the simulation are as follows: $\gamma_{i1} = 50$, $\gamma_{2i} = \gamma_{3i} = \gamma_{4i} = 0.5$, $x(0) = [\pi/6, 0]^T$, $x_m(0) = [\pi/9, 0]^T$, $b_{1,u_0}^* = 0.1$, $b_{2,u_0}^* = 0.003$, $\hat{a}_1(0) = [15, 0]^T$, $\hat{a}_2(0) = [8, 0]^T$, $[\hat{b}_1(0), \hat{b}_2(0)] = [-0.15, -0.005]$, $[\hat{b}_{1,u_0}(0), \hat{b}_{2,u_0}(0)] = [-0.3, -0.01]$, and $[\hat{b}_{11,d}(0), \hat{b}_{12,d}(0), \hat{b}_{21,d}(0), \hat{b}_{22,d}(0)] = [0, 0, 0, 0]$.

The control law is applied to both the T–S fuzzy system and the original plant. The state tracking responses under a stuck fault are shown in Fig. 10.2. It can be observed that both the T–S fuzzy system and the original plant can track the states of the reference model well despite the actuator fault. The adaptive compensation control signal adjusts itself when the actuator fault occurs at $t = 30\,\text{sec}$. The parameter adaptation of $\hat{a}_i$ is shown in Fig. 10.3.

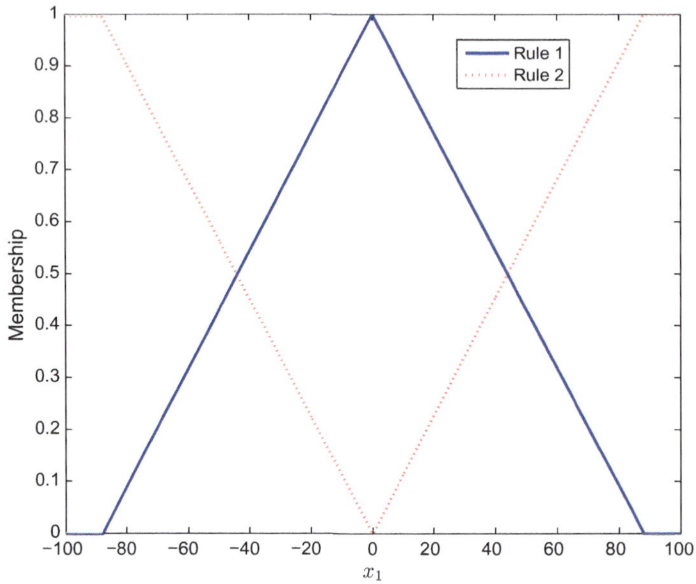

Fig. 10.1 Membership functions for x_1

The simulation results verify that all the closed-loop signals are bounded under the proposed adaptive compensation control scheme and the asymptotical tracking can still be achieved under actuator faults. The parameters converge to constant values. The tracking responses of the original plant demonstrate that if the approximation error is small, the proposed control scheme can be effectively applied to tracking control of the original plant and produces desirable control performance.

10.3 Adaptive Fault Compensation Control Using Output Feedback

In this section,[1] we consider adaptive fault compensation control using output feedback. Consider a discrete-time nonlinear system with redundant actuators described by the following difference equation:

$$\begin{aligned} y(t) &= f(y(t-1), \ldots, y(t-n), u_1(t-d), \ldots, u_1(t-n), \\ &\quad \ldots, u_m(t-d), \ldots, u_m(t-n)), \end{aligned} \tag{10.44}$$

[1]©[2012] IEEE. Parts of Sect. 10.3 are reprinted, with permission, from Qi et al. (2012).

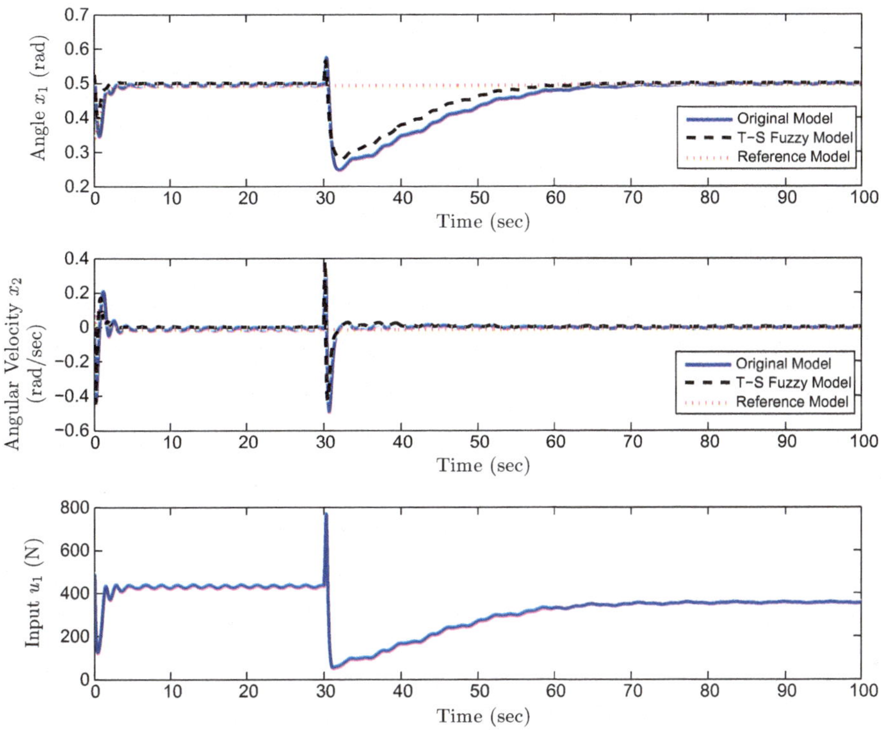

Fig. 10.2 State tracking responses under constant fault $\bar{u}_2(t) = 1$

where $y \in R$ is the output, $u_i \in R, i = 1, 2, \ldots, m$ represent m inputs generated by m actuators.

In Chap. 7, an input–output discrete-time T–S fuzzy system (7.4) is employed to approximate the nonlinear system (10.44) for the case $m = 1$ and an adaptive control scheme is developed. In this section, we study the adaptive fault compensation control design of T–S fuzzy systems with redundant inputs ($m > 1$).

10.3.1 Problem Statement

Similar to (7.4) for the case when $m = 1$ (that is, the nonredundant actuator case), the following discrete-time T–S fuzzy system model can be used for the case when $m > 1$:

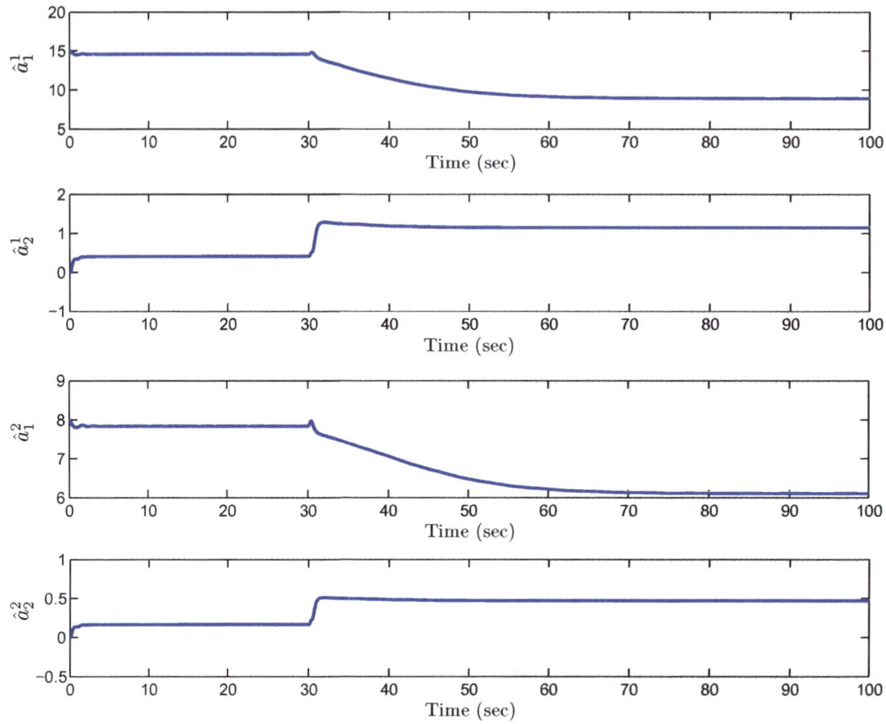

Fig. 10.3 Parameter estimation

$$
\begin{aligned}
R^i:\ &\text{IF}\ \ \xi_1\ \text{is}\ F_1^i\ \text{and}\ \xi_2\ \text{is}\ F_2^i\ \text{and}\ \ldots\ \text{and}\ \xi_L\ \text{is}\ F_L^i\\
&\text{THEN}\ \ y(t)+a_{i,1}y(t-1)+\cdots+a_{i,n}y(t-n)=b_{1,i,0}u_1(t-d)\\
&+b_{1,i,1}u_1(t-d-1)+\cdots+b_{1,i,\,n-d}u_1(t-n)+\cdots+b_{m,i,0}u_m(t-d)\\
&\ \ +b_{m,i,1}u_m(t-d-1)+\cdots+b_{m,i,\,n-d}u_m(t-n),
\end{aligned}
\tag{10.45}
$$

with $b_{j,i,0} \neq 0$, $j = 1, 2, \ldots, m$, $i = 1, 2, \ldots, N$.

The actuator faults can be described by

$$
\bar{u}_j(t) = \bar{u}_{j,0} + \sum_{l=1}^{n_j} \bar{u}_{j,l} f_{j,l}(t),\ \ t \geq t_j,
\tag{10.46}
$$

for some unknown index $j \in \{1, 2, \ldots, m\}$, unknown time instant t_j and unknown constants $\bar{u}_{j,0}$ and $\bar{u}_{j,l}$, $l = 1, \ldots, n_j$, and known bounded signals $f_{j,l}(t)$ and $n_j \geq 1$ (Tao et al. 2004). This parameterized actuator fault model can be used to closely approximate a large class of practical fault, by a proper selection of these basis functions $f_{j,l}(t)$. The selection of the basis functions of the actuator faults depends on individual applications; for example, a common fault is that the actuator is stuck at an unknown constant value $\bar{u}_{j,0}$, for which those $f_{j,l}(t)$ can be set to zero.

The control objective is to find an adaptive control scheme for the global version of the fuzzy system (10.45) with unknown parameters and subject to failures belonging to a fault pattern set with unknown fault patterns, values, and time instants, to ensure closed-loop signal boundedness and asymptotic tracking of a reference output $y_m(t)$ by the system output $y(t)$.

10.3.2 Nominal Control Design

In this section, we develop the solution to the adaptive actuator fault compensation problem. For simplicity of presentation and without loss of generality, we consider the fuzzy system model (10.45) with two actuators (that is, with $m = 2$), to construct a nominal control law for actuator fault compensation and develop an adaptive fault compensation scheme.

We first derive a global fuzzy system model for the fuzzy system model (10.45) with two actuators ($m = 2$). With $A_i(z^{-1})$ used in (7.50) and

$$\bar{B}_{j,i}(z^{-1}) = b_{j,i,0} + b_{j,i,1}z^{-1} + \cdots + b_{j,i,n-d}z^{-n+d}, \tag{10.47}$$

$j = 1, 2$, we denote the local system (10.45) as

$$A_i(z^{-1})[y](t) = z^{-d}\bar{B}_{1,i}(z^{-1})[u_1](t) + z^{-d}\bar{B}_{2,i}(z^{-1})[u_2](t). \tag{10.48}$$

With (7.8): $y(t+d) = z^d F_i(z^{-1})A_i(z^{-1})\,[y](t) + G_i(z^{-1})[y](t)$, and using (10.48), we obtain

$$\begin{aligned} y(t+d) &= F_i(z^{-1})\bar{B}_{1,i}(z^{-1})[u_1](t) + F_i(z^{-1})\bar{B}_{2,i}(z^{-1})[u_2](t) + G_i(z^{-1})[y](t) \\ &= \beta_{1,i}(z^{-1})[u_1](t) + \beta_{2,i}(z^{-1})[u_2](t) + \alpha_i(z^{-1})[y](t), \end{aligned} \tag{10.49}$$

where $\alpha_i(z^{-1}) = G_i(z^{-1})$ as in (7.11) and $\beta_{j,i}(z^{-1}) = F_i(z^{-1})\bar{B}_{j,i}(z^{-1})$, $j = 1, 2$:

$$\begin{aligned} \beta_{j,i}(z^{-1}) &= \beta_{j,i,0} + \beta_{j,i,1}z^{-1} + \cdots + \beta_{j,i,n-1}z^{-n+1}, \\ \beta_{j,i,0} &= b_{j,i,0} \neq 0, \quad j = 1, 2. \end{aligned} \tag{10.50}$$

Similar to that described in Proposition 7.2, we have the global fuzzy prediction system model for the two actuators case ($m = 2$):

$$y(t+d) = \sum_{i=1}^{N} \mu_i \alpha_i(z^{-1})[y](t) + \sum_{i=1}^{N} \mu_i \beta_{1,i}(z^{-1})[u_1](t) + \sum_{i=1}^{N} \mu_i \beta_{2,i}(z^{-1})[u_2](t). \tag{10.51}$$

For this system, there are three possible situations which we need to deal with: (i) both actuators u_1 and u_2 are healthy, (ii) the actuator u_1 is healthy while u_2 fails (that is, $u_2 = \bar{u}_2(t)$), and (iii) the actuator u_2 is healthy while u_1 fails (that is, $u_1 = \bar{u}_1(t)$). Our goal is to develop one controller structure which is suitable for all the three situations.

With $\bar{\beta}_{j,i}(z^{-1}) = \beta_{j,i,1} z^{-1} + \cdots + \beta_{j,i,n-1} z^{-n+1}$, $j = 1, 2$, we first consider the second case with $u_2 = \bar{u}_2$). In this case, the nominal control law for $u_1(t)$ is

$$u_1(t) = \frac{1}{\sum_{i=1}^{N} \mu_i \beta_{1,i,0}} \left[-\sum_{i=1}^{N} \mu_i \alpha_i(z^{-1})[y](t) - \sum_{i=1}^{N} \mu_i \bar{\beta}_{1,i}(z^{-1})[u_1](t) - \sum_{i=1}^{N} \mu_i \beta_{2,i}(z^{-1})[\bar{u}_2](t) + y_m(t+d) \right] \tag{10.52}$$

which, when applied to the system (10.51) with $u_2(t) = \bar{u}_2(t)$, brings $y(t+d)$ to $y_m(t+d)$ in one step and leads the closed-loop system stability if the subsystem (u_1, y) is minimum phase.

The third case can be similarly handled if the subsystem (u_2, y) is minimum phase. Hence, for actuator fault compensation, we need

Assumption 10.3 The individual subsystems (u_1, y) and (u_2, y) are minimum phase.

Finally, for the first case with no fault, that is, when both actuators u_1 and u_2 are healthy, there is a need of actuator coordination to meet a desired system output performance (otherwise, for example, $u_1(t)$ and $u_2(t)$ are against to each other, there would be a problem). Such an actuator coordination is characterized by the chosen actuation scheme

$$u_i(t) = \delta_i v_0(t), \;\; \delta_i > 0, \;\; i = 1, 2 \tag{10.53}$$

for an applied input signal $v_0(t)$ to be designed. Then the system (10.51) becomes

$$y(t+d) = \sum_{i=1}^{N} \mu_i \alpha_i(z^{-1})[y](t) + \sum_{i=1}^{N} \mu_i \beta_i(z^{-1})[v_0](t), \tag{10.54}$$

where $\beta_i(z^{-1}) = \delta_1 \beta_{1,i}(z^{-1}) + \delta_2 \beta_{2,i}(z^{-1}) \triangleq \beta_{i,0} + \beta_{i,1} z^{-1} + \cdots + \beta_{i,n-1} z^{-n+1}$.

To ensure the controllability of (10.54) using v_0, the following assumption is used.

Assumption 10.4 $\sum_{i=1}^{N} \mu_i(\xi(t)) \beta_{i,0} \neq 0$, for all $t \geq 0$.

The control signal $v_0(t)$ can be designed as $u(t)$ in (9.49), under Assumptions 10.3 and 10.4 for the system (10.54). This brings up the assumption for the redundant actuator system (10.54):

Assumption 10.5 The coordinated system (10.54) is minimum phase for the chosen $\delta_i > 0, i = 1, 2,$.

Nominal control law. Under a chosen actuation scheme (10.53), we propose the following nominal controller structure for $v_0(t)$ to accommodate the above all three cases:

$$\begin{aligned} v_0(t) &= \frac{1}{\sum_{i=1}^N \mu_i \beta_{i,0}^*}\left[-\sum_{i=1}^N \mu_i \alpha_i(z^{-1})[y](t) - \sum_{i=1}^N \mu_i \bar{\beta}_i^*(z^{-1})[v_0](t)\right. \\ &\left. -\sum_{i=1}^N \mu_i \beta_{1,i}^*(z^{-1})[\bar{u}_1](t) - \sum_{i=1}^N \mu_i \beta_{2,i}^*(z^{-1})[\bar{u}_2](t) + y_m(t+d)\right], \end{aligned} \tag{10.55}$$

where

$$\begin{aligned} &(\beta_i^*(z^{-1}),\ \beta_{1,i}^*(z^{-1}),\ \beta_{2,i}^*(z^{-1})) \\ &= \begin{cases} (\delta_1\beta_{1,i}(z^{-1}) + \delta_2\beta_{2,i}(z^{-1}), 0, 0) & \text{for no failure} \\ (\delta_1\beta_{1,i}(z^{-1}),\ 0,\ \beta_{2,i}(z^{-1})) & \text{for } u_2 = \bar{u}_2(t) \\ (\delta_2\beta_{2,i}(z^{-1}),\ \beta_{1,i}(z^{-1}),\ 0) & \text{for } u_1 = \bar{u}_1(t) \end{cases} \end{aligned} \tag{10.56}$$

which are all piecewise time-invariant polynomials which change when an actuator fault occurs.

Although in the nominal control case $\beta_{j,i}^*(z^{-1})$ may be zero for some $j = 1, 2$, for adaptive control when the actuator fault pattern (which actuator fails) is uncertain, they are both treated as parameterized polynomials whose parameters are to be adaptively estimated.

10.3.3 Adaptive Control Scheme

To develop an adaptive actuator fault compensation scheme, we need to obtain the estimates of $\alpha_i(z^{-1})$, $\beta_i^*(z^{-1})$, $\beta_{1,i}^*(z^{-1})$ and $\beta_{2,i}^*(z^{-1})$. In view of (10.56), we express the system (10.51) as

$$\begin{aligned} y(t+d) &= \sum_{i=1}^N \mu_i \alpha_i(z^{-1})[y](t) + \sum_{i=1}^N \mu_i \beta_i^*(z^{-1})[v_0](t) \\ &+ \sum_{i=1}^N \mu_i \beta_{1,i}^*(z^{-1})[\bar{u}_1](t) + \sum_{i=1}^N \mu_i \beta_{2,i}^*(z^{-1})[\bar{u}_2](t). \end{aligned} \tag{10.57}$$

In this model, $\beta_i^*(z^{-1})$ is parameterized in the same structure as that of $\delta_1\beta_{1,i}(z^{-1}) + \delta_2\beta_{2,i}(z^{-1})$, $\delta_1\beta_{1,i}(z^{-1})$ and $\delta_2\beta_{2,i}(z^{-1})$ (they all have the same structure), and $\beta_{1,i}^*(z^{-1})$ and $\beta_{2,i}^*(z^{-1})$ are parameterized in the structures of $\beta_{1,i}(z^{-1})$ and $\beta_{2,i}(z^{-1})$, respectively. For example,

$$\beta_i^*(z^{-1}) = \beta_{i,0}^* + \beta_{i,1}^* z^{-1} + \cdots + \beta_{i,n-1}^* z^{-n+1}. \tag{10.58}$$

The parametrization of $\beta_{j,i}^*(z^{-1})[\bar{u}_1](t)$ involves the parameters of $\beta_{j,i}^*(z^{-1})$ and that of $\bar{u}_j(t)$, $j = 1, 2$, for which an augmented parametrization using the Kronecker product of the two sets of parameters is to be employed. Moreover, the parametrization of $\sum_{i=1}^N \mu_i \beta_{j,i}^*(z^{-1})[\bar{u}_j](t)$ is based on combining the known signals μ_i with the known basis functions $f_{j,l}(t)$, $j = 1, 2$, $l = 1, \ldots, n_j$, of the actuator fault model (10.46).

Parameter estimation. The system (10.57) can be expressed as

$$y(t+d) = \theta_a^T \phi_a(t) + \sum_{j=1}^{2} \sum_{i=1}^{N} \mu_i \beta_{j,i}^*(z^{-1})[\bar{u}_j](t), \tag{10.59}$$

where

$$\begin{aligned}
\theta_a &= [\theta_{a1}^T, \theta_{a2}^T, \ldots, \theta_{aN}^T]^T \\
\phi_a(t) &= [\phi_{a1}^T, \phi_{a2}^T, \ldots, \phi_{aN}^T]^T \\
\phi_{ai}(t) &= [\mu_i y(t), \mu_i y(t-1), \ldots, \mu_i y(t-n+1), \mu_i v_0(t), \ldots, \mu_i v_0(t-n+1)]^T \\
\theta_{ai} &= [\alpha_{i,0}, \alpha_{i,1}, \ldots, \alpha_{i,n-1}, \beta_{i,0}^*, \beta_{i,1}^*, \ldots, \beta_{i,n-1}^*]^T.
\end{aligned}$$

The actuator fault $\bar{u}_j(t)$ in (10.46) can be parameterized as

$$\bar{u}_j(t) = p_{j,0} + p_j^T f_j(t), \quad j = 1, 2, \tag{10.60}$$

where $p_{j,0} = \bar{u}_{j,0}$, $p_j = [\bar{u}_{j,1}, \bar{u}_{j,2}, \ldots, \bar{u}_{j,n_j}]^T$ and $f_j(t) = [f_{j,1}(t), f_{j,2}(t), \ldots, f_{j,n_j}(t)]^T$.

Using (10.60), the second term in (10.59) can be written as

$$\sum_{i=1}^{N} \mu_i \beta_{j,i}^*(z^{-1})[\bar{u}_j](t) = \bar{p}_{j,0}^T \bar{\mu}(t) + \bar{p}_j^T \phi_{f_j}(t), \tag{10.61}$$

where

$$\bar{p}_{j,0} = [\sum_{k=0}^{n-1} \beta^*_{j,1,k} p_{j,0}, \ldots, \sum_{k=0}^{n-1} \beta^*_{j,N,k} p_{j,0}]^T \in R^N$$
$$\bar{\mu}(t) = [\mu_1(t), \mu_2(t), \ldots, \mu_N(t)]^T \in R^N$$
$$\bar{p}_j = [\beta^{*T}_{j,1} \otimes p_j^T, \ldots, \beta^{*T}_{j,N} \otimes p_j^T]^T \in R^{n \times n_j \times N}$$
$$\phi_{f_j}(t) = [\mu_1 \bar{f}_j^T(t), \ldots, \mu_N \bar{f}_j^T(t)]^T \in R^{n \times n_j \times N}$$
$$\bar{f}_j(t) = M_j(z^{-1})[f_j](t) \in R^{n \times n_j}$$
$$M_j(z^{-1}) = [I_{n_j}, z^{-1} I_{n_j}, \ldots, z^{-n+1} I_{n_j}]^T \in R^{(n \times n_j) \times n_j}$$

with $\otimes$ denoting the Kronecker product.

Substituting (10.61) into (10.59) yields

$$\begin{aligned} y(t+d) &= \theta_a^T \phi_a(t) + \sum_{j=1}^{2} (\bar{p}_{j,0}^T \bar{\mu}(t) + \bar{p}_j^T \phi_{f_j}(t)) \\ &= \theta^T \phi(t), \end{aligned} \tag{10.62}$$

where $\theta = [\theta_a^T, \bar{p}_0^T, \bar{p}_1^T, \bar{p}_2^T]^T$ with $\bar{p}_0 = \bar{p}_{1,0} + \bar{p}_{2,0}$, and $\phi(t) = [\phi_a^T(t), \bar{\mu}^T(t), \phi_{f_1}^T(t), \phi_{f_2}^T(t)]^T$.

With the closed-loop system in the form (10.62), the estimation of θ can be obtained by using the similar parameter adaptation scheme as (7.26):

$$\hat{\theta}(t) = \hat{\theta}(t-1) + \frac{\gamma(t)\phi(t-d)\varepsilon(t)}{c + \phi^T(t-d)\phi(t-d)}, \tag{10.63}$$

where $\gamma(t) \in (\gamma_0, 2-\gamma_0)$ is an adaptation gain for some constant $\gamma_0 \in (0, 1)$, $c > 0$ is a small design parameter, and $\varepsilon(t) = y(t) - \hat{\theta}^T(t-1)\phi(t-d)$.

The parameter adaptation law (10.63) has the similar desired stability properties as that for (7.26), which have been summarized in Lemma 7.1.

Adaptive control law. With the parameter estimation, the nominal control law (10.55) can be implemented as

$$\begin{aligned} v_0(t) = \frac{1}{\sum_{i=1}^N \mu_i \hat{\beta}^*_{i,0}} \Bigg[&- \sum_{i=1}^{N} \mu_i \hat{\alpha}_i(z^{-1})[y](t) - \sum_{i=1}^{N} \mu_i \widehat{\bar{\beta}}{}^*_i(z^{-1})[v_0](t) \\ &- \sum_{i=1}^{N} \mu_i \hat{\beta}^*_{1,i}(z^{-1})[\bar{u}_1](t) - \sum_{i=1}^{N} \mu_i \hat{\beta}^*_{2,i}(z^{-1})[\bar{u}_2](t) + y_m(t+d) \Bigg], \end{aligned} \tag{10.64}$$

where the parameters of $\hat{\alpha}_i(z^{-1})$ and $\hat{\beta}_i^*(z^{-1})$ can be obtained directly from corresponding terms in $\hat{\theta}$ and the remaining fault-related parts can be calculated as

$$\sum_{i=1}^{N} \mu_i \hat{\beta}_{1,i}^*(z^{-1})[\bar{u}_1](t) + \sum_{i=1}^{N} \mu_i \hat{\beta}_{2,i}^*(z^{-1})[\bar{u}_2](t)$$
$$= \widehat{\bar{p}}_0^T \bar{\mu}(t) + \widehat{\bar{p}}_1^T \phi_{f_1}(t) + \widehat{\bar{p}}_2^T \phi_{f_2}(t) \tag{10.65}$$

with $\widehat{\bar{p}}_0$, $\widehat{\bar{p}}_1$ and $\widehat{\bar{p}}_2$ being the corresponding elements in the parameter estimate

$$\hat{\theta} = [\hat{\theta}_a^T, \widehat{\bar{p}}_0^T, \widehat{\bar{p}}_1^T, \widehat{\bar{p}}_2^T]^T. \tag{10.66}$$

Based on the desired properties of the parameter adaptive law (10.63), the following closed-loop stability and asymptotic tracking results can be proved (Qi et al. 2012).

Theorem 10.2 *The adaptive controller (10.64) with the parameter adaptive law (10.63), applied to the system (10.57) under Assumptions 10.3–10.5 and with actuator faults (10.46), guarantees that all closed-loop system signals are bounded and* $\lim_{t\to\infty} e(t) = 0$ *for the tracking error* $e(t) = y(t) - y_m(t)$.

Proof The proof of Theorem 10.2 is similar to that of Theorem 7.1 and is sketched here.

For $\phi(t) = [\phi_a^T(t), \bar{\mu}^T(t), \phi_{f_{j_1}}^T(t), \ldots, \phi_{f_{j_q}}^T(t)]^T$ in (10.62), $\phi_a(t)$ has the same signals as that in $\phi(t)$ in (7.25) and $\bar{\mu}(t)$, $\phi_{f_1}(t)$ and $\phi_{f_2}(t)$ are all bounded, so that we also have the result of Lemma 7.2: $\|\phi(t-d)\| \le \rho_1 + \rho_2 \max_{0\le\tau\le t} |e(\tau+d)|$, for the new signal $\phi(t)$ in (10.62). We can also derive the feedback structure of the closed-loop adaptive control system (as given in (7.106)):

$$|e(t)| \le c_2 + c_3 c_5 |\bar{\varepsilon}(t)| \max_{0\le\tau\le t} |e(\tau)|$$
$$+ c_4 \|\hat{\theta}(t-d) - \hat{\theta}(t-1)\| \max_{0\le\tau\le t} |e(\tau)|, \tag{10.67}$$

in which $\bar{\varepsilon}(t) \in L_2 \cap L_\infty$ and $\hat{\theta}(t-d) - \hat{\theta}(t-1) \in L_2 \cap L_\infty$ as from the properties of the adaptive law (10.63) (these L_2 properties ensure that the above feedback system has a small gain). Then, we can conclude that $e(t)$ is bounded, and so is $u(t)$ from (10.64) (the minimum phase property). From (7.40), we have $e(t) \in L_2$, so that $\lim_{t\to\infty} e(t) = 0$. ∇

The above adaptive fault compensation scheme has been designed and analyzed for fuzzy systems with two redundant actuators. The results obtained can be readily extended to systems with more actuators.

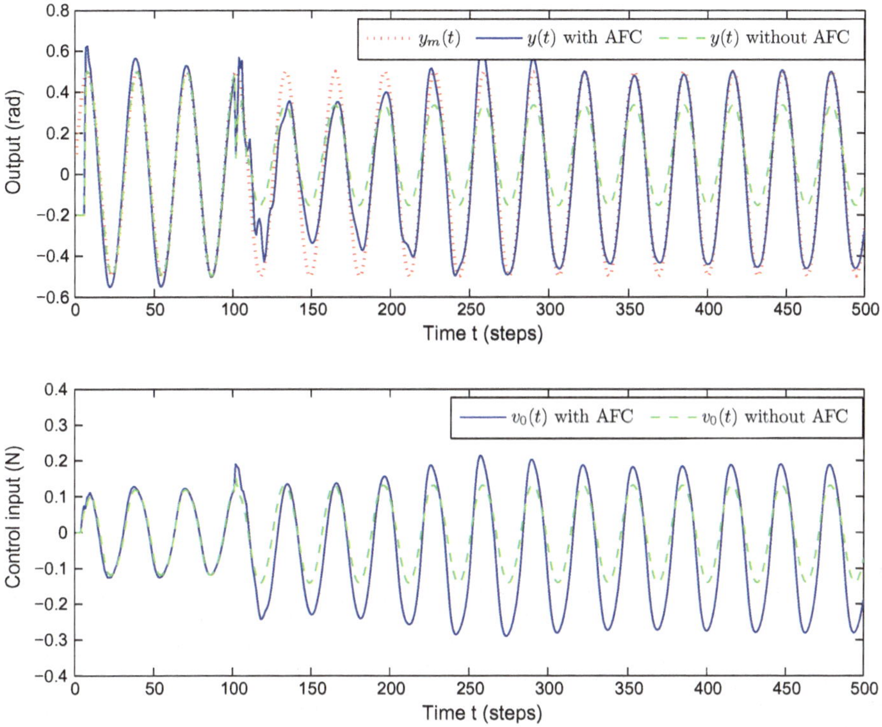

Fig. 10.4 System responses with and without adaptive fault compensation (AFC) for $\bar{u}_2(t) = 0.05$. ©[2012] IEEE. Reprinted, with permission, from Qi et al. (2012)

10.3.4 Simulation Study

In this section, we demonstrate the effectiveness of the adaptive actuator fault compensation design through a simulation example.

Simulation system. In Sect. 7.2.4, we use a flexible-joint manipulator system described by (7.43) as the simulation system to show the performance of the adaptive control scheme. We have constructed the discrete-time input–output T–S fuzzy model for this system, as given in Sect. 7.2.4.

Here, we use the same plant but with 2 actuators. Correspondingly, a redundant actuator is added to the fuzzy model so that when one fails, the other can be used to adaptively compensate the effect of the failed actuator. We start our design from the following model:

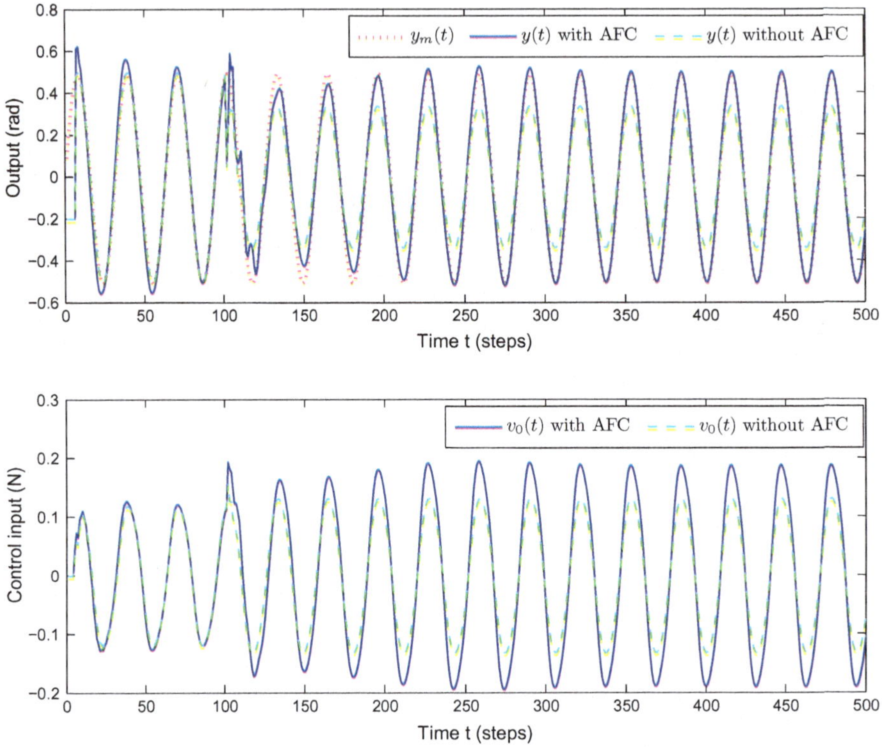

Fig. 10.5 System responses with and without adaptive fault compensation (AFC) for $\bar{u}_2(t) = 0.05\sin(0.2t)$. ©[2012] IEEE. Reprinted, with permission, from Qi et al. (2012)

$$
\begin{aligned}
R^i :\ & \text{IF} \quad x_1(t) \quad \text{is} \quad F_1^i \quad \text{and} \quad x_3(t) \quad \text{is} \quad F_2^i, \quad \text{THEN} \\
& y(t) + a_{i,1}y(t-1) + a_{i,2}y(t-2) + a_{i,3}y(t-3) + a_{i,4}y(t-4) \\
& = b_{1,i,0}u_1(t-2) + b_{1,i,1}u_1(t-3) + b_{1,i,2}u_1(t-4) + b_{1,i,3}u_1(t-5) + \\
& \quad b_{2,i,0}u_2(t-2) + b_{2,i,1}u_2(t-3) + b_{2,i,2}u_2(t-4) + b_{2,i,3}u_2(t-5),
\end{aligned}
$$

where $b_{j,i,k} = b_{i,k}$, $j = 1, 2$; $k = 0, 1, \ldots, 3$. The global fuzzy system model for the two actuator case ($m = 2$) can be derived as follows:

$$
y(t+2) = \sum_{i=1}^{5} \mu_i \alpha_i(z^{-1})[y](t) + \sum_{i=1}^{5} \mu_i \beta_{1,i}(z^{-1})[u_1](t) + \sum_{i=1}^{5} \mu_i \beta_{2,i}(z^{-1})[u_2](t),
$$

where, for $j = 1, 2$,

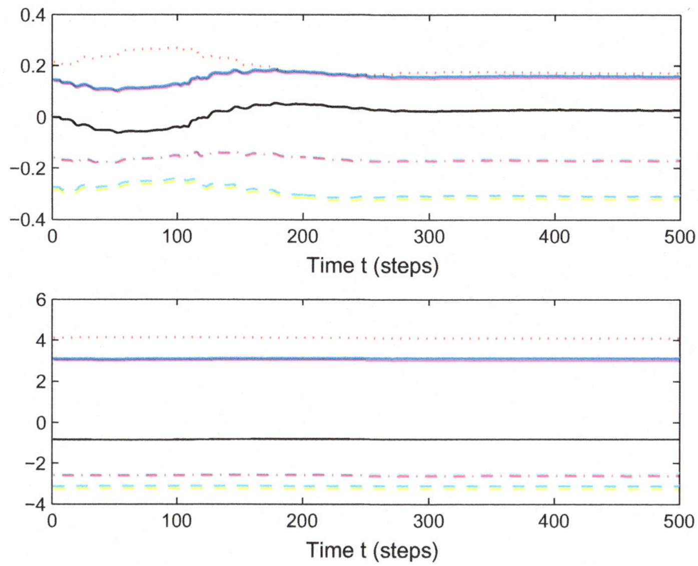

Fig. 10.6 Parameter estimation for $\bar{u}_2(t) = 0.05\sin(0.2t)$

$$\begin{aligned}\beta_{j,i}(z^{-1}) &= \beta_{j,i,0} + \beta_{j,i,1}z^{-1} + \cdots + \beta_{j,i,4}z^{-4} \\ &= b_{j,i,0} + (b_{j,i,1} - a_{i,1}b_{j,i,0})z^{-1} - a_{i,1}b_{j,i,3}z^{-4} \\ &\quad +(b_{j,i,2} - a_{i,1}b_{j,i,1})z^{-2} + (b_{j,i,3} - a_{i,1}b_{j,i,2})z^{-3}.\end{aligned}$$

If the jth actuator fails at $t = t_j$, it can be described by $u_j(t) = \bar{u}_j(t), \quad t \geq t_j$, where the fault $\bar{u}_j(t)$ is characterized by (10.46). In the simulation, we assume $u_2(t) = \bar{u}_2(t)$ and consider two classes of failures:

- a constant fault: $\bar{u}_2(t) = 0.05$
- a sinusoidal fault: $\bar{u}_2(t) = 0.05\sin(0.2t)$.

Simulation results. The initial parameter values are set as 80% of their true values. Other parameters are chosen as $\gamma(t) = 0.8$ and $c = 0.01$. The fault is added at $t = 100$s. We consider output tracking of a sinusoidal signal $y_m(t) = 0.5\sin(0.2t)$ under actuator fault. Figure 10.4 shows the system responses for a constant fault with and without adaptive fault compensation and Fig. 10.5 for a time-varying fault. It shows that for both cases, with adaptive fault compensation (AFC), the asymptotic tracking is achieved, while without adaptive actuator fault compensation, the closed-loop system performance degrades significantly, leading to large tracking errors. For the AFC case, the parameter estimation is stable and the parameter estimates tend to converge to some constant values, as shown in Fig. 10.6.

10.4 Summary

In this chapter, we have developed two adaptive compensation schemes for T–S fuzzy systems subject to actuator faults, which are applicable to solve the fault-tolerant control problems for a class of nonlinear systems approximated by T–S fuzzy systems. To use adaptive control techniques to compensate for the effects of actuator faults, the key step is to develop a parametrization model containing unknown system parameters and uncertain fault parameters. Then parameter adaptive laws are derived to estimate the system parameters and fault parameters together. Detailed design procedures and rigorous stability and tracking performance analysis have been provided. Simulation results have demonstrated the effectiveness of the presented adaptive actuator fault compensation schemes.

10.5 Concluding Remarks

Takagi–Sugeno (T–S) fuzzy modeling and adaptive control has proven to be a powerful tool to deal with the identification and control problems for uncertain nonlinear systems, which is not only conceptually easy to understand but also owns a sound analytical basis for carrying out systematic control designs and rigorous stability analysis. This book has provided readers with a systematic and unified framework for identification and adaptive control of T–S fuzzy systems, helping readers improve their understanding and gain enough design techniques to apply this powerful tool to solve challenging practical control problems.

For different nonlinear systems and based on different modeling methods, T–S fuzzy systems identified may have different forms: state-space form or input–output form, continuous-time form or discrete-time form, single-input-single-output (SISO) form or multi-input-multi-output (MIMO) form, canonical form or noncanonical form. From the feedback control design perspective, T–S fuzzy system control can be designed using state feedback or output feedback. From the adaptive control perspective, adaptive T–S fuzzy control can be direct adaptive control or indirect adaptive control. From the parameter estimation perspective, T–S fuzzy systems have both linear parameters and nonlinear parameters, requiring different parameter adaptation algorithms. In a clear, balanced and unified way, this book has offered a unique text describing T–S fuzzy identification and adaptive control methodologies for T–S fuzzy systems in various forms and employing different feedback control and parameter estimation techniques, aiming at reflecting the state of the art in adaptive T–S fuzzy control as well as at presenting its fundamentals.

Focused on fuzzy system theory and topics, our book has provided a systematic study of fuzzy system modeling and control problems, using fuzzy system tools for both function approximation and for feedback control of nonlinear systems. Our book has shown how fuzzy approximation methods can be used to represent various dynamic systems, especially, in parameterized forms which are suitable for adaptive control to deal with system uncertainties. Our book has explicitly developed the design and analysis of various system identification and feedback control schemes using fuzzy approximators, such as those for state-space system models and input–output models, for state feedback control to achieve either state tracking or output tracking, for output feedback control to achieve output tracking, in the presence of system parametric uncertainties and fault uncertainties, and for continuous-time systems and for discrete-time systems as well. Our book has built a general framework of adaptive fuzzy system based identification and control for various cases as mentioned above. The book has developed a set of adaptive fuzzy system based theoretical tools parallel to those applicable to ordinary (non-fuzzy) systems to make essential additions to the literature.

References

Ballé P (1999) Fuzzy-model-based parity equations for fault isolation. Control Eng Pract 7(2):261–270

Chen J, Patton RJ (1999) Robust model-based fault diagnosis for dynamic system. Kluwer, London

Ichtev A, Hellendoorn J, Babuska R (2002) Fault detection and isolation using multiple Takagi–Sugeno fuzzy models. Proceedings of the IEEE international fuzzy system conference, Melbourne, Australia, pp 1498–1502

Jiang L, Qi R (2016) An adaptive actuator failure compensation scheme for discrete-time T–S fuzzy systems. Int J Innov Comput Info Contr 12(4):1043–1058

Jiang B, Gao Z, Shi P, Xu Y (2010) Adaptive fault-tolerant tracking control of near-space vehicle using Takagi–Sugeno fuzzy models. IEEE Trans Fuzzy Syst 18(5):1000–1007

Jiang B, Zhang K, Shi P (2011) Integrated fault estimation and accommodation design for discrete-time Takagi–Sugeno fuzzy systems with actuator faults. IEEE Trans Fuzzy Syst 19(2):291–304

Nguang SK, Shi P, Ding S (2007) Fault detection for uncertain fuzzy systems: an LMI approach. IEEE Trans Fuzzy Syst 15(6):1251–1262

Qi R, Tao G, Jiang B, Tan C (2012) Adaptive control schemes for discrete-time T–S fuzzy systems with unknown parameters and actuator failures. IEEE Trans Fuzzy Syst 20:471–486

Qi R, Zhu L, Jiang B (2013) Fault-tolerant reconfigurable control for MIMO systems using online fuzzy identification. Int J Innov Comput, Inf Control 9(10):3915–3928

Shen Q, Jiang B, Cocquempot V (2012) Fault-tolerant control for T-S fuzzy systems with application to near-space hypersonic vehicle with actuator faults. IEEE Trans. Fuzzy Syst. 20(4):652–665

Tao G, Chen S, Tang X, Joshi SM (2004) Adaptive control of systems with actuator failures. Springer, London

Wang HO, Tanaka K, Griffin MF (1996) An approach to fuzzy control of nonlinear systems: stability and design issues. IEEE Trans Fuzzy Syst 4:14–23

Zhang Y, Jiang J (2008) Bibliographical review on reconfigurable fault-tolerant control systems. Annu Rev Control 32:229–252

Zheng Y, Fang H, Wang HO (2006) Takagi–Sugeno fuzzy-model-based fault detection for networked control systems with Markov delays. IEEE Trans Syst, Man, Cybern Part B, Cybern 36(4):924–929

Zhang K, Jiang B, Shi P (2010) Dynamic output feedback fault tolerant controller design for Takagi–Sugeno fuzzy systems with actuator faults. IEEE Trans Fuzzy Syst 18(1):194–201

Appendix A
Proof of Lemma 8.1

Proof (i) Consider the following Lyapunov candidate function

$$V = \mathrm{tr}[\tilde{\Theta}^T(t)\tilde{\Theta}(t)], \ \tilde{\Theta}(t) = \hat{\Theta}(t) - \Theta. \tag{A.1}$$

With (8.48), we have

$$\begin{aligned} & V(t) - V(t-1) \\ &= \gamma(t)\left(-2 + \frac{\gamma(t)\phi^T(t-d)\phi(t-d)}{c + \phi^T(t-d)\phi(t-d)}\right) \\ &\cdot \frac{\varepsilon^T(t)\varepsilon(t)}{c + \phi^T(t-d)\phi(t-d)}. \end{aligned} \tag{A.2}$$

Since $\gamma_0 < \gamma(t) < 2 - \gamma_0$ for some constant $\gamma_0 \in (0, 1)$, we have

$$\gamma(t)\left(-2 + \frac{\gamma(t)\phi^T(t-d)\phi(t-d)}{c + \phi^T(t-d)\phi(t-d)}\right) < -a_0, \tag{A.3}$$

where a_0 is a positive constant. Then we obtain

$$V(t) - V(t-1) \leq -\frac{a_0\varepsilon^T(t)\varepsilon(t)}{c + \phi^T(t-d)\phi(t-d)}, \tag{A.4}$$

which means $\tilde{\Theta}(t) \in L_\infty$ and $\|\hat{\Theta}(t) - \Theta\| \leq \|\hat{\Theta}(t-1) - \Theta\| \leq \cdots \leq \|\hat{\Theta}(0) - \Theta\|$, for the matrix norm $\|\hat{\Theta}(t) - \Theta\| = \sqrt{\mathrm{tr}[(\hat{\Theta}(t) - \Theta)^T(\hat{\Theta}(t) - \Theta)]}$.

(ii) From (A.4) we also have

$$\frac{\varepsilon^T(t)\varepsilon(t)}{c + \phi^T(t-d)\phi(t-d)} \leq \frac{V(t-1) - V(t)}{a_0} < \infty, \tag{A.5}$$

R. Qi et al., *Fuzzy System Identification and Adaptive Control*, Communications and Control Engineering,
https://doi.org/10.1007/978-3-030-19882-4

which means $\frac{\varepsilon(t)}{\sqrt{c+\phi^T(t-d)\phi(t-d)}} \in L_\infty$.

Summing (A.4) from 1 to N, we have

$$V(N) = V(0) - \sum_{t=1}^{N} \frac{a_0 \varepsilon^T(t)\varepsilon(t)}{c + \phi^T(t-d)\phi(t-d)}. \tag{A.6}$$

Since V is a nonincreasing bounded function, we can conclude from (A.6) that

$$\sum_{t=1}^{N} \frac{\varepsilon^T(t)\varepsilon(t)}{c + \phi^T(t-d)\phi(t-d)} < \infty, \tag{A.7}$$

and this implies $\frac{\varepsilon(t)}{\sqrt{c+\phi^T(t-d)\phi(t-d)}} \in L_2$.

(iii) Follows immediately from (A.7).

(iv) From (8.48), we have

$$\begin{aligned}
\|\hat{\Theta}(t) - \hat{\Theta}(t-1)\| &= \|\frac{\gamma(t)\phi(t-d)\varepsilon(t)}{c + \phi^T(t-d)\phi(t-d)}\| \\
&\leq \frac{(2-\gamma_0)\|\phi(t-d)\|}{\sqrt{c + \phi^T(t-d)\phi(t-d)}} \cdot \frac{\|\varepsilon(t)\|}{\sqrt{c + \phi^T(t-d)\phi(t-d)}}.
\end{aligned}$$

It is easy to see $\|\hat{\Theta}(t) - \hat{\Theta}(t-1)\| < \infty$. Consequently, $\hat{\Theta}(t) - \hat{\Theta}(t-t_1) \in L_\infty$ for finite t_1.

Furthermore,

$$\begin{aligned}
&\|\hat{\Theta}(t) - \hat{\Theta}(t-1)\|^2 \\
&= \|\frac{\gamma^2(t)\phi(t-d)\varepsilon(t)}{c + \phi^T(t-d)\phi(t-d)}\|^2 \\
&\leq \frac{(2-\gamma_0)^2\|\phi(t-d)\|^2}{c + \phi^T(t-d)\phi(t-d)} \cdot \frac{\|\varepsilon(t)\|^2}{c + \phi^T(t-d)\phi(t-d)} \\
&\leq \frac{(2-\gamma_0)^2\|\varepsilon(t)\|^2}{c + \phi^T(t-d)\phi(t-d)}.
\end{aligned} \tag{A.8}$$

Summing (A.8) from 1 to N and with (A.7), we have

$$\begin{aligned}
&\sum_{t=1}^{N} \|\hat{\Theta}(t) - \hat{\Theta}(t-1)\|^2 \\
&\leq \sum_{t=1}^{N} \frac{(2-\gamma_0)^2\|\varepsilon(t)\|^2}{c + \phi^T(t-d)\phi(t-d)} < \infty,
\end{aligned} \tag{A.9}$$

which means $\hat{\Theta}(t) - \hat{\Theta}(t-1) \in L_2$.

Using the Schwarz inequality, we have

$$\begin{aligned}
&\|\hat{\Theta}(t) - \hat{\Theta}(t - t_1)\|^2 \\
&\le k(\|\hat{\Theta}(t) - \hat{\Theta}(t - 1)\|^2 + \|\hat{\Theta}(t - 1) - \hat{\Theta}(t - 2)\|^2 \\
&\quad + \cdots + \|\hat{\Theta}(t - t_1 + 1) - \hat{\Theta}(t - t_1)\|^2),
\end{aligned} \tag{A.10}$$

where k is a positive constant. Using (A.9) and (A.10), we obtain

$$\sum_{t=1}^{N} \|\Theta(t) - \Theta(t - t_1)\|^2 < \infty \tag{A.11}$$

since t_1 is finite. Hence, $\hat{\Theta}(t - 1) - \hat{\Theta}(t - d) \in L^2$.

(v) Follows immediately from (A.11).

Index

R. Qi et al., *Fuzzy System Identification and Adaptive Control*, Communications and Control Engineering,
https://doi.org/10.1007/978-3-030-19882-4

T

U

W

Z

Printed by Printforce, the Netherlands